T0188629

Finite Elements
⇕
Computational
Engineering Sciences

Finite Elements
↕
Computational
Engineering Sciences

A. J. Baker

Professor Emeritus
The University of Tennessee
USA

A John Wiley & Sons, Ltd., Publication

Library of Congress Cataloguing-in-Publication Data

Baker, A. J., 1936-
   Finite elements : computational engineering sciences / A.J. Baker.
      p. cm.
   Includes bibliographical references and index.
   ISBN 978-1-119-94050-0 (cloth)
1.   Finite element method.   I. Title.
   TA347.F5B3423 2012
   620.001'51825–dc23

                        2012011745

A catalogue record for this book is available from the British Library.

ISBN: 9781119940500

Set in 10/12.5 pt, Palatino-Roman by Thomson Digital, Noida, India
Printed in Singapore by Ho Printing Singapore Pte Ltd

Yogi Berra is quoted,
"If you come to a fork in the road, take it,"
so I did and ended up here.

Contents

Preface

The computer revolution has profoundly impacted how engineers and scientists conduct professional activities. In the early 1960s, a computer fully occupied and amply heated (!) a space the size of a classroom. The PC, introduced in the mid-1970s, was a "toy." Yet by the millennia, Linux clusters of cheap gigahertz–gigabyte PCs could execute truly large-scale computational simulations. Indeed, the "desktop Cray," fantasized in ∼1980, was here and was truly inexpensive!

The companion maturation of theory and practice in the computational engineering sciences has been an evolutionary (not revolutionary!) process. It remains highly fragmented by discipline, even though computational fluid dynamics (CFD) and computational structural mechanics (CSM) emerged simultaneously from the research laboratory in the late 1950s. The former relied on finite difference (FD) methods to convert theory to computable form. Conversely, the latter's classical *virtual work* foundation enabled a calculus-based finite element (FE) theory implementation of the underlying variational principle extremum. Finally, in chemical engineering collocation methods were developed for process simulations, and at first glance these theories appear absolutely "linearly independent."

Research now completed has proven that practically all developments supporting the *computational engineering sciences* can be formulated from the *extremum* of the mathematician's *weak form theory* termed a *weak statement* (WS). The weak form process enables *theorization* to be completed in the *continuum*, using calculus, vector field theory, and modern approximation concepts. When finished, the discrete implementation of the theory extremum can be formed using FE, FD, and/or finite volume (FV) procedures. The FE implementation is typically guaranteed *optimal* in its performance, that is, accuracy, asymptotic convergence rate, and so on. Furthermore, FE methodology leads to *precise constructions* devoid of heurism, since integral–differential calculus is used rather than difference algebra to generate the algebraic statement amenable to computing.

This text develops discrete implementations of WS theory for a diverse variety of problem statements in the *computational engineering sciences*. Unique to the FE discrete development, the resulting *algorithms* are immediately stated in *computable* form via a transparent, object-oriented programming syntax. The engineering science problem classes developed herein include

- heat conduction
- structural mechanics
- mechanical vibrations
- heat transfer, with convection and radiation
- fluid mechanics
- heat/mass convective transport

The text is organized into twelve chapters. Following an introduction, and some very pertinent overview material, an elementary heat conduction *tutorial* clearly illustrates all element matrix constructs, the "famous" assembly algorithm and the concept of error estimation and measurement. Subsequent chapter pairs develop expository one-dimensional, then general n-dimensional FE WS implementations in each continuum engineering sciences discipline.

The sequence of developments serves to illustrate, examine, and generalize the available theoretical error estimates, with the concept of a *norm* central to this process. In moving to the convection–diffusion problem class, a sequence of Taylor series manipulations leads to *modified* conservation principle expressions, expressed in the *continuum*, which collectively improve asymptotic convergence rate coupled with annihilation of significant order discretization-induced *phase lag* and *dispersive* error mechanisms.

Incisive computer lab experiments complement each development, with principle focus to gain a firm usable understanding of approximation error mechanisms as influenced by data *nonsmoothness*, problem *nonlinearity*, stability, dispersion error and boundary conditions, each impacted by the selected FE basis completeness degree. The n-dimensional computer experiments focus on refinements for error nuances associated with nonconvex boundaries, phase lag, and artificial *numerical diffusion*. An intervening brief chapter clearly identifies the connections between FD, FV, and FE discrete implementations for a Poisson equation in n-dimensions.

Engineers are clearly of the opinion that, "theory is fine, but show me the numbers!," which requires theory conversion to code practice. Since the FE-implemented WS theory is highly organized, the algorithm statement in any discipline ends up constituted of six, and *only* six, types of data to convert theory to practice. Capitalizing on *object-oriented* concepts, these six data types are organized into a *template* such that the *computing statement*, including explicit nonlinearity, is unambiguously expressible.

In summary, this text fully develops modern FE discrete algorithms for the computational engineering sciences with applications aimed to available and emergent *problem solving environments* (PSEs). Its organization and content has evolved from two decades of teaching the subject at UT. This text fully obsoletes the predecessor 1991 text *Finite Elements* 1-2-3, marketed with a "spaghetti" Fortran PC code on a 5.25 inch floppy disk.

All computer lab exercise MATLAB® .m files along with the specifically written MATLAB® toolbox *FEmPSE* are available for download from www.wiley.com/go/baker/finite. The .mph files for the COMSOL design studies may be downloaded from their user community web site www.comsol.com/community/exchange/?page=2. University faculty interested in presenting the internet-enabled academic course from which this text was generated will find complete support materials available at www.wiley.com/go/baker/finite.

Many colleagues and graduate students have contributed to the creation and refinement of text content. My thinking formality on the subject has benefited from a multi-decade collegial association with Prof. J. Tinsley Oden. I owe a deep debt of gratitude to my Computational Mechanics Corp. co-founders Paul Manhardt, who invented the template concept, and Joe Orzechowski, who assimilated templates into reliable computational syntax for mainly CFD applications.

The dissertation research of Dr. Jin Kim, Dr. Subrata Roy, Dr. David Chaffin, Dr. Alexy Kolesnikov, and Dr. Sunil Sahu collectively formalized the improved theoretical and practical understanding of FE algorithm performance nuances detailed herein. Dr. Zac Chambers and Dr. Marcel Grubert along with Messrs. Mike Taylor and Shawn Ericson contributed significantly to polishing these fundamental underlying precepts to pedagogical acceptability.

A. J. Baker
Knoxville, TN
January 2012

Note: All color originals are accessible at www.wiley.com/go/baker/finite.

About the Author

A. J. Baker, PhD, PE, left commercial aerospace research to join the University of Tennessee College of Engineering in 1975, to lead academic research in the exciting new field of CFD (computational fluid dynamics). Now Professor Emeritus and still Director, UT CFD Laboratory (http://cfdlab.utk.edu), his professional career started as a mechanical engineer with Union Carbide Corp. The challenges there prompted resigning after 5 years to enter graduate school full time in 1963 with the goal to "learn what a computer was and could do." The introduction involved *driving* an IBM 1620 with 5 kB memory and no disk pack! A 1967 summer job with Bell Aerospace Company required assessing the *first* publication claiming unsteady heat conduction was amenable to finite element analysis. This led to the 1968 Bell Aerospace technical memorandum, "A Numerical Solution Technique for a Class of Two-dimensional Problems in Fluid Dynamics Formulated via Discrete Elements," a truly pioneering expose in the fledgling FE CFD field. Finishing his dissertation in 1970, he joined Bell Aerospace as Principal Research Scientist to pursue full-time finite element methods in CFD. NASA Langley contracts with summer appointments at ICASE led to a visiting professorship at Old Dominion University, 1974–1975, from which he moved directly to UT forming Computational Mechanics Consultants, Inc., with two Bell colleagues, to assist converting academic FE CFD research progress into computing practice.

FE ⟷ Computational Engineering Sciences with hands-on computing:
This is the first *introductory* level text to fully integrate the underlying *theory* with *hands-on* computer experiments supported by the MATLAB® and COMSOL® Problem Solving Environments (PSEs). You may download all .m and .mph files supporting each suggested computer experiment, also eight topical lectures for video-streaming on your PC available from www.wiley.com/go/baker/finite. The academic course engendering the text technical content became totally distance-enabled on Internet in 2005. Academics interested in presenting this course at their institution may acquire the complete academic support material at www.wiley.com/go/baker/finite.

Notations

a	expansion coefficient
A	plane area; one-dimensional FE matrix prefix; coefficient
A	generic square matrix
[A]	factored global matrix
b	coefficient; boundary condition subscript; body force component, generic column matrix
{b}	global data matrix
B	two-dimensional FE matrix prefix
B	body force, structural FE matrix
c	coefficient; specific heat
C	three-dimensional FE matrix prefix, constant, Courant number
d	coefficient; FE matrix indicator
D	diagonal matrix, diffusion coefficient
[DIFF]	global diffusion matrix
DOF	approximation degrees-of-freedom
e	element-dependent; unit vector component, error
$e(\cdot)$	error, a function of $(\cdot)$
e^N	approximation error
e^h	discrete approximation error
eta$_{ji}$	coordinate transformation data
E	energy seminorm (subscript), elastic modulus
E	Hooke's law matrix
F	radiation viewfactor
F	applied force, flux on $\partial\Omega$
f	kinetic flux vector
FD	finite difference
FE	finite element
FV	finite volume
{F}	homogeneous form of a discretized weak statement
g	gravity magnitude
g	gravity

G	elastic shear modulus, amplification factor, Gebhart factor
Gr	Grashoff number
GWS	Galerkin weak statement
h	discretization (superscript), heat transfer coefficient, measure
H	Gauss quadrature weight; Hilbert space
[BC]	boundary condition matrix
i	summation index, mesh node, imaginary unit
$\hat{\mathbf{i}}$	unit vector parallel to x
I	moment of inertia; element matrix summation index
[I]	identity (diagonal) matrix
j	summation index, mesh node
$\hat{\mathbf{j}}$	unit vector parallel to y
J	template summation index
[J]	coordinate transformation jacobian
[JAC]	jacobian
k_{ij}	element of the [DIFF] and/or [STIFF] matrix
k	thermal conductivity, basis degree, index, diffusion coefficient,
k	spring constant
$\bar{k}$	average value of conductivity
$\hat{\mathbf{k}}$	unit vector parallel to z
K	template matrix summation index, viewfactor kernel
ℓ	element length; summation index
$\ell(\cdot)$	differential equation on $\partial\Omega$
L	domain span, length measure, lower triangular matrix, lagrangian
$\mathcal{L}(\cdot)$	differential equation on Ω
m	integer
m_i	point mass
mGWS	Taylor series-modified Galerkin weak statement
mPDE	Taylor series-modified conservation principle PDE
M	elements in Ω^h; moment; matrix prefix; particle system mass
M	iteration matrix
[MASS]	global mass matrix
n	index; normal subscript; dimension of domain Ω; integers, normal coordinate, time index (subscript)
$\hat{\mathbf{n}}$	outward pointing unit vector normal to $\partial\Omega$
N	matrix prefix
N	summation termination; approximation (superscript), iteration matrix
NC	natural coordinate basis
$\{N_k\}$	finite element basis of degree k non-D non-dimensional
p	load (data); pressure, iteration index
P	point load; Gauss quadrature order
$\{P\}$	computational matrix, distributed load DOF
Pa	non-D parameter on Ω
Pb	non-D parameter on $\partial\Omega$

Pr	Prandtl number
q	generalized dependent variable
Q	discretized dependent variable; heat added
$\{Q\}$:	approximation DOF matrix
r	reference state subscript; radius
Re	Reynolds number
R^+	the positive real axis
$\Re^n$	Euclidean space
$\{RES\}$	global matrix statement residual
s	source term on Ω; heat added, tangent coordinate
$\mathbf{s}$	unit vector tangent to $\partial\Omega$
S	finite element assembly operator; entropy
SOR	successive over-relaxation
$\{S\}$	computational matrix
t	time
T	temperature, kinetic energy
T_c	convection heat transfer exchange temperature
T_r	radiation heat transfer exchange temperature
$\mathbf{T}$	surface traction vector
T^N	approximate temperature solution
TE	truncation error
TP	tensor product basis
TS	Taylor series
$\mathbf{u}$	displacement vector; velocity vector
U	upper triangular matrix
u	velocity x component; speed
U	discretized speed DOF, phase velocity (speed)
[VEL]	global fluid convection matrix
v	velocity y component
V	shear force; volume; potential energy
$\mathbf{V}$	velocity
w	weight function; fin thickness; velocity z component
W	weight; work done by system
WF	weak form
WS	weak statement
x	generic unknown
x, x_i	cartesian coordinate, coordinate system $1 \leq i \leq n$
$\bar{x}$	transformed local coordinate
X	discrete cartesian coordinate
y	displacement; cartesian coordinate
Y	discrete cartesian coordinate
z	cartesian coordinate
Z	thickness ratio; discrete cartesian coordinate
$(\cdot)$	scalar (number)

$\{\cdot\}$	column matrix
$\{\cdot\}^T$	row matrix
$[\cdot]$	square matrix
$\|\cdot\|$	norm
$\cup$	union (non-overlapping sum)
$\cap$	intersection
$\det[\cdot]$	matrix determinant
sym	symmetric
α	coefficient
β	coefficient
γ	shear strain, coefficient
δ_{ij}	Kronecker delta
δQ	iterate
Δ	discrete increment
ε	normal strain, emissivity
ϕ	electric potential, flow potential
$\phi(\cdot)$	trial space function; potential function
Φ	potential function
$\Phi_\beta(\mathbf{x})$	test space
$\Psi_\alpha(\mathbf{x})$	trial space
$\boldsymbol{\eta}$	coordinate system in transform space
η_i	tensor product coordinate system
κ	thermal diffusivity, wave number
$\boldsymbol{\kappa}_{\alpha\beta}$	element of a square matrix
λ	Lamé parameter, wavelength
μ	Lamé parameter, dynamic viscosity
υ	Poisson ratio, kinematic viscosity
$O(\cdot)$	order of $(\cdot)$
π	pi (3.1415926 . . .)
θ	time integration implicitness factor
Θ	potential temperature
ρ	density, absorbtivity
$\mathrm{d}\sigma$	differential element on $\partial\Omega$
$\mathrm{d}\tau$	differential element on Ω
τ	normal stress
ω	frequency
Ω	domain of differential equation
Ω_e	finite element domain
Ω^h	discretization of Ω
$\partial\Omega$	boundary of Ω
ζ_α	natural coordinate system
$\mathrm{d}(\cdot)/\mathrm{d}x$	ordinary derivative
$\partial(\cdot)/\partial x$	partial derivative
∇	vector derivative
∇^2	laplacian derivative operator

1

The Computational Engineering Sciences:
an introduction

1.1 Engineering Simulation

The digital computer, coupled with engineering and computer science plus modern approximation theory, have coalesced to render computational simulation via math modeling an alternative modality supporting *design optimization* in engineering. Design has historically been conducted in the *physical laboratory* (Figure 1.1). The test device is a miniature of reality and the laboratory process sequence is:

- *model* the geometry (similitude)
- *determine* desired data (cost)
- *acquire* the data
- *interpret* the data
- *draw* conclusions

The *computational engineering sciences laboratory* has emerged as the complement to, or replacement of, the legacy modality (Figure 1.2). The computational laboratory process sequence is:

- *model* the mathematics (fidelity)
- *model* the physics (cost)
- *compute* the data
- *interpret* the data
- *draw* conclusions

Finite Elements ⇔ Computational Engineering Sciences, First Edition. A. J. Baker.
© 2012 John Wiley & Sons, Ltd. Published 2012 by John Wiley & Sons, Ltd.

Figure 1.1   Classic wind tunnel test

The first two components of the computational engineering sciences (CES) laboratory place a significant *new burden* on the engineer/scientist. Aspects of calculus and vector field theory, the *language* for expressing conservation principles in the engineering sciences, must be recalled. Additionally, dexterity with *constitutive closure* approximations, that is, the *"physics model,"* must be understood on a fidelity/mathematics as well as cost/benefit basis.

The identical calculus and vector field topics underpin modern approximation theory guidance for generating a conservation principle *approximate solution* based on a *weak formulation* (WF) [1]. The mathematicians, in developing this approach to solution approximation, have endowed it with an elegant theory on *optimal* construction and error estimation. A WF, *always* completed in the *continuum*, theoretically *transforms* the

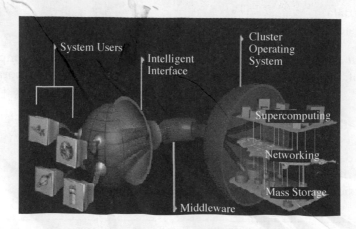

Figure 1.2   Cloud-computing visualization

solution of the partial differential equation (PDE) into a *computable* large-order algebraic equation system.

Once the continuum weak form theory is completed, the sole remaining decision is implementation. Herein this is accomplished by replacing the trial and test spaces with *finite element* (FE) *trial/test space bases* defined for a spatial *discretization* of the PDE domain of dependence. This identification directly enables WF integral evaluations using the calculus. Detailing this process for a diverse spectrum in the *engineering sciences* is the content of this text.

1.2 A Problem-Solving Environment

Historically, a frustrating aspect of computational simulation was interacting with *the computer code*! User interface and computer science issues dominate this facet, and the engineer/scientist interested in analysis is typically not well founded in the required skills. This issue is compounded by the tradition in *olden times*, that is, a decade or so ago, for the individual to code his/her own computer program.

This incredible dissipation of time and effort has been superceded by the emergence of *component-based software* leading to the concept of a *problem-solving environment* (PSE). Commercial code systems now exist throughout the engineering sciences possessing very powerful advances in user interfaces. Maturation of *grid computing* concepts will lead, in the not too distant future, to Internet-enabled *just-in-time* capabilities using remotely accessible *high-performance computing and communications* (HPCC) constructs [2].

Figure 1.3 illustrates this emergent scenario. The practicing design engineer possesses knowledge about his/her problem statement, and after absorbing this text's content will be thoroughly comfortable with the associated mathematics/physics issues with seeking an *optimal* approximate solution. From that point on only casual

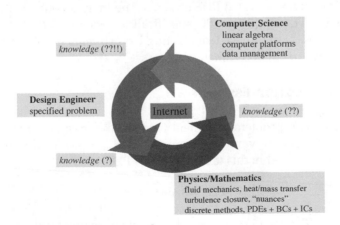

Figure 1.3   The problem-solving environment

knowledge about the subsequent computer science issues will be required, as the Internet modality exists to complete the loop.

An added historical aspect is that computational codes were irrevocably tailored to the specific discrete theory, for example, finite difference (FD), finite volume (FV), FE for a given engineering science problem class. This is now moot as completed research confirms these apparently very distinct computational constructs can be interpreted as specific decisions in implementing a *weak statement* (WS), the *extremum* of a WF for a PDE. Invariably the FE discrete implementation generates the *optimal* construction, the consequence of WF theory, and the use of calculus rather than difference algebra to form the algebraic statement.

The computational practice of FE methods is rapidly maturing, as academics in math, engineering, and computer science collectively resolve key theoretical issues. A by-product, developed in thoroughness in this text, is the *object-oriented* FE algorithm construct that directly communicates "compute desire" to a PSE via a *template*.

This approach recognizes a code is but a data-handling system, and the FE implementation of a WS generates *only* six data types for each and every (!) FE domain Ω_e specifically including nonlinearity. The objects for *all* element-level matrix contributions $\{WS\}_e$ to a WS algebraic statement are thus organized as:

$$\{WS\}_e \equiv \begin{pmatrix} \text{global} \\ \text{constant} \end{pmatrix} \begin{pmatrix} \text{element} \\ \text{average} \end{pmatrix}_e \left\{ \begin{matrix} \text{element} \\ \text{variable} \end{matrix} \right\}_e \begin{pmatrix} \text{metric} \\ \text{data} \end{pmatrix}_e \begin{bmatrix} \text{master} \\ \text{matrix} \end{bmatrix} \left\{ \begin{matrix} \text{unknown} \\ \text{or data} \end{matrix} \right\}_e .$$

Coding of a FE WS discrete implementation is thus reduced to data identification in these six object categories.

Herein, the progression of a WS algorithm for an engineering science topic, FE discrete-implemented, leads to the object-oriented *template* transparently converting theory to executable code. Template generation occurs in a word-processing environment, and the result precisely encompasses all complexities, specifically including nonlinearity, in coupled PDE systems. The template-enabled computing PSE herein employs MATLAB® [3], via the specifically written *FEmPSE toolbox* for expository computing labs. Design-based computing experiments employ COMSOL [4], an FE-implemented *multiphysics* commercial PSE.

1.3 Weak Formulation Essence

An engineering design problem statement is invariably cast as a PDE written on the *state variable* (the dependent variable), herein labeled $q = q(\mathbf{x})$ for the steady definition. The compact notation used in this text to denote a PDE is

$$\mathcal{L}(q) = 0, \text{ on } \Omega \subset \Re^n. \tag{1.1}$$

In equation (1.1), $\mathcal{L}$ is the PDE placeholder and its domain of influence is symbolized as Ω, a region lying on an n-dimensional euclidean space $\Re^n$.

To "connect" the PDE to the specific problem statement requires boundary conditions (BCs) communicating this given information, that is, the *data*. The text-utilized BC compact notation is

$$\ell(q) = 0, \text{ on } \Omega \subset \Re^{n-1}, \tag{1.2}$$

where $\partial\Omega$ is the $(n-1)$-dimensional bounding enclosure of Ω. Figure 1.4 illustrates these formalisms.

The exact solution $q(\mathbf{x})$ satisfying a *genuine* problem equations (1.1) and (1.2) can never (!) be found analytically. Consequently, the *key* WF theory requirement is to formally define an (any!) *approximation* to $q(\mathbf{x})$. Herein this requirement is expressed as

$$q(\mathbf{x}) \approx q^N(\mathbf{x}) \equiv \sum_{\alpha=1}^{N} \Psi_\alpha(\mathbf{x}) Q_\alpha. \tag{1.3}$$

The assumption in equation (1.3) is that one can identify a suitable *trial space* $\Psi_\alpha(\mathbf{x})$, a set of functions on $\Re^n$, to *support* any approximate solution. The summation therein couples each trial space member to an *unknown* expansion coefficient Q_α, called a *degree-of-freedom* (DOF) of the approximation, the set of which is to become determined in the algebraic computing process.

Unless equations (1.1) and (1.2) define a trivial problem, q^N cannot be identical with q. The difference between q and q^N is the *approximation error*, herein denoted e^N. Since everything is a function, obviously

$$e^N(\mathbf{x}) \equiv q(\mathbf{x}) - q^N(\mathbf{x}). \tag{1.4}$$

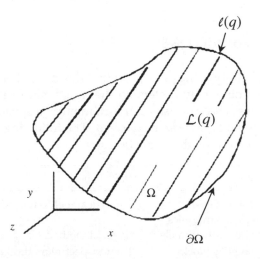

Figure 1.4 Engineering problem statement notation

The singular goal is to seek the *best* approximation q^N, hence to *constrain* in some sense the "size" of $e^N(\mathbf{x})$. This is elegantly accomplished via the mathematicians' *WF* that requires the available measure of error

$$\mathcal{L}(q^N) \neq 0!$$

be made *orthogonal* (mathematically "perpendicular") to an *arbitrarily* chosen *test function* $w(\mathbf{x})$. The *weak form* (WF^N) expression of this constraint on the approximation (1.3) is

$$WF^N \equiv \int_\Omega w(\mathbf{x})\mathcal{L}(q^N)\mathrm{d}\tau \equiv 0. \tag{1.5}$$

The requirement of *any* function $w(\mathbf{x})$ is cleanly handled via an interpolation, ([5], Chapter 2.2), followed by forming the *extremum* of rearranged equation (1.5) which identifies the *test space* $\Phi_\beta(\mathbf{x})$ companion to the trial space $\Psi_\alpha(\mathbf{x})$. The scalar equation (1.5) thus becomes a *set* of equations of the order N defined in equation (1.3), which is termed the *weak statement* (WS^N). All $\mathbf{x}$-dependence vanishes in evaluating the defined integrals, hence WS^N generates the *algebraic* equation system

$$WS^N \Rightarrow [\text{Matrix}]\{Q\} = \{b\}. \tag{1.6}$$

As the final caveat, inserting equation (1.2) into equation (1.5) moves the BC-constrained DOF in the set Q_α, into the *data* matrix $\{b\}$ in equation (1.6). The remaining equation (1.2)-defined DOF populate the column matrix $\{Q\}$ in equation (1.6), the *exactly* correct order algebraic equation system for determination of the *unknown* DOF defined in equation (1.3).

1.4 Decisions on Forming WS^N

The key to weak form utility is the *assumption* that the integrals in WS^N, equation (1.6), can be evaluated. This obviously centers on the functional form selected for the test and trial function spaces. These decisions in turn identify a specific algorithm from the wide range of WS^N methods that can be derived. The following table provides a WS^N summary *essence* categorized on these function sets.

$\Phi_\beta(\mathbf{x}),\Psi_\alpha(\mathbf{x})$	Examples	WS^N Label
Global	Sine, cosine, Bessel, spherical harmonics	Analytical methodology (separation of variables)
	Chebyshev polynomials	Spectral methods
Global–local	Chebyshev by blocks	Pseudospectral methods
Local	Lagrange polynomials	Spatially discrete methods (FE, FV, FD)

For $\Phi_\beta(\mathbf{x})$ and $\Psi_\alpha(\mathbf{x})$ spanning the entirety of Ω generates formulations closely associated with analytical PDE-solution methodology. However, this choice precludes geometric flexibility, as closures $\partial\Omega$ of the domain Ω must be coordinate surfaces. The singular key attribute is that these spaces contain functions that are indeed orthogonal on Ω. Hence, [Matrix] in equation (1.6) is typically diagonal, which renders the algebraic solution process trivial (recall *separation of variables* in your sophomore calculus class?).

Spectral methodology retains the definition and use of global span function spaces. Pseudospectral methods lie halfway between spectral and spatially discrete algorithms, and both typically inherit the liability that closure segments be coincident with global coordinate surfaces.

For domains with absolutely arbitrarily geometric closure, that is, essentially *all* practical problems, the FE discrete implementation of WS^N, hereon denoted WS^h, *guarantees* the extremum of WF^N, equation (1.5), generates integrals that can be evaluated. This is accomplished by subdividing Ω into the *union* (nonoverlapping sum, symbol $\cup$) of small subdomains, see Figure 1.5. Each subdomain is called a FE, denoted Ω_e, and their union can be manipulated to fit any geometrical shape of the domain closure $\partial\Omega$.

A WS^N can be manipulated to interpret FD and FV methodology as will be illustrated. The formulation distinctions include integrals not being generated via calculus and the resultant algorithms are not predictable *optimal*, the key attribute of a specific FE implementation WS^N, detailed shortly.

This process of subdividing Ω into the *union* of small subdomains is called spatial *discretization*, symbolized in the literature by superscript h. Unambiguously then

$$\Omega \approx \Omega^h \equiv \cup_e \Omega_e \qquad (1.7)$$

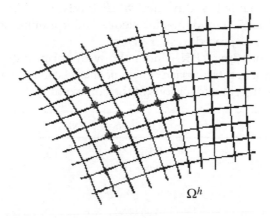

Ω^h

Figure 1.5 Domain discretization

and the region Ω_e is called a *FE*. The resultant FE solution approximation form (1.3) transitions to

$$q^N(\mathbf{x}) \equiv q^h(\mathbf{x}) = \cup_e q_e(\mathbf{x})\}. \tag{1.8}$$

Constructing the required discrete equivalents of $\Psi_\alpha(\mathbf{x})$ and $\Phi_\beta(\mathbf{x})$, equations (1.3) and (1.5), generates the trial and test space *basis functions*. The theoretical foundation is typically Lagrange or Hermite interpolation polynomials, and FE basis functions are herein symbolized as the column matrix $\{N(\mathbf{x})\}$. Hence, for equation (1.8)

$$q_e(x) = \{N(x)\}^T \{Q\}_e. \tag{1.9}$$

With equations (1.7)–(1.9) the WS^h-generated [Matrix] in equation (1.6) is *never* diagonal. Hence, one must find an algebraic solution replacement for Cramer's rule, which introduces iterative matrix *linear algebra* methodology.

A *fundamentally significant* solution facet results upon making the discrete approximation decision. In the FE implementation, the DOF $\{Q\}_e$ in the element-level approximation equation (1.9), that is, select DOF Q_α in equation (1.3), are usually generated *only* at mesh intersections on Ω^h. Illustrated in Figure 1.5 as dots (•) they are called the *nodes* of the mesh.

The union of the local solutions $q_e(\mathbf{x})$ forms $q^h(\mathbf{x})$, equation (1.8), with the resultant *spectral resolution* controlled by node-separation distance. Specifically, for a mesh of *measure* Δx any $2\Delta x$ wavelength information *cannot* be resolved. This is clearly illustrated in Figure 1.6; on the left the DOF $\{Q\}$ for the $2\Delta x$ sine waves are all zero (!) while those on the right are nonzero. Hence *mesh resolution* is central to accuracy; a too coarse mesh can produce totally wrong solutions, as will be illustrated.

1.5 Discrete WS^h Implementations

Legacy FD and FV methods also employ a domain *discretization* $\Omega^h = \cup_c \Omega_c$, where Ω_c is a computational *cell*. Further, mathematicians and chemical engineers (in particular) have developed many node-based numerical methods, for example, *collocation, least squares, weighted residuals*. Do fundamental theory underpinnings exist for these apparently very diverse discrete procedures for PDEs?

The answer is a resounding *YES!!* Under the weak form umbrella, the distinctions reside strictly in the test and trial space *basis functions* chosen to form WS^h. The following table summarizes algorithm decisions in the context of WS^h.

Name	Trial space, $\Psi_\alpha(x)$	Test space, $\Phi_\beta(x)$
Galerkin (FE)	Basis $\{N\}$	Basis $\{N\}$
Collocation	Basis $\{N\}$	Kronecker δ
Finite difference (FD)	?	None
Finite volume (FV)	?	Unity
Least squares	Basis $\{N\}$	$\mathcal{L}(\{N\})$
Boundary element (BEM)	Basis $\{N\}$	Green's function

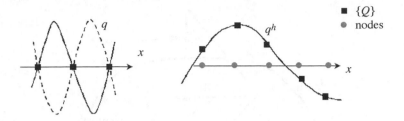

Figure 1.6 Resolution illustrations on a mesh of measure Δx

The fact that myriad choices exist, and have been computer implemented, immediately raises the *fundamental* question:

Does an optimal choice for the WS^N trial and test space function sets $\Psi_\alpha(x)$ and $\Phi_\beta(x)$ exist?

One must first define *optimal* to answer this. Mathematicians will work this to the point of distraction but engineers are not so burdened. Their obvious choice is *the* selection that produces the absolute *minimum* approximation error $e^N(x)$, equation (1.4).

Importantly, this answer must be and *is* absolutely independent of the particular choice for discrete implementation! For a wide range of PDEs describing problem statements in the engineering sciences, in the *continuum* the answer to the fundamental question is:

The WS^N approximation error $e^N(x)$ is minimized, in a suitable norm, when the trial and test spaces are identical.

Now moving to the WS^N spatially discrete implementation WS^h, on a mesh Ω^h, the approximation error becomes $e^h(x) \equiv q(x) - q^h(x)$. Thereby, the WS *discrete* implementation corollary for optimal performance is:

The WS^h approximation error $e^h(x)$ is minimized, in a suitable norm, when the trial and test space replacements contain the identical FE trial space basis functions.

The name historically attached to identical trial and test space functions is *Galerkin*; herein this form of WS^N is denoted GWS^N. The Galerkin criterion is *optimal in theory and in the FE discrete implementation.*

The vector "cartoon" in Figure 1.7 serves to illustrate that the exact solution q cannot lie in the "plane" containing the trial function set $\Psi_\alpha(x)$ supporting q^N unless they are identical (not likely!). The "distance" between q and q^N is the error e^N and its "magnitude" is the smallest when e^N is *orthogonal* (mathematically perpendicular) to the plane, as induced by the Galerkin criterion $q^N(GWS^N)$. The solution $q^N(WS^N)$ generated by any other trial/test function criterion produces the error $\bar{e}^N$ which is not

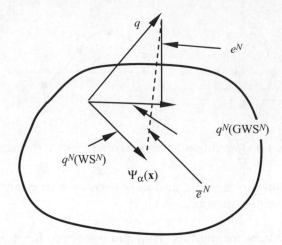

Figure 1.7   Vector cartoon illustrating GWS^N optimality

orthogonal to the trial/test function plane, hence the (dashed) error vector $\bar{e}^N$ possesses a larger "magnitude."

1.6 Chapter Summary

The FE discrete implementation of GWS^N, that is, GWS^h, is the *optimal* decision for a wide class of problem statements in the CES. FE trial space basis availability guarantees accurate evaluation of the integrals defined in equation (1.5) for domains Ω enclosed by an arbitrarily nonregular shaped boundary $\partial\Omega$. With this theory in place, the key topic becomes identification of the *trial space basis functions* spanning FE domains Ω_e on n-dimensions. Their identification and performance quantification across the *CES* is the subject of this text.

In summary, the GWS^N ⇒ GWS^h process *essence* for designing optimally performing algorithms is:

$$\text{approximation: } q(\mathbf{x}) \cong q^N(\mathbf{x}) \equiv \sum_{\alpha}^{N} \Psi_\alpha(\mathbf{x}) Q_\alpha$$

$$\text{error extremization: } \text{GWS}^N \equiv \int_{\Omega} \Psi_\beta(\mathbf{x}) \mathcal{L}(q^N) d\tau \equiv 0$$

$$\text{discretization: } \Omega \Rightarrow \Omega^h = \cup_e \Omega_e$$

$$\text{FE construction: } q^N \Rightarrow q^h \equiv \cup_e \{N(\mathbf{x})\}^T \{Q\}_e$$

$$\text{FE implementation: } \text{GWS}^h \Rightarrow \sum_e \{WS\}_e = \{0\}$$

$$\text{linear algebra: } [\text{Matrix}]\{Q\} = \{b\}$$

$$\text{error estimation: solution-adapted } \Omega^h \text{ refinement}$$

References

[1] Oden, J.T. and Reddy, J.N. (1976) *An Introduction to the Mathematical Theory of Finite Elements*, Wiley, New York.
[2] Moore, S., Baker, A.J., Dongarra, J., Halloy, C. and Ng, C. (2002) Active Netlib: An Active mathematical software collection for inquiry-based computational science and engineering education. *J. Digit. Inf.*, **2**.
[3] www.mathworks.com/Matlab.
[4] www.comsol.com.
[5] Baker, A.J. (2013) *Optimal Modified Continuous Galerkin CFD*, Wiley, London.

2

Problem Statements:
in the engineering sciences

2.1 Engineering Simulation

The engineer or scientist interested in computer simulations must acquire some dexterity with fundamental mechanics principles to be a *cogent* and competent analyst. This chapter provides a basic refresher on the range of pertinent academic materials traditionally included in undergraduate engineering curricula, which the author assumes is the reader's background.

Fundamentally, computational simulation involves seeking a *solution* to one or more usually *nonlinear* partial differential equations (PDEs) that stem from basic conservation principles. In the lagrangian point mass perspective, the mathematical statement of these mechanics principles is

$$\text{conservation of mass}: \quad dM = 0, M = \Sigma m_i \tag{2.1}$$

$$\text{Newton's 2nd law}: \quad d\mathbf{P} = \Sigma \mathbf{F}, \mathbf{P} = M\mathbf{V} \tag{2.2}$$

$$\text{thermodynamics}: \quad dE = dQ - dW \tag{2.3}$$

$$\text{thermodynamic process}: \quad dS \geq 0 \tag{2.4}$$

In equations (2.1) and (2.2), m_i denotes the ith point mass, M is total mass of a particle system, $\mathbf{V}$ the system velocity, and $\mathbf{F}$ denotes the applied (external) forces. Equations (2.3) and (2.4) state the first and second laws of thermodynamics where E is the total system energy, Q the heat added, W the work done by the system, and S the entropy. The undergraduate academic courses in statics, dynamics, strength of materials, thermodynamics, fluid mechanics, thoroughly cover the pertinent developments [1–3]. Do you recall your exposure to this material?

Finite Elements ⇔ Computational Engineering Sciences, First Edition. A. J. Baker.
© 2012 John Wiley & Sons, Ltd. Published 2012 by John Wiley & Sons, Ltd.

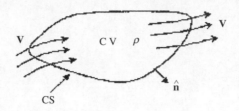

CS

Figure 2.1   Control volume for Reynolds transport theorem

2.2 Continuum Mechanics Viewpoint

In practice, engineers almost never deal with the conservation principles written in the Lagrangian form. Instead, transition to the *continuum* (eulerian) description is made, wherein one assumes there are so many mass points per characteristic volume V (not boldface!) that a *density* function ρ can be defined, that is,

$$\rho(x,t) = \lim_{V \to 0} \frac{1}{V} \sum_i m_i. \tag{2.5}$$

One then defines a control volume CV, with encompassing control surface CS, see Figure 2.1, and transforms the conservation principles from the lagrangian to the eulerian viewpoint via *Reynolds transport theorem* which states, [4]

$$d() \Rightarrow D() \equiv \frac{\partial}{\partial t} \int_{cv} (\cdot)d\tau + \oint_{cs} (\cdot)\mathbf{V} \cdot \hat{n}\, d\sigma. \tag{2.6}$$

The bracket $(\cdot)$ in equation (2.6) contains an appropriate conservation principle variable, $\mathbf{V}$ is the *velocity* vector field traversing CV, $\mathbf{V} \cdot \hat{n}$ signifies the net *efflux* of $(\cdot)$ across CS, and $d\tau$ and $d\sigma$ are the volume and surface differential elements.

One immediately notes the eulerian viewpoint involves transition to multidimensional vector differential and integral calculus! You will be called upon to regain dexterity with these fundamental undergraduate academic subjects.

2.3 Continuum Conservation Principle Forms

Reynolds transport theorem enables deriving very precise conservation principle statements for *continuum* systems, for example, structural mechanics, fluid dynamics, heat/mass transport, electromagnetics, mechanical vibrations, and so on. For the generic control volume CV with enclosing surface CS, mass conservation, Newton's law and the first law of thermodynamics convert to [4]

$$DM = \frac{\partial}{\partial t} \int_{cv} \rho d\tau + \oint_{cs} \rho \mathbf{V} \cdot \hat{n} d\sigma = 0. \tag{2.7}$$

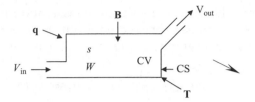

Figure 2.2 Unidirectional CV: CS example

$$\text{DP} \Rightarrow \frac{\partial}{\partial t} \int_{cv} \rho V d\tau + \oint_{cs} V \rho V \cdot \hat{n} d\sigma = \int_{cv} \rho B d\tau + \int_{cs} T d\sigma. \tag{2.8}$$

$$\text{DE} \Rightarrow \frac{\partial}{\partial t} \int_{cv} \rho e d\tau + \oint_{cs} (e + p/\rho) \rho V \cdot \hat{n} d\sigma = \int_{cv} s d\tau + \int_{cs} (W - q \cdot \hat{n}) d\sigma. \tag{2.9}$$

Observe the eulerian "fill in" of the right-hand sides of DP and DE with $\sum F \Rightarrow$ body force **B** (gravity) + surface tractions **T**, and $dQ - dW \Rightarrow$ heat added s, heat efflux $q \cdot \hat{n}$, and work done W.

The continuum expressions (2.7) and (2.9) are the most general eulerian frame integrodifferential equations. However, they are rarely used except for elementary unidirectional control volume analyses, as developed in the first fluid mechanics course, see Figure 2.2.

One can readily develop the more practical expressions for computational simulation by assuming the control volume CV is *stationary* (not moving in time) and invoking the *divergence theorem* for the surface integrals [4]. For example, in equation (2.7)

$$\oint_{cs} \rho V \cdot \hat{n} d\sigma = \int_{cv} \nabla \cdot \rho V d\tau \tag{2.10}$$

which introduces the gradient (vector) derivative ∇. The expanded form in rectangular cartesian coordinates with unit vector triad $\hat{i}, \hat{j}, \hat{k}$ is

$$\nabla = \hat{i} \frac{\partial}{\partial x} + \hat{j} \frac{\partial}{\partial y} + \hat{k} \frac{\partial}{\partial z}. \tag{2.11}$$

The vector inner (*dot*) product in equation (2.10) thus generates the directional derivative of the velocity vector. Assuming resolution of **V** into the scalar cartesian components u, v, w parallel to $\hat{i}, \hat{j}, \hat{k}$, expanding this term in equation (2.10) produces the *partial derivatives*

$$\nabla \cdot \rho V = \frac{\partial \rho u}{\partial x} + \frac{\partial \rho v}{\partial y} + \frac{\partial \rho w}{\partial z}. \tag{2.12}$$

This serves to illustrate that dexterity with vector differential calculus is an asset in developing computational engineering sciences problem statement definitions.

Using the divergence theorem, the eulerian conservation law integrodifferential equation system (2.7)–(2.9) can be expressed uniformly as integrals on (stationary) CV which vanish identically. For example, conservation of mass becomes

$$DM \Rightarrow \int_{cv} \left(\frac{\partial \rho}{\partial t} + \nabla \cdot \rho \mathbf{V} \right) d\tau = 0 \tag{2.13}$$

and the only way this expression can hold in general for stationary CV is that the *integrand* must identically vanish. Using this process, the conservation principles DM, DP, and DE become re-expressed as the *coupled nonlinear* PDE system

$$DM: \quad \frac{\partial \rho}{\partial t} + \nabla \cdot \rho \mathbf{V} = 0. \tag{2.14}$$

$$DP: \quad \frac{\partial \rho \mathbf{V}}{\partial t} + \nabla \cdot \rho \mathbf{V} \mathbf{V} = \rho \mathbf{g} + \nabla \mathbf{T}. \tag{2.15}$$

$$DE: \quad \frac{\partial \rho e}{\partial t} + \nabla \cdot (\rho e + p) \mathbf{V} = s - \nabla \cdot \mathbf{q}. \tag{2.16}$$

2.4 Constitutive Closure for Conservation Principle PDEs

The eulerian continuum viewpoint PDE system (2.14)–(2.16) is valid for any medium with mass density (distribution) ρ. So how does one get expressions specific to structures, fluids, heat transfer, reacting systems, electromagnetics, and so on? The answer lies in definition of *constitutive* closure models for the terms residing on the right-hand sides of equations (2.15) and (2.16).

The common example of a constitutive closure model is the *Fourier* conduction law that defines the heat flux vector $\mathbf{q}$ in terms of thermal conductivity k, a material property, and the gradient of temperature T. The form is

$$\mathbf{q} = -k \nabla T \tag{2.17}$$

where the minus sign recognizes that heat "flows" in the direction opposite to the temperature gradient. Therein, k depends on the material present, for example, aluminum, asbestos, and is usually at least mildly temperature-dependent.

Next, assume the continuum is a solid hence $\mathbf{V} = 0$ in equation (2.16). Then $e \equiv c_v T$ defines mass-specific internal energy in terms of material-dependent volumetric specific heat c_v and temperature T. Substituting yields the highly recognized *unsteady heat conduction* PDE form of DE

$$DE: \quad \frac{\partial T}{\partial t} = \kappa \nabla^2 T + s. \tag{2.18}$$

In forming equation (2.18), the assumed-constant continuum material properties are combined into the *thermal diffusivity* $\kappa = k / \rho c_v$, a function only of the material in the

PDE domain. If temperature dependence is important then $\kappa\nabla^2 T$ must be replaced with the vector derivative dot product $\nabla \cdot (\kappa\nabla T)$ to handle the induced thermal diffusivity nonlinearity.

The constitutive closure model (2.17) has introduced the higher order *laplacian* differential operator ∇^2 into DE, equation (2.18). This is the common occurrence for eulerian conservation principles $D(\cdot)$, since closure models are usually of force–flux form.

The companion formulation for the surface traction vector **T** in DP is more detailed, since $\nabla\mathbf{T}$ in equation (2.15) defines a *tensor* product calculus operation. For both fluids and solids, **T** is replaced by the dot product of the local *stress tensor* with the outwards pointing unit vector $\hat{\mathbf{n}}$, Figure 2.3.

In solid mechanics, stress is then related to strain (displacement/unit length) via *Hooke's law*. For a linear elastic material this introduces the *elastic modulus* E and *Poisson ratio* v for the material present, cf. [5].

Assuming $n=2$ (two dimensional, to keep it simple), stationary and steady state ($\mathbf{V}=0=\partial/\partial t$), the *left-hand* side of equation (2.15) vanishes identically and DP becomes the PDE pair

$$\text{DP}: \quad \frac{\partial\sigma_x}{\partial x} + \frac{\partial\tau_{xy}}{\partial y} + \rho g_x = 0, \quad \frac{\partial\tau_{xy}}{\partial x} + \frac{\partial\sigma_y}{\partial y} + \rho g_y = 0. \tag{2.19}$$

Further assuming the material is homogeneous linear elastic, the $n=2$ matrix form for Hooke's law relating stress and strain is

$$\begin{pmatrix} \sigma_x \\ \sigma_y \\ \tau_{xy} \end{pmatrix} = \frac{E}{1-v^2} \begin{bmatrix} 1 & v & 0 \\ v & v & 0 \\ 0 & 0 & (1-v)/2 \end{bmatrix} \begin{pmatrix} \varepsilon_x \\ \varepsilon_y \\ \gamma_{xy} \end{pmatrix}. \tag{2.20}$$

Now invoking the strain-displacement (kinematics) definitions $\varepsilon_x = \partial u/\partial x$, $\gamma_{xy} = \frac{1}{2}(\partial u/\partial y + \partial v/\partial x)$, equations (2.19) and (2.20) transform to the coupled laplacian

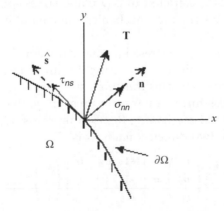

Figure 2.3 Surface traction vector resolution

PDE system on the cartesian resolution of the displacement vector $\mathbf{u} = u(x,y)\hat{\mathbf{i}} + v(x,y)\hat{\mathbf{j}}$ as

$$
\text{DP}: \qquad
\begin{aligned}
\nabla^2 u - \frac{1}{1-2\upsilon}\frac{\partial}{\partial x}\left(\frac{\partial u}{\partial x} + \frac{\partial v}{\partial y}\right) + \frac{\rho g_x}{G} = 0 \\[2mm]
\nabla^2 v - \frac{1}{1-2\upsilon}\frac{\partial}{\partial y}\left(\frac{\partial u}{\partial x} + \frac{\partial v}{\partial y}\right) + \frac{\rho g_y}{G} = 0
\end{aligned}
\qquad (2.21)
$$

where $G = E/2(1-\upsilon)$ is the (material-dependent) *shear modulus*.

The n-dimensional *vector* PDE for linear plane stress/plane strain is

$$
\text{DP}: \qquad \mathcal{L}(u) = -\nabla^2 \mathbf{u} - f(\upsilon)\nabla(\nabla \cdot \mathbf{u}) - \mathbf{b} = 0 \qquad (2.22)
$$

for $f(\upsilon) \equiv (1+\upsilon)/(1-\upsilon)$ *or* $1/(1-2\upsilon)$, respectively, and $\mathbf{b}$ contains $\mathbf{g}$, ρ, and G. The key point is that the laplacian operator again dominates DP via the constitutive closure model.

An alternative approach, more conventional for setting up structural mechanics analyses, balances the internal and external work terms W in the *energy principle DE*, equation (2.9). Assuming all other terms vanish identically yields the *Principle of Virtual Work*, [5], an elementary form of which traditionally appears as an appendix in a deformable solid mechanics text, for example [2].

The resultant n-dimensional form of DE is

$$
\text{DE}: \qquad \Pi = \int de = \frac{1}{2}\int_\Omega d\,\mathrm{vol}\int_0^\varepsilon \sigma_{ij}d\varepsilon_{ij} - \int_{\partial\Omega} \mathbf{u}\cdot\mathbf{T}d\mathrm{surf}. \qquad (2.23)
$$

Assuming validity of Hooke's law, the equivalent matrix integrodifferential form is

$$
\text{DE}: \qquad \Pi = \int_\Omega \left(\frac{1}{2}\{\varepsilon\}^T[\mathbf{E}]\{\varepsilon\} - \{\mathbf{u}\}^T\{\mathbf{B}\}\right)d\mathrm{vol} - \int_{\partial\Omega}\{\mathbf{u}\}^T\{\mathbf{T}\}d\mathrm{surf} \qquad (2.24)
$$

where $[\mathbf{E}]$ is the n-dimensional generalization of the linear Hooke's law matrix (2.20), and $\{\mathbf{B}\}$ and $\{\mathbf{T}\}$ contain the body force and surface traction contributions. Indeed, equations (2.22) and (2.24), restricted to 2-D plane stress/plane strain, are formally equivalent expressions for structural mechanics static equilibrium! It's not very obvious is it?

Switching to *fluid mechanics*, the stress kinematic relationship to strain transcends to *strain-rate*, that is, *velocity* $\mathbf{V}$, since a fluid cannot withstand a static shear stress. The counterpart to the Hooke's law *constitutive* closure model is Stokes "viscosity" law, which segregates the stress formulation from pressure p, the (negative) equivalent of the normal stress (σ) of solid mechanics. Reinserting pressure, the matrix form analogous to equation (2.20) for two-dimensional flow is

$$
\begin{pmatrix} \sigma_x \\ \sigma_y \\ \tau_{xy} \end{pmatrix} =
\begin{bmatrix} -1 & \upsilon\frac{\partial}{\partial x} & 0 \\ -1 & 0 & \upsilon\frac{\partial}{\partial y} \\ 0 & \upsilon\frac{\partial}{\partial y} & \upsilon\frac{\partial}{\partial x} \end{bmatrix}
\begin{pmatrix} p \\ u \\ v \end{pmatrix}, \qquad (2.25)
$$

where υ is now the *kinematic viscosity* of the fluid.

For the *incompressible* flow assumption ρ is a constant. Hence DM, equation (2.14), states that the velocity vector **V** must be divergence-free

$$DM: \quad \nabla \cdot \mathbf{V} = 0. \tag{2.26}$$

Then assuming steady flow and neglecting gravity, the *incompressible n = 2* form for DP becomes the explicitly *nonlinear* laplacian PDE system

$$DP: \quad v\nabla^2 u - \frac{\partial uu}{\partial x} - \frac{\partial uv}{\partial y} - \frac{1}{\rho_0}\frac{\partial p}{\partial x} = 0 \tag{2.27}$$

$$v\nabla^2 v - \frac{\partial uv}{\partial x} - \frac{\partial vv}{\partial y} - \frac{1}{\rho_0}\frac{\partial p}{\partial x} = 0, \tag{2.28}$$

where u, v is the cartesian scalar resolution of velocity vector **V**.

Equations (2.26)–(2.28) constitute the "famous" *Navier–Stokes* PDE system addressing laminar 2-D incompressible steady flow. The *differential constraint* of DM, equation (2.26), renders this nonlinear PDE system very challenging to "solve" computationally.

This is totally eased upon the further assumption that the velocity field is *irrotational* which requires that the curl of the velocity vector **V** also vanish. Therefore, for $\nabla \times \mathbf{V} \equiv 0$ then $\mathbf{V} = -\nabla\Phi$ is guaranteed via vector field theory, [6], and Φ is termed a *potential function*.

Substituting the potential definition into DM, equation (2.26) produces

$$DM: \quad \nabla \cdot \mathbf{V} = 0 = -\nabla \cdot \nabla\Phi = -\nabla^2\Phi \tag{2.29}$$

a *linear* laplacian PDE. Note that no material property is involved, hence conservation of mass DM for *all* incompressible fluids undergoing irrotational flow is independent of the fluid present.

In summary, a laplacian PDE results throughout engineering science problem statements when a constitutive closure model is coupled with a conservative principle. As a further example, *Maxwell's equations* express conservation-constitutive principles in continuum *electromagnetics*. The simplified case of plane wave propagation in an *optically perfect* medium results in DE appearing as

$$DE: \quad \nabla^2\phi - \omega^2\phi = 0, \tag{2.30}$$

where ϕ is the electric potential (volts) and ω the frequency (rad/s). For *mass transport*, for example, grain drying, dialysis, and so on, DM for the distribution of a mass fraction ϕ for an (implicitly nonlinear) source s is

$$DM: \quad \nabla^2\phi + s(\phi) = 0. \tag{2.31}$$

The *creeping flow* of fluid through a saturated aquifer is characterized by the explicitly nonlinear laplacian PDE

$$DM: \quad \nabla \cdot \phi \nabla \phi = 0, \tag{2.32}$$

where ϕ is the flow potential function.

2.5 Engineering Science Continuum Mechanics

From this brief reprise, you hopefully realize that the union of conservation principles with constitutive modeling defining embedded continuum material behavior confirms:

- engineering problem statements are PDEs
- the highest derivative is typically the laplacian ∇^2
- the closure model coefficients are material dependent
- the PDE may have nonlinear terms
- the PDE may have an unsteady and/or source term

Perhaps you recall from your second-year calculus course that solving a laplacian type PDE requires one to define conditions for the unknown *everywhere* on the bounding surface (CS $\Rightarrow \partial\Omega$) enclosing the region of space (domain, CV $\Rightarrow \Omega$) where the PDE holds, Figure 2.4. It is this *problem-specific* boundary condition (BC) *data* that communicates the "outside world" to the problem at hand. Admissible imposed BCs, $l(q)$ in Figure 2.4, for a laplacian PDE include

- fixed level for unknown (Dirichlet)
- fixed flux of unknown (Neumann)
- solution-dependent flux of unknown (Robin)

and the three types may be variously mixed on the entirety of $\partial\Omega$.

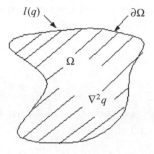

Figure 2.4 Engineering science PDE + BCs statement cartoon

To summarize, the *computational engineering sciences* constitute establishing and implementing mathematical constructions to generate *approximate solutions* to PDE systems with BCs. The legacy discrete methodology was based on *finite difference* (FD) constructions, wherein terms in the PDE were *replaced* by divided differences via *Taylor series*. This constituted a return to the historical *delta* process to express "*change.*" Sir Isaac Newton's invention of *the calculus*, via the *limit*, formalized the mathematics of change centuries ago.

The finite element (FE) spatially discrete implementation of the mathematician's *weak form* extremum (the weak statement, WS) returns calculus to the computational engineering sciences formulation process. More importantly, appropriate discrete WS implementations encompass essentially *all* historical theorizations (FD and *finite volume*, FV) in an elegant, theoretically rich formal process.

The purpose of this text is to concisely develop WS discrete FE implementations for the engineering sciences with ample exposure to coding practice and performance testing against theory. The resultant assimilated knowledge base will well serve the reader in design of optimally performing algorithms for *genuine real-world* applications.

References

[1] Shames, I.H. (1966) *Engineering mechanics*, 2nd edn, vol. **I**, *Statics*, vol. II, *Dynamics*, Prentice-Hall, Englewood Cliffs, NJ.

[2] Shames, I.H. (1964) *Mechanics of Deformable Solids*, Prentice-Hall, Englewood Cliffs, NJ.

[3] White, F.M. (1999) *Fluid Mechanics*, 3rd edn, McGraw-Hill, New York, NY.

[4] Shames, I.H. (1982) *Mechanics of Fluids*, 2nd edn, McGraw-Hill, New York, NY.

[5] Gere, J.M. and Timoshenko, S.P. (1997) *Mechanics of Materials*, DWS Publishing, Boston, MA.

[6] Kreyszig, E. (1967) *Advanced Engineering Mathematics*, Wiley, New York, NY.

3

Some Introductory Material: PDEs, BCs, solutions, discrete concepts

3.1 Example Linear Heat Conduction Solutions

The laplacian partial differential equation (PDE), including a time and source term, typifies a problem statement in the computational engineering sciences. The classic example is linear heat conduction with temperature distribution the solution to the PDE

$$\frac{\partial T}{\partial t} = \kappa \nabla^2 T + s \tag{3.1}$$

for T the temperature, κ the material-dependent thermal diffusivity, and s a source term.

The $n = 1$ steady form of equation (3.1) provides the chance to easily construct some example solutions. Hence, find the function $T = T(x,s,\kappa)$ that is the solution to

$$\frac{d^2 T}{dx^2} + \frac{s}{\kappa} = 0 \tag{3.2}$$

for the three source definitions

$$s/\kappa = \text{constant} = C$$
$$s/\kappa = \text{polynomial} = \sum_i c_i x^i$$
$$s/\kappa = \text{Fourier expansion} = \sum_n c_n \sin\left(\frac{n\pi x}{l}\right)$$

Finite Elements ⇔ Computational Engineering Sciences, First Edition. A. J. Baker.
© 2012 John Wiley & Sons, Ltd. Published 2012 by John Wiley & Sons, Ltd.

for the Dirichlet (fixed) boundary conditions (BCs)

$$T(x = x_L) = T_L$$
$$T(x = x_R) = T_R$$

(3.3)

imposed at the Left and Right ends of the solution domain Ω of span $l = x_R - x_L$.

For the constant source C simply integrate equation (3.2) twice producing

$$T(x) = \frac{Cx^2}{2} + ax + b$$

(3.4)

where a and b are the unknown *constants of integration*. These are then determined (an exercise) by inserting equation (3.3) into equation (3.4) and solving the resultant 2×2 algebraic equation system for $a,b = f(T_L, T_R, l)$.

For the polynomial series source, again simply integrate equation (3.2) twice which yields

$$T(x) = \frac{1}{(i+1)(i+2)} \sum_i c_i x^{i+2} + ax + b.$$

(3.5)

The integration constants a, b are again evaluated from the BCs, equation (3.3), a process made easier by translating the x-axis origin in equation (3.5) to x_L (an exercise).

For the Fourier series source, and realizing how trigonometric functions differentiate, the direct approach is to assume a *trial solution* for $T(x)$ of the form

$$T(x) = \sum_m A_m \sin\left(\frac{m\pi\bar{x}}{l}\right) + a\bar{x} + b$$

(3.6)

in the translated origin coordinate system $\bar{x} = x - x_L$. Evaluating a,b in terms of the BCs, equation (3.3), and the coefficient set A_m in terms of c_n, is a suggested exercise, whence you will determine that m must be equal to n.

3.2 Multidimensional PDEs, Separation of Variables

The linear, steady one-dimensional (degenerate) laplacian (3.1) is indeed quite easy to solve analytically using classical integral calculus. For equation (3.1) containing dependence on more than one independent variable the classical analytical solution process is called *separation of variables* (SOV) [1]. This fundamentally involves assuming some form of a *trial solution*, as in equation (3.6), then matching coefficients of the expansion, recall equation (1.3).

Let us seek the solution to the *unsteady* $n = 1$ form of equation (3.1)

$$\frac{1}{\kappa}\frac{\partial T}{\partial t} = \frac{\partial^2 T}{\partial x^2}$$

(3.7)

for the Dirichlet BCs $T(x=0, t)=0=T(x=l, t)$ and the initial condition (IC) $T(x, t=0) \equiv f(x)$. The SOV process starts with a trial solution of the form given in the first equality.

$$T(x,t) \equiv F(x)G(t) = \sum_\alpha \Psi_\alpha(x,t)Q_\alpha. \tag{3.8}$$

The second equality in equation (3.8) connects the SOV construction to the weak form fundamental assumption (1.3), with the goal to support exposure of the intrinsic duality.

Substituting the first expression in equation (3.8) into equation (3.7), and then dividing through by FG, as it cannot be identically zero (!) yields

$$\frac{1}{\kappa G}\frac{dG}{dt} = \frac{1}{F}\frac{d^2F}{dx^2} \equiv \pm\beta^2. \tag{3.9}$$

Equating to β^2, an (unknown) constant, is the *only* way arbitrary functions of the independent variables t and x can equal each other. Thus, the SOV assumption modifies the PDE (3.7) into the pair of ordinary differential equations (ODEs) in equation (3.9).

Hence, one returns to the familiar 1-D integral calculus but with the *added twist* that the "source term" involves the unknown, for example,

$$\frac{d^2F}{dx^2} \mp \beta^2 F = 0. \tag{3.10}$$

The example (3.6) reminds one that trigonometric functions sin $(\cdot)$ and cos $(\cdot)$ return themselves when differentiated twice. Hence, a candidate solution *trial function* for F in equation (3.8), for *any* constants A and B, is

$$F(x) = A \sin (\beta x) + B \cos (\beta x) \tag{3.11}$$

provided that the $+$ sign is selected in equation (3.9). Choosing the minus sign replaces sin, cos, in equation (3.11) with sinh, cosh, which leads to the failure of the process (try it!). For the $+$ sign the second ODE in equation (3.9) possesses the analytical solution

$$G(t) = C \exp (-\kappa\beta^2 t) \tag{3.12}$$

for C another arbitrary constant. The negative exponent argument in equation (3.12) additionally confirms that $+\beta^2$ is appropriate, as otherwise $T(x,t)$ would grow unbounded (*blow up*) for large t.

Combining equations (3.11) and (3.12) and folding C into A and B, the *generic* (α^{th}) member of the SOV solution process *trial space* is

$$\Psi_\alpha(x,t)Q_\alpha = [A_\alpha\sin (\beta x) + B_\alpha\cos (\beta x)]\exp (-\kappa\beta^2 t). \tag{3.13}$$

The equality in equation (3.13) indicates the arbitrary constants A_α, B_α, and β must somehow be coalesced into the weak form coefficient set Q_α, equation (3.8) Since every term in equation (3.11) must satisfy the BCs for all time t, these two requirements are

$$A_\alpha \sin(0) + B_\alpha \cos(0) = 0$$
$$A_\alpha \sin(\beta l) + B_\alpha \cos(\beta l) = 0 \tag{3.14}$$

and can be satisfied *only* for $B_\alpha \equiv 0$. Since A_α cannot also vanish (the trivial solution), equation (3.14) generates the requirement $\beta \Rightarrow \beta_n \equiv n\pi/l$ for all *integers* n. Then switching the summation index from α to n, hence A_α to Q_n, the trial function duality defined in equation (3.8) becomes

$$T(x,t) \equiv \sum_\alpha \Psi_\alpha(x,t)Q_\alpha \equiv \sum_n^\infty Q_n \sin\left(\frac{n\pi x}{l}\right)\exp\left(-\kappa\beta_n^2 t\right). \tag{3.15}$$

The solution trial space identification (3.15) worked out this cleanly only because the BCs are homogeneous (equal to zero) Dirichlet. The SOV process "will work" for other BCs, taking guidance from the weak form trial space example (3.6), which is another exercise.

The expansion coefficient set Q_n in equation (3.15) becomes quantified by requiring it to satisfy the given IC. At $t = 0$, the exp $(\cdot)$ term therein becomes unity yielding

$$T(x, t = 0) \equiv f(x) = \sum_n^\infty Q_n \sin\left(\frac{n\pi x}{l}\right). \tag{3.16}$$

Interestingly, note that equation (3.16) is *one* equation for the determination of an *infinite* (!) number of coefficients Q_n. The resolution to this dilemma is the construction of a *Galerkin weak statement* (GWS) for equation (3.16) using identical test and trial function spaces.

Since n therein is a summation index, the test function set must have a unique index, hence select $\sin(m\pi x/l)$ for $1 \leq m \leq \infty$. Thereby, the GWS written in equation (3.16)

$$\text{GWS} \equiv \int_\Omega \sin\left(\frac{m\pi x}{l}\right)f(x)dx - \sum_n^\infty \int_\Omega Q_n \sin\left(\frac{m\pi x}{l}\right)\sin\left(\frac{n\pi x}{l}\right)dx \equiv 0 \tag{3.17}$$

must vanish for *all* integers $m = 1, 2, 3, \ldots, M, \ldots, n = 1, 2, 3, \ldots, N, \ldots$. Hence, equation (3.17) is an algebraic equation *system* defining the determination of the coefficient set Q_n for all n and m.

The practical dilemma remains however, as equation (3.17) defines a doubly-infinite number of equations! But the resolution is again at hand, as asserted in Chapter 1, as *optimal* GWS requires the approximation error to be *orthogonal* to the trial space. This property manifests itself in equation (3.17) in the fact that trigonometric functions are indeed orthogonal on Ω. Thereby, every integral in the second term in equation (3.17) reduces to $\delta_{mn}(l/2)$, with δ_{mn} the *Kronecker delta* which equals zero for all $m \neq n$.

Hence [Matrix] in equation (1.6) for the algebraic equation system (3.17) is *diagonal* and the resultant fully segregated solution process for the range $n \equiv m = 0, 1, 2, \ldots$ is

$$Q_n \equiv \frac{2}{l} \int_0^l \sin\left(\frac{n\pi x}{l}\right) f(x) dx. \tag{3.18}$$

Continuing the illustration, the $n = 3$ steady form of the multidimensional PDE (3.1) is widely present in engineering sciences problem statements. Consider the SOV process for generating a solution to the homogenous laplacian PDE

$$\nabla^2 T = \frac{\partial^2 T}{\partial x^2} + \frac{\partial^2 T}{\partial y^2} + \frac{\partial^2 T}{\partial z^2} = 0 \tag{3.19}$$

for solution domain Ω the unit cube and with Dirichlet BCs on the union of cartesian closure segments $\partial\Omega$ as

$$T(0, y, z) = 0 = T(a, y, z)$$
$$T(x, 0, z) = 0 = T(x, b, z) \tag{3.20}$$
$$T(x, y, z) = 0 \text{ and } T(x, y, c) = f(x, y)$$

The SOV assumption is $T(x,y,z) = X(x)Y(y)Z(z)$. Proceeding as illustrated in equations (3.9) and (3.10), and then enforcing the homogenous BCs leads to the solution *trial space* companion to equation (3.15)

$$\Psi_{n,m}(x) \equiv \sin\left(\frac{n\pi x}{a}\right) \sin\left(\frac{m\pi y}{b}\right) \sinh\left(\pi \beta_{n,m} z\right). \tag{3.21}$$

In equation (3.21) $\beta_{nm} = \text{sqrt}\left[(n/a)^2 + (m/b)^2\right]$ with n, m each the (infinite) set of integers. Restricting equation (3.21) to satisfy the remaining nonhomogeneous BC, then forming the GWS and using orthogonality again generates a diagonal [Matrix] statement with solution

$$Q_{n,m} \equiv \frac{4}{ab} \int_0^a \int_0^b f(x, y) \sin\left(\frac{n\pi x}{a}\right) \sin\left(\frac{m\pi y}{b}\right) dx \, dy. \tag{3.22}$$

These examples clearly illustrate how the *analytical* SOV process leads to the identification of a *solution trial space* $\Psi_\alpha(\mathbf{x}, t)$, the fundamental starting point of a weak formulation. Then, via GWS *optimally*, that is, error orthogonal to $\Psi_\alpha(\mathbf{x}, t)$, a computable determination of the DOF array Q_α identified in equation (1.3) results.

Unfortunately(?), the *analytical* solutions (3.15) with (3.18), (3.21), and (3.22), are intractable since each involves an *infinite* sum, never achievable in practice. However, this is another nonissue easily resolved using the *completeness* attribute of the orthogonal solution trial space functions. This property enables forming an *approximation* T^N to the exact solution T possessing an *estimable error* when truncating the exact solution infinite series at some $n = N$.

Thus for equation (3.15), the approximate solution with estimable error is

$$T(x,t) \approx T^N(x,t) \equiv \sum_{n=1}^{N} Q_n \sin\left(\frac{n\pi x}{l}\right) \exp\left(-\kappa\beta_n^2 t\right). \tag{3.23}$$

As $\sin(n\pi x/l)$ forms a *complete* set, one can *a posteriori* determine the N producing a verifiable *accuracy* level for T^N, cf. [1]. This is the consequence of the monotone convergence property of the sum in equation (3.15). Similarly, the truncations $n \leq N$ and $m \leq M$ in equations (3.21) and (3.22) lead to a verifiably accurate approximation $T^N(\mathbf{x}) \equiv \sum_{\alpha}^{N} \Psi_\alpha(\mathbf{x}) Q_\alpha$ for α identified with n,m and some determinable M,N.

3.3 Mathematical Foundation Essence for GWS^N

The restrictions for the SOV analytical solution process "to work" are severe, that is,

- the PDE must be linear
- the BCs must be separable on coordinate surfaces
- the source term is at least quasilinear

Unfortunately, none of these restrictions hold for *genuine* problem statements in the engineering sciences.

As intimated, the weak form process does embody SOV process *mathematical principles* of merit, specifically:

- solution is a function
- formulation using calculus
- trial space orthogonality/completeness
- verifiable approximation error

and the finite element discrete implementation of the (extremized WF) GWS opens the reach of SOV process mathematical elegance to:

- systems of explicitly nonlinear PDEs
- BCs on arbitrary noncoordinate domain closures
- asymptotic error estimates, verifiable accuracy
- source, time, first derivative, and constitutive nonlinearities
- complete geometric flexibility

and the computationally practical attributes of the GWS^N ⇒ GWS^h process for "solving" such PDEs + BCs systems results from:

defining the trial space supporting the approximation, forming an expansion with the approximation DOF $Q\alpha$, then generating their optimal determination via GWS^N ⇒ GWS^h

Once into the GWS^h, the process of *mesh refinement* corresponds essentially with adding more functions (N larger) to the trial space $\Psi_\alpha(\mathbf{x})$. Theory exists to predict FE solution convergence in terms of *asymptotic error estimates* in PDE-appropriate *norms* (integrals over the domain Ω) as a function of the mesh density. The rate at which convergence occurs typically responds to the completeness (degree) of the FE *trial space basis*. In combination, these weak form theory attributes enable achieving a *verifiable* specified level of accuracy for nonlinear problem statements in the computational engineering sciences.

3.4 A Legacy FD Construction

A brief introduction to finite difference (FD) constructions with their (i,j,k) pictogram *stencils* is appropriate. The FD approach involves constructing *discrete approximations* to terms in the PDE, rather than seeking the solution approximation. A domain discrete geometry supporting basic $n=2$ Taylor series (TS) operations is given as Figure 3.1.

For the first derivative and *mesh measure* (size) Δx, a *forward* TS for $\phi(x,y)$ written parallel to the x-axis is

$$\phi(x+\Delta x, y) = \phi(x,y) + \Delta x \frac{\partial \phi}{\partial x}\bigg|_{x,y} + \frac{\Delta x^2}{2}\frac{\partial^2 \phi}{\partial x^2}\bigg|_{x,y} + O(\Delta x^3)$$

$$\Rightarrow \frac{\partial \phi}{\partial x}\bigg|_{x,y} \cong \frac{\partial \phi}{\partial x}\bigg|_{i,j} = \frac{\phi_{i+1,j} - \phi_{i,j}}{\Delta x} + O(\Delta x) \qquad (3.24)$$

This is a 1st order accurate approximation since the *truncation error* (TE) is (order) $O(\Delta x)$. The companion *backward* TS generates an alternative 1st order approximation

$$\phi(x-\Delta x, y) = \phi(x,y) - \Delta x \frac{\partial \phi}{\partial x}\bigg|_{x,y} + \frac{\Delta x^2}{2}\frac{\partial^2 \phi}{\partial x^2}\bigg|_{x,y} + O(\Delta x^3)$$

$$\Rightarrow \frac{\partial \phi}{\partial x}\bigg|_{i,j} = \frac{\phi_{i-1,j} - \phi_{i,j}}{\Delta x} + O(\Delta x) \qquad (3.25)$$

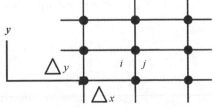

Figure 3.1   Generic FD mesh on Ω, $n=2$

Subtracting equation (3.25) from equation (3.24) will eliminate the $O(\Delta x)$ TE, leading to the 2nd order accurate FD approximation to the first derivative $\frac{\partial \phi}{\partial x}\big|_{i,j}$ on a uniform mesh

$$\frac{\partial \phi}{\partial y}\bigg|_{i,j} = \frac{\phi_{i+1,j} - \phi_{i-1,j}}{2\Delta x} + O(\Delta x^2). \tag{3.26}$$

Hence, by cartesian extension

$$\frac{\partial \phi}{\partial y}\bigg|_{i,j} = \frac{\phi_{i,j+1} - \phi_{i,j-1}}{2\Delta y} + O(\Delta y^2). \tag{3.27}$$

Multidimensional FD derivative approximations such as equations (3.26) and (3.27) are often expressed as a *stencil*, a geometric *pictogram* of an FD expression, for example,

$$\frac{\partial \phi}{\partial x}\bigg|_{i,j} = \frac{1}{2h}(-1 \quad 0 \quad 1) + O(h^2), \quad \frac{\partial \phi}{\partial y}\bigg|_{i,j} = \frac{1}{2k}\begin{pmatrix} -1 \\ 0 \\ 1 \end{pmatrix} + O(k^2) \tag{3.28}$$

using h, k, to replace $\Delta x, \Delta y$.

Second derivative TS approximations, as required for the laplacian, are generated by similar operations. A suggested exercise is to verify that

$$\frac{\partial^2 \phi}{\partial x^2}\bigg|_{i,j} = \frac{1}{h^2}(1 \quad -2 \quad 1) + O(h^2), \frac{\partial^2 \phi}{\partial y^2}\bigg|_{i,j} = \frac{1}{k^2}\begin{pmatrix} 1 \\ -2 \\ 1 \end{pmatrix} + O(k^2). \tag{3.29}$$

Thereby, the uniform ($h = k$) cartesian mesh FD stencil for the $n = 2$ laplacian $\nabla^2 \phi$ sums these two TS expressions yielding

$$\nabla^2\big|_{i,j} = \frac{1}{h^2}\begin{bmatrix} & 1 & \\ 1 & -4 & 1 \\ & 1 & \end{bmatrix} + O(h^2). \tag{3.30}$$

The stencil for $n = 3$ laplacian FD approximation is the obvious extension (an exercise).

In summary, the FD discrete algorithm for "solving" the $n = 2$ laplacian PDE $\mathcal{L}(\phi) = -\nabla^2 \phi - s = 0$ on a *uniform mesh* directly generates an algebraic equation system *stencil*

$$(\nabla^2 \phi + s)_{\text{FD}} \Rightarrow \frac{1}{h^2}\begin{bmatrix} & 1 & \\ 1 & -4 & 1 \\ & 1 & \end{bmatrix}\Phi_{i,j} + S_{i,j} + O(h^2) = 0. \tag{3.31}$$

The placeholder (capital) $\Phi_{i,j}$ is the set of *numbers* sought, located at mesh *nodes* (x_i, y_j), the dots in Figure 3.1, while the $S_{i,j}$ are source term specified data similarly evaluated as $s(x_i, y_j)$.

FD *stencils* provide an accurate, understandable visualization of the interior node discrete algebraic process, but do possess *caveats*. The stencil (3.31) is incorrect for BC-constrained nodes on the laplacian domain Ω boundary, which requires alteration using one-sided operators and/or phantom nodes. For a nonuniform mesh the stencil is much less transparent, loosing its visual appeal, and the TE order invariably increases (a suggested exercise).

The FE discrete implementation of the weak form extremum, in constructing an approximation solution for the laplacian PDE, does not inherit such liabilities. Specifically, the theoretical error estimate replacing TE order holds for arbitrary nonuniform meshing in n-dimensions, and all gradient BC forms are intrinsically embedded, not appended. Both are significant attributes of the weak formulation process to become exposed and validated.

3.5 An FD Approximate Solution

An FD algorithm generates the set of numbers $\Phi_{i,j}$ located at the nodes of the mesh on Ω. Conversely, the FE discrete implementation GWS^h generates a distributed function, the approximate solution. However, both processes generate an algebraic equation, that is,

$$\frac{1}{h^2} \begin{bmatrix} & 1 & \\ 1 & -4 & 1 \\ & 1 & \end{bmatrix} \Phi_{i,j} + S_{i,j} + \text{TE} \Leftrightarrow [\text{Matrix}]\{\Phi\} + \{b\} = \{0\} \qquad (3.32)$$

hence a correspondence must certainly exist.

The "solution" connection is via *interpolation polynomials*. FE trial space basis functions are indeed thus formed, and the FD algorithm numbers $\Phi_{i,j}$ are admissible as polynomial *knot* data. Returning to $n=1$ for visualization, Figure 3.2 illustrates the FD solution coefficient set Φ_k on a partition of the x axis with knots $1 \leq k \leq K$.

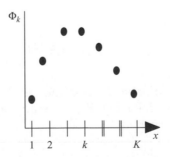

Figure 3.2   FD knot coefficient set Φ_k

A candidate N^{th} degree interpolation polynomial

$$f(x) = a_0 + a_1 x + a_2 x^2 + a_3 x^3 + \ldots = \sum_{\alpha=0}^{N} a_\alpha x^\alpha \tag{3.33}$$

is associated with $N+1$ coefficients a_α. Substituting the K FD algorithm node data Φ_k equation (3.33) becomes

$$f(x_k) = \Phi_k = a_0 + a_1 x_k + a_2 x_k^2 + \ldots a_N x_k^N \text{ for } 1 \leq k \leq K \tag{3.34}$$

which is a matrix statement for determination of the a_α as

$$[\text{Matrix}(x_k^\alpha)]\{a\} = \{\Phi_k\}. \tag{3.35}$$

The elements of $[\text{Matrix}(\cdot)]$ contain α-exponentiations of the knot coordinates x_k and to be *solvable* the matrix rank K must equal $N+1$. Hence, the admissible interpolation polynomial degree must be one less than the number of expansion coefficients a_α. Substituting the solution to equation (3.35) into equation (3.33) generates the FD approximate solution, a distributed function of x, in terms of the algorithm discrete data Φ_k as

$$\phi_{FD}^h(x) = \sum_{\substack{\alpha=0, k-1 \\ k=1, K}} a_\alpha(x_k, \Phi_k) x^\alpha. \tag{3.36}$$

3.6 Lagrange Interpolation Polynomials

While informative, equation (3.36) is an inappropriate construction. The simple power series interpolation (3.33) will generate spurious knot-to-knot oscillations in equation (3.36) for any $K \geq 6$. This constitutes an unacceptable mesh restriction for generation of accurate $\phi_{FD}^h(x)$.

The resolution is replacing the global-span coordinate x with local coordinate support of Lagrange *piecewise-continuous* interpolation polynomials, [2]. In essence, a Lagrange polynomial interchanges the functional locations of Φ_k and x in equation (3.36), that is,

$$\phi^h(x) \equiv \sum_{k=1}^{K} L_K(x, x_k) \Phi_k. \tag{3.37}$$

The defined Lagrange polynomial L_K is of degree K–1, hence requires data at K knots. Recalling the Kronecker delta definition, the functional expression of L_K is easily derived starting with the observation

$$\phi^h(x_j) \equiv \Phi_j = \sum_{k=1}^{K} L_K(x_j, x_k) \Phi_k = \delta_{jk} \Phi_k. \tag{3.38}$$

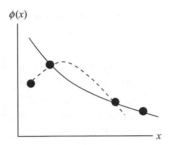

Figure 3.3 Lagrange polynomial nonuniqueness, $K=2$

Hence for the K–1 zeros of δ_{jk} and C an arbitrary constant

$$L_K(x, x_k) = C(x - x_1)(x - x_2)\ldots(x - x_{k-1})(x - x_{k+1})\ldots(x - x_K). \tag{3.39}$$

The single instance of δ_{jk} equal to unity generates

$$C = 1/(x_k - x_1)(x_k - x_2)\ldots(x_k - x_{k-1})(x_k - x_{k+1})\ldots(x_k - x_K). \tag{3.40}$$

Coalescing equations (3.38)–(3.40), the *Lagrange polynomial* of degree K–1

$$L_K(x, x_k) \equiv \frac{(x - x_1)(x - x_2)\ldots(x - x_{k-1})(x - x_{k+1})\ldots(x - x_K)}{(x_k - x_1)(x_k - x_2)\ldots(x_k - x_{k-1})(x_k - x_{k+1})\ldots(x_k - x_K)} \tag{3.41}$$

is amenable to interpolating any function $\phi(x)$ via data Φ_k located at the K *knots* x_k.

Thus equation (3.37) is a preferred functional form supporting an FD algorithm approximate solution, even though the TS process never identified its existence. It must be emphasized this representation is only *piecewise-continuous*. For the simplest linear definition, $K=1$ in equation (3.41), the resultant interpolation will be everywhere unique. For $K > 1$, the interpolation may not be *unique*, as illustrated in Figure 3.3 for $K=2$.

3.7 Chapter Summary

The goal of this chapter was to introduce formalisms about solution processes for PDEs with boundary/initial (BC/IC) condition data. The exposure starts with classical analytical calculus leading to a SOV review to introduce mathematical attributes that underlay weak form elegance. Indeed, the classical concepts of trial space, orthogonality, completeness, and predictable accuracy approximation are each assimilated into the discrete implementation of a weak form extremum.

In distinction to FD methodology, existence of a continuum (*nondiscrete*) approximate solution with trial space universally initializes the weak form process. The FE *trial space bases* utilized in the subsequent *discrete* implementation of (in particular) a GWS^h possess commonality with *multidimensional* Lagrange polynomials of completeness degree K.

The literature notation for an FE basis is the column matrix $\{N_k(\boldsymbol{\eta})\}$, where k (rather than K) indicates completeness degree and $\boldsymbol{\eta}$ is the n-dimensional *local* coordinate system spanning the FE domain Ω_e. Thus, a coordinate transformation from the global $\mathbf{x}$ to local $\boldsymbol{\eta}$ is always required in the discrete implementation process.

Summarily, the FE approximate solution $q^h(\mathbf{x})$ is constituted of the nonoverlapping sum (*union*, denoted $\cup$) of the local *piecewise-continuous* trial space basis construction for each FE domain. Symbolically, the transition from WF extremum to GWS^h is

$$q^N(x) \equiv q^h(x) = \cup_e q_e(x), \quad and \quad q_e(x) \equiv \{N_k(\eta)\}^T \{Q\}_e. \tag{3.42}$$

Analytical completion of the GWS^h calculus process generates the [Matrix] statement, the algebraic solution of which determines the approximation *degrees-of-freedom* (DOF) $\{Q\} = \cup_e \{Q\}_e$ that ultimately define the function $q^h(\mathbf{x})$ via the trial space basis.

Exercises

3.1 Verify the heat-conduction example ODE solutions for given BCs, for the given heat source forms (3.1).

3.2 (Optional, brain-teaser): Solve the 1D unsteady heat conduction PDE, equation (3.7) to determine the temperature distribution in a rod initially at uniform temperature T_1, with the "right" end suddenly increased to T_2 at time t_0.
(Hint: cast the solution process on the difference between the steady-state temperature distribution (linear in x) and the time-evolution of temperature.)

3.3 (Optional, straight forward): Solve the 2D laplacian for ϕ on the rectangular domain $0 < x < a$, $0 < y < b$ for the BCs $\phi(0,y) = 0$, grad $\phi(x,0) \cdot \mathbf{n} = 0 =$ grad $\phi(x,b) \cdot \mathbf{n}$ and $\phi(a,y) = \cos(3\pi y/b)$.

3.4 Complete derivation details for the second order accurate FD approximation for a first derivative, equation (3.25).

3.5 Derive the FD stencil for the first derivative on a nonuniform mesh, hence determine the specific form of the truncation error.

3.6 Derive the uniform mesh FD stencil for the Laplacian in 2D and 3D, cf. (3.26).

3.7 Work through, hence verify equation (3.31) as the Lagrange interpolation polynomial of degree K. Comment on the potential for interpolation nonuniqueness for $K > 1$.

References

[1] Kreyszig, E. (1967) *Advanced Engineering Mathematics*, Wiley, New York, NY.
[2] Jeffrey, A. (1969) *Mathematics for Engineers and Scientists*, Barnes & Noble, Boca Raton, FL.

4

Heat Conduction:
an FE weak statement tutorial

4.1 A Steady Heat Conduction Example

A *weak formulation* (WF^N) organizes construction of a computer-enabled algorithm for *any* problem statement in the engineering sciences via the *systematic* set of decisions identified in Chapter 1. This chapter thoroughly details the process $\text{WF}^N \Rightarrow \text{WS}^N \Rightarrow \text{GWS}^N \Rightarrow \text{GWS}^h \Rightarrow e^h$ for an elementary steady-state one-dimensional $\text{DE} + \text{BCs}$ statement.

Consider the slab spanning $a \leq x \leq b$, Figure 4.1. The continuum DE conservation principle statement for *state variable* $T(x)$ is

$$\text{DE}: \mathcal{L}(T) = -\frac{\mathrm{d}}{\mathrm{d}x}\left(k\frac{\mathrm{d}T}{\mathrm{d}x}\right) - s = 0, \quad \text{on } a < x < b. \tag{4.1}$$

$$\text{BCs}: \ell(T) = -k\frac{\mathrm{d}T}{\mathrm{d}x} - f_n = 0, \quad \text{at } x = a. \tag{4.2}$$

$$T = T_b, \quad \text{at } x = b. \tag{4.3}$$

In equation (4.1) $\mathcal{L}(\cdot)$ is the placeholder for the homogeneous form ($= 0$) of the *differential* equation. Similarly, $\ell(\cdot)$ in equation (4.2) is the (Neumann) boundary condition (BC) constraining the state variable *efflux* from the domain of $\mathcal{L}(\cdot)$. Finally, equation (4.3) is a fixed (Dirichlet) BC for the state variable.

The problem statement *data* is everything defined beforehand. In equations (4.1)–(4.3) the data are a, b, thermal conductivity k, source s, efflux f_n, and fixed temperature T_b.

Finite Elements $\Leftrightarrow$ Computational Engineering Sciences, First Edition. A. J. Baker.
© 2012 John Wiley & Sons, Ltd. Published 2012 by John Wiley & Sons, Ltd.

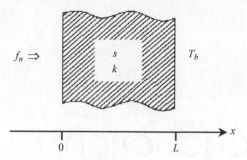

Figure 4.1 DE + BCs tutorial definition

This DE + BCs definition is so elementary that, for data uniformly constant and recalling Chapter 3.1 the analytical solution is readily determined to be (an exercise)

$$T(x, \text{data}) = \frac{sL^2}{2k}\left[1 - \left(\frac{x}{L}\right)^2\right] + \frac{f_n L}{k}\left(1 - \frac{x}{L}\right) + T_b, \tag{4.4}$$

where $L \equiv b–a$ is domain span.

The weak form (WF) theory assertion that equation (1.3), specifically,

$$T^N(x) = \sum_{\alpha=1}^{N} \Psi_\alpha(x) Q_\alpha = Q_1 \Psi_1(x) + Q_2 \Psi_2(x) + \cdots + Q_N \Psi_N(x) \tag{4.5}$$

is a *viable* representation of a (*any!*) solution is clearly illustrated by observing that equation (4.5) contains the analytical solution for $N = 3$ with expansion coefficients and trial space members

$$\begin{aligned} Q_1 &= \frac{sL^2}{2k}, & \Psi_1(x) &\equiv 1 - \left(\frac{x}{L}\right)^2 \\ Q_2 &= \frac{f_n L}{k}, & \Psi_2(x) &\equiv 1 - \left(\frac{x}{L}\right) \\ Q_3 &= T_b, & \Psi_3(x) &\equiv 1 \end{aligned} \tag{4.6}$$

Note that trial space members are functions *only* of (normalized) spatial coordinate, the independent variable, while the expansion coefficients contain all given *data!*

This chapter exercise does not select the Ψ_α in equation (4.6), hence the GWS^N solution $T^N(x)$ cannot be exact. This admits theory development for estimation of the GWS^N *approximation error* e^N, which following FE discrete implementation becomes e^h.

4.2 Weak Form Approximation, Error Extremization

Simply stated, one seeks to generate the *solution approximation* (4.5) possessing the *minimum error* possible. Assuming $T^N(x)$ cannot match the exact solution, approximation *error* $e^N(x)$ is introduced which, recalling equation (1.4), is distributed over the solution domain as is $T^N(x)$. The mathematical statement is

$$T(x) = T^N(x) + e^N(x) \tag{4.7}$$

and no *a priori* knowledge exists about $e^N(x)$ except the desire for it to be an *extremum*.

A measure of error can be determined upon substitution of equation (4.7) into equation (4.1), This confirms $e^N(x)$ satisfies the same differential equation as does $T^N(x)$ with a minus sign. Thus, equation (4.1) written on $T^N(x)$

$$\mathcal{L}(T^N) = -\frac{d}{dx}\left(k\frac{dT^N}{dx}\right) - s \neq 0 \tag{4.8}$$

is the available statement of approximation error.

Since equation (4.8) is not homogeneous, that is, does not equate to zero, a weak form WF written thereon can control the "*size*" of the error $e^N(x)$, recall equation (1.5). Upon handling the WF "for any test function" requirement, recall Chapter 1.3, the resultant weak statement

$$WS^N = \int_\Omega \Phi_\beta(x)\mathcal{L}(T^N)dx \equiv 0, \text{ for all } \Phi_\beta \tag{4.9}$$

defines an equation *system* holding for *any and all* $\Phi_\beta(x)$, the set of functions populating the *test space*.

Recalling Figure 1.5, the *optimal* choice for the $\Phi_\beta(x)$ is identically $\Psi_\alpha(x)$ which renders the error $e^N(x)$ *orthogonal* to the trial space. This is the *Galerkin criterion* for a weak statement, the *optimal* construction herein denoted GWS^N. Inserting this definition into equation (4.9), substituting equation (4.8), and then performing an integration-by-parts (recall this?), which introduces the boundary fluxes, yields

$$GWS^N = \int_\Omega \frac{d\Psi_\beta}{dx} k \frac{dT^N}{dx} dx - \int_\Omega \Psi_\beta s dx - \Psi_\beta k \frac{dT^N}{dx}\bigg|_a^b = 0, \quad \text{for } 1 \leq \beta \leq N \tag{4.10}$$

Now substituting the approximation (4.5) and inserting the Neumann efflux BC (4.2), the WF *theorization* process is *complete* in establishment of the *continuum* Galerkin weak statement

$$GWS^N = \sum_{\alpha=1}^N \left(Q_\alpha \int_\Omega \frac{d\Psi_\beta}{dx} k \frac{d\Psi_\alpha}{dx} dx \right) - \int_\Omega \Psi_\beta s dx$$

$$- \Psi_{Nk}\frac{dT^N}{dx}\bigg|_a^b - \Psi_1 f_n\bigg|_a = 0, \text{ for } 1 \leq \beta \leq N \tag{4.11}$$

4.3 GWS^N Discrete Implementation, FE Trial Space Basis

The GWS^N theory is complete in equation (4.11); all one needs do is identify a *trial space* $\Psi_\alpha(x)$, $1 \le \alpha \le N$, that admits evaluation of the integrals therein to generate the algebraic equation system for computing. Viewing equation (4.11), the only requirement is that every member $\Psi_\alpha(x)$ reside in H^1, the *Hilbert space* of all functions with first-derivative products that are square integrable.

True practical and geometric versatility in GWS^N completion accrues to selecting relatively low-degree polynomial subsets lying in the trial space $\Psi_\alpha(x)$. The essential foundation is interpolation theory, specifically Lagrange and/or Hermite piecewise-continuous polynomials, recall (3.41), and the underlying mathematical rigor is rich [1]. Figure 4.2 clearly illustrates how the *"closeness"* of an interpolation to a smooth function is readily controlled by appropriate spacing of the interpolation polynomial *knots*, denoted X1, X2, ...

These low-degree polynomial subsets in $1 \le n \le 3$ dimensions are manipulated into finite element (FE) *trial space bases*, with the *knots* transitioning to *nodes* of the FE domain where are colocated the approximation degrees of freedom (DOF), the Q_α defined in equation (4.5). Full geometric versatility results and the transition GWS^N ⇒ GWS^h involves precise evaluation of the integrals in equation (4.11) via calculus.

Keeping the development thoroughly elementary, define a uniform two-element discretization of the domain $a < x < b$, Figure 4.3. This generates three *nodes*, labeled X1, X2, X3, hence the solution approximation (4.5) requires three DOF coefficients

$$T^N(x) \equiv \sum_{\alpha=1}^{N=3} \Psi_\alpha(x)Q_\alpha = \Psi_1(x)Q_1 + \Psi_2(x)Q_2 + \Psi_3(x)Q_3. \qquad (4.12)$$

Viewing equation (4.6), the $\Psi_\alpha(x)$ should contain only coordinate data, hence temperature and the effects of *data* lie in the DOF Q_α. Further, as implied by the Kronecker

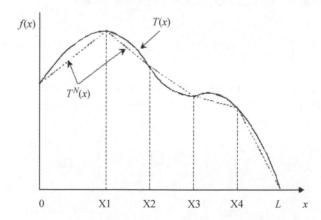

Figure 4.2 Lagrange polynomial interpolation

Figure 4.3 Two-element domain discretization

delta, these Q_α must reduce to the temperature at each node, for example, $T^N(x \Rightarrow a)$ must equate to $Q_{\alpha=1}$, denoted as DOF Q1 in Figure 4.3. Thereby, $\Psi_1(x \Rightarrow a)$ must be unity while $\Psi_2(a)$ and $\Psi_3(a)$ must each vanish.

Development transparency results upon choosing the simplest Lagrange linear interpolation polynomial. Figure 4.4 graphs the associated piecewise-continuous distributions for each $\Psi_\alpha(x)$, $1 \le \alpha \le 3$, on $a \le x \le b$. You can visually verify that for these $\Psi_\alpha(x)$ definitions the Q_α correspond precisely to nodal temperature.

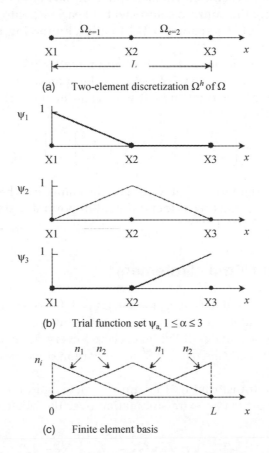

Figure 4.4 Lagrange linear trial space basis illustration, (a) two element discretization of Ω, (b) trial function set, Ψ_α, $1 \le \alpha \le 3$, (c) finite element basis functions

Reverting to the subscript notation of Chapter 3.6, the Lagrange piecewise-continuous linear polynomial defining the generic member of the set Ψ_α centered at node X_j is

$$\Psi(x) = \begin{cases} \dfrac{x - X_{j-1}}{X_j - X_{j-1}}, & \text{for } X_{j-1} \leq x \leq X_j \\[2mm] \dfrac{X_{j+1} - x}{X_{j+1} - X_j}, & \text{for } X_j \leq x \leq X_{j+1} \\[2mm] 0, & \text{for } x < X_{j-1}, x > X_{j+1} \end{cases} \tag{4.13}$$

The last line in equation (4.13) emphasizes that each $\Psi_\alpha(x)$ is of global span, that is, defined on the entirety of $a \leq x \leq b$. The resultant FE linear trial space basis $\{N\} = \{n_1, n_2\}^T$ spanning each element is graphed at the bottom of Figure 4.4, as generated by superposing the nonzero components of each Ψ_α onto each FE domain Ω_e. The matrix elements of $\{N\}$ span *only* Ω_e, totally eliminating the burden of $\Psi_\alpha(x)$ being global.

First observe that the elements n_i of the column matrix $\{N\}$ are independent of Ω_e, that is, they are identical on every Ω_e! Defining the span of (any!) element Ω_e as $l_e \equiv$ $(XR-XL)_e$, for XR and XL the right and left-node coordinates of Ω_e, the universal linear trial space basis matrix is

$$\{N\} \equiv \begin{Bmatrix} n_1 \\ n_2 \end{Bmatrix} = \frac{1}{(XR - XL)_e} \begin{Bmatrix} XR - x \\ x - XL \end{Bmatrix}_e = \begin{Bmatrix} 1 - \bar{x}/l_e \\ \bar{x}/l_e \end{Bmatrix} \tag{4.14}$$

for the $\bar{x}$ coordinate origin located at XL of Ω_e. The similarities between equation (4.6) and equation (4.14) are patently evident; the all-important distinguishing feature is local $\bar{x}/l_e$ replacing global x/L!

4.4 Finite Element Matrix Statements

Since equation (4.14) is local to Ω_e *only*, the integrals defined in global GWS^N, equation (4.11), can now be evaluated via a FOR loop over the elements Ω_e, $1 \leq e \leq M$, of the discretization. In fact, *universal* $\{N\}$ on each and every domain Ωe renders integral evaluations for the *spatially discrete* GWS^h an operation done *once only* on the generic (*master*) element.

The first step is transforming the approximate solution global expression, equation (4.12), to local via an extra summation over the M elements of the domain discretization.

$$T^N(x) \equiv \sum_{\alpha=1}^{N} \Psi_\alpha(x)Q_\alpha = \sum_{e=1}^{M}\sum_{j=1}^{2} n_j(\bar{x})Q_j^e = \sum_{e=1}^{M} \{N\}^T \{Q\}_e. \tag{4.15}$$

This also introduces the matrix nomenclature preferred by the FE community. The resulting operation $GWS^N \Rightarrow GWS^h$ generates

$$GWS^h = \sum_{e=1}^{M} \left[\sum_{j=1}^{2} \int_{\Omega_e} \frac{dn_i}{dx} k \frac{dn_j}{dx} dx Q_j^e - \int_{\Omega_e} n_i s dx \right] - k \frac{dT}{dx} n_i \Big|_a^b = 0 \qquad (4.16)$$

for the *free* index range $1 \le i \le 2$. For the lead term in equation (4.16), the transition from summation index to matrix notation is

$$\sum_{j=1}^{2} \int_{\Omega_e} \frac{dn_i}{dx} k \frac{dn_j}{dx} dx Q_j^e \Big|_{i=1,2} \Rightarrow \int_{\Omega_e} \frac{d\{N\}}{dx} k \frac{d\{N\}^T}{dx} dx \{Q\}_e. \qquad (4.17)$$

Here, the $1 \le j \le 2$ summation loop is the row-column matrix product $\{N\}^T\{Q\}_e$, while the (free) index range $1 \le i \le 2$ corresponds to rows in the lead column matrix $\{N\}$.

The FE basis derivative required for equation (4.17) is easy to form via the definition (4.14) and the chain rule

$$\frac{d\{N\}}{dx} = \frac{d\{N\}}{d\bar{x}} \frac{d\bar{x}}{dx} = \frac{1}{l_e} \left\{ \begin{array}{c} -1 \\ 1 \end{array} \right\} \qquad (4.18)$$

noting the coordinate transformation $\bar{x} = \bar{x}(x)$ is affine (only a translation). Hence, equation (4.17) evaluates to

$$\int_{\Omega_e} \frac{d\{N\}}{dx} k \frac{d\{N\}^T}{dx} dx = \int_{\Omega_e} \frac{1}{l_e} \left\{ \begin{array}{c} -1 \\ 1 \end{array} \right\} \frac{1}{l_e} \{ -1 \quad 1 \} k dx$$

$$= \frac{1}{l_e^2} \left[\begin{array}{cc} 1 & -1 \\ -1 & 1 \end{array} \right] \int_{\Omega_e} k dx = \frac{\bar{k}_e}{l_e} \left[\begin{array}{cc} 1 & -1 \\ -1 & 1 \end{array} \right] \qquad (4.19)$$

The terminal integral evaluation leads to the definition of $\bar{k}_e$, the average value of conductivity on Ω_e (if it is indeed variable).

The second term in GWS^h, equation (4.16), requires a function integration on Ω_e. Via equation (4.14) and affine $\bar{x} = \bar{x}(x)$ the result is

$$\int_{\Omega_e} n_i s dx \Big|_{i=1,2} \Rightarrow \int_{\Omega_e} \{N\} s dx = s \int_{\Omega_e} \left\{ \begin{array}{c} 1 - \bar{x}/l_e \\ \bar{x}/l_e \end{array} \right\} d\bar{x} = \frac{s l_e}{2} \left\{ \begin{array}{c} 1 \\ 1 \end{array} \right\} \qquad (4.20)$$

with verification a suggested exercise. The last term in equation (4.16) requires no integration, only an evaluation at the termini (nodes) of Ω_e. This term contribution

$$-k \frac{dT}{dx} n_i \Big|_{i=1,2} \Rightarrow -k \frac{dT}{dx} \{N\} \Big|_{XL}^{XR} = -k \frac{dT}{dx} \left\{ \begin{array}{c} 0 \\ 1 \end{array} \right\}_{e=M} + k \frac{dT}{dx} \left\{ \begin{array}{c} 1 \\ 0 \end{array} \right\}_{e=1} \qquad (4.21)$$

recognizes elements n_i of $\{N\}$ take on $(0,1)$ values at the nodes of Ω_e. Hence equation (4.21) enforces that efflux BCs are admissible *only* at the right node of the last element $\Omega_{e=M}$ or the left node of first element $\Omega_{e=1}$.

The generic Ω_e GWS^h algorithm integrated statement, as enabled via the FE linear trial space basis $\{N\}$, equation (4.16), is precisely

$$\text{GWS}^h = \sum_{e=1}^{M} \{WS\}_e$$
$$= \sum_{e=1}^{M} \left(\frac{\bar{k}_e}{l_e} \begin{bmatrix} 1 & -1 \\ -1 & 1 \end{bmatrix} \{Q\}_e - \frac{sl_e}{2} \begin{pmatrix} 1 \\ 1 \end{pmatrix} - k\frac{dT}{dx} \begin{Bmatrix} -\delta_{e1} \\ \delta_{eM} \end{Bmatrix} \right) = \{0\}$$

(4.22)

In equation (4.22) everything subscripted "e" is element-*dependent* data. Further, the Kronecker delta "switches" δ_{e1} and δ_{eM} clearly define imposition of efflux BCs with the correct sign.

For the uniform $M=2$ mesh, elements Ω_e have length $l_e = L/2$. Completing the scalar-matrix products in equation (4.22) clears element data, hence generates the terminal $M=2$ contributions to GWS^h. Substituting equation (4.2) the two $\{WS\}_e$ contributions to GWS^h, equation (4.22), are

$$\text{for } e=1: \quad \{WS\}_1 = \frac{\bar{k}_1}{l_1} \begin{bmatrix} 1 & -1 \\ -1 & 1 \end{bmatrix} \{Q\}_{e=1} - \frac{sl_1}{2} \begin{Bmatrix} 1 \\ 1 \end{Bmatrix} - k\frac{dT}{dx} \begin{Bmatrix} -\delta_{11} \\ 0 \end{Bmatrix}$$
$$= \frac{\bar{k}_1}{L/2} \begin{bmatrix} 1 & -1 \\ -1 & 1 \end{bmatrix} \begin{Bmatrix} Q1 \\ Q2 \end{Bmatrix} - \frac{sL/2}{2} \begin{Bmatrix} 1 \\ 1 \end{Bmatrix} - \begin{Bmatrix} f_n \\ 0 \end{Bmatrix}$$

(4.23a)

$$\text{for } e=2: \quad \{WS\}_2 = \frac{\bar{k}_2}{l_2} \begin{bmatrix} 1 & -1 \\ -1 & 1 \end{bmatrix} \{Q\}_2 - \frac{sl_2}{2} \begin{Bmatrix} 1 \\ 1 \end{Bmatrix} - k\frac{dT}{dx} \begin{Bmatrix} 0 \\ \delta_{22} \end{Bmatrix}$$
$$= \frac{2\bar{k}_2}{L} \begin{bmatrix} 1 & -1 \\ -1 & 1 \end{bmatrix} \begin{Bmatrix} Q2 \\ Q3 \end{Bmatrix} - \frac{sL}{4} \begin{Bmatrix} 1 \\ 1 \end{Bmatrix} + \begin{Bmatrix} 0 \\ F3 \end{Bmatrix}$$

(4.23b)

introducing $F3$, the *unknown* efflux $-k\,dT/dx$ at $x = b$.

4.5 Assembly of $\{WS\}_e$ to form Algebraic GWS^h

The contributions (4.23) sum to form equation (4.22), *the* sought algebraic statement. The nodal DOF number order, equation (4.12), is left to right, Figure 4.3, hence $\{Q\} = \{Q1,Q2,Q3\}^T$ in equation (4.5). Indicating matrix order via subscripts, the terminal $M=2$ GWS^h algebraic statement (4.22) is

$$\text{GWS}^h \Rightarrow [\text{Matrix}]_{3\times3} \begin{Bmatrix} Q1 \\ Q2 \\ Q3 \end{Bmatrix}_{3\times1} - \{b\}_{3\times1} = \{0\}_{3\times1}.$$

(4.24)

The key completion issue is entering the 2×2 element matrices (4.23) into the 3×3 global [Matrix] (4.24), also the 3×1 *data* matrix $\{b\}$. This is readily accomplished by simply adding a row–column pair of zeros into each matrix in (4.23) appropriate for the missing DOF entry in $\{Q\}_e$. Thus modified (4.23) inserted into equation (4.24) yields

$$[\text{Matrix}] = \sum_{e=1}^{M} [\text{Matrix}]_e$$

$$= \frac{2\bar{k}_1}{L} \begin{bmatrix} 1 & -1 & 0 \\ -1 & 1 & 0 \\ 0 & 0 & 0 \end{bmatrix} + \frac{2\bar{k}_2}{L} \begin{bmatrix} 0 & 0 & 0 \\ 0 & 1 & -1 \\ 0 & -1 & 1 \end{bmatrix} \equiv \frac{2k}{L} \begin{bmatrix} 1 & -1 & 0 \\ -1 & 2 & -1 \\ 0 & -1 & 1 \end{bmatrix} \quad (4.25)$$

$$\{b\} = \sum_{e=1}^{2} \{b\}_e = \frac{sL}{4} \begin{Bmatrix} 1 \\ 1 \\ 0 \end{Bmatrix} + \begin{Bmatrix} f_n \\ 0 \\ 0 \end{Bmatrix} + \frac{sL}{4} \begin{Bmatrix} 0 \\ 1 \\ 1 \end{Bmatrix} + \begin{Bmatrix} 0 \\ 0 \\ -F3 \end{Bmatrix}$$

$$= \frac{sL}{4} \begin{Bmatrix} 1 \\ 2 \\ 1 \end{Bmatrix} + \begin{Bmatrix} f_n \\ 0 \\ -F3 \end{Bmatrix} \quad (4.26)$$

where $\bar{k}_1 \equiv k \equiv \bar{k}_2$ has been assumed in equation (4.25). Then substituting equations (4.25) and (4.26) into equation (4.24), dividing through by $2k/L$ and moving the data $\{b\}$ across the equal sign produces the terminal *assembled* global-order GWS^h algorithm algebraic statement

$$\text{GWS}^h \Rightarrow \begin{bmatrix} 1 & -1 & 0 \\ -1 & 2 & -1 \\ 0 & -1 & 1 \end{bmatrix} \begin{Bmatrix} Q1 \\ Q2 \\ Q3 \end{Bmatrix} = \frac{sL^2}{8k} \begin{Bmatrix} 1 \\ 2 \\ 1 \end{Bmatrix} + \frac{L}{2k} \begin{Bmatrix} f_n \\ 0 \\ -F3 \end{Bmatrix}. \quad (4.27)$$

The final step enabling solution of equation (4.27) is enforcing the Dirichlet BC $Q3 = T_b$. One thus notes the *unknown* in the last matrix row of equation (4.27) is the energy efflux $F3$ not $Q3$! Moving $F3$ to the column matrix of unknowns and substituting T_b for $Q3$ *decouples* the equations for DOF $Q1$ and $Q2$ from that for $F3$.

Thereby, the final form of matrix statement (4.27) for solution is

$$\begin{bmatrix} 1 & -1 \\ -1 & 2 \end{bmatrix} \begin{Bmatrix} Q1 \\ Q2 \end{Bmatrix} = \begin{Bmatrix} \dfrac{L}{2k}\left(\dfrac{sL}{4} + f_n\right) \\ \dfrac{sL^2}{4k} + T_b \end{Bmatrix} \quad (4.28)$$

which via Cramer's rule results in

$$Q1 = \frac{sL^2}{2k} + \frac{f_n L}{k} + T_b$$

$$Q2 = \frac{3sL^2}{8k} + \frac{f_n L}{2k} + T_b \quad (4.29)$$

Of course the third entry in $\{Q\}$, equation (4.24), is $Q3 = T_b$. Solving the decoupled algebraic equation in the rearrangement of equation (4.27) produces the GWS^h algorithm-identified DOF solution $F3 = sL + f_n$, with verification a suggested exercise.

4.6 Solution Accuracy, Error Distribution

The GWS^h algorithm DOF algebraic solution is valid for *any data* specification, that is, conductivity k (including variable, e.g., $\bar{k}_1 \equiv 2\bar{k}_2$), domain span L, source s, boundary efflux f_n, and fixed temperature T_b. Note specifically that DOF (4.29), also $Q3 = T_b$, are *all* functions of the specified data, as in the analytical solution, equation (4.6). One thus observes that the FE discrete implemented GWS^h solution appears to possess mathematical attributes intrinsic to the classical procedure.

Several summary observations are appropriate. First note the GWS^h solution $T^h(x)$, as a construction replacement for continuum $T^N(x)$, is a function continuous on x. Starting with equation (4.12), which becomes equation (4.15) in the FE implementation, the generated linear piecewise-continuous GWS^h approximate solution is

$$T^N(x) \equiv \sum_{\alpha=1}^{N=3} \Psi_\alpha(x)Q_\alpha = \Psi_1(x)Q_1 + \Psi_2(x)Q_2 + \Psi_3(x)Q_3$$

$$\equiv T^h(x) = \cup_e \{N\}^T \{Q\}_e \tag{4.30}$$

$$\equiv \cup_e \left(\{n_1, n_2\} \begin{Bmatrix} Q1 \\ Q2 \end{Bmatrix}_{e=1}, \{n_1, n_2\} \begin{Bmatrix} Q2 \\ Q3 \end{Bmatrix}_{e=2} \right)$$

recalling that union ($\cup$) denotes nonoverlapping sum.

Secondly, even though the GWS^h solution $T^h(x)$ possesses classical attributes, it does not generate the exact solution! Comparing equation (4.30) to equation (4.6), the distinction is in the selected trial space with equation (4.30) missing the quadratic member in equation (4.6). Therefore, $T^h(x)$ has associated the approximation error $e^h(x)$, the distribution of which can be quantified.

Figure 4.5 graphs the solution (4.30) for assumed positive and negative source s, compared to the analytical solution (4.6). Factually, this tutorial problem is so simple that the $T^h(x)$ DOF coincide *exactly* with the analytical solution at mesh node coordinates. Indeed, the solution $T^h(x)$ is precisely its Lagrange piecewise linear interpolation, an approximation as close to the exact solution as possible for the linear restriction!

This observation gives testimony to the richness inherent with the GWS^h process. The approximation error is indeed a *minimum*, but it exists and is nonzero almost everywhere. Viewing Figure 4.5, it is obvious that nowhere else but at the mesh nodes does $T^h(x)$ agree with $T(x)$! Indeed, equation (4.7) transitions to

$$T(x) = \bar{T}^h(x) + e^h(x) \tag{4.31}$$

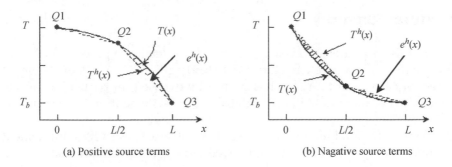

(a) Positive source terms   (b) Nagative source terms

Figure 4.5 Graphs of $T(x)$, $T^h(x)$, and $e^h(x)$, (a) positive source s, (b) negative source s

with the discrete approximation error $e^h(x)$ *distribution* illustrated by the shaded regions in the figure. It is hard to imagine an $e^h(x)$ that could be smaller than that produced by the GWS^h algorithm!

4.7 Convergence, Boundary Heat Flux

Viewing Figure 4.5, the *size* and distribution of $e^h(x)$ will certainly respond favorably to use of FE meshes having more DOF than for $M=2$. In fact, $T^h(x)$ will approach $T(x)$ under mesh refinement in a very regular and predictable manner. Derivation of the GWS^h algorithm *asymptotic convergence* (rate) of error annihilation is detailed in the next chapter, for a much broader class of FE trial space bases and diverse examples in the computational engineering sciences.

Additionally, and unanticipated, the GWS^h process identified the $M=2$ mesh right-end heat efflux $F3$ as the true DOF at this boundary node. Factually, Dirichlet BCs applied to the DE principle always engender the associated *key* heat flux an unknown. The GWS^h solution for $F3$ is identical with the *exact* solution (a suggested exercise via equation (4.6)). A similar exercise will verify that the GWS^h left-end efflux $F1$ *exactly* reproduces the imposed data, specifically

$$F1 = -k\frac{dT^N}{dx}\bigg|_{x=\alpha} \Rightarrow f_n.$$
(4.32)

These exact efflux DOF result, even though the approximate solution $T^h(x)$ obviously does not possess the slope of the exact solution at either end, $x=a$ or $x=b$! The alternative approach to boundary flux determination is via a finite difference (FD) approximation to $-kdT^N/dx$ which generates a significantly less accurate solution (another exercise). Generalizing to n-dimensions, boundary flux derivatives can always be post-processed with an accuracy equivalent to that of the FE trial space basis.

4.8 Chapter Summary

The goal of this chapter is to develop in painstaking detail the process $\mathrm{WF}^N \Rightarrow \mathrm{WS}^N \Rightarrow$ $\mathrm{GWS}^N \Rightarrow \mathrm{GWS}^h \Rightarrow e^h$ for an elementary $n = 1$ steady $DE + \mathrm{BCs}$ problem statement. The key aspect of each algorithm step has been fully exposed, in particular the *assembly* process loading element data into the global matrix statement and direct insertion of derivative BCs.

The generated DOF solution for the linear FE trial space basis GWS^h implementation clearly anticipates that the process is endowed with aspects of mathematical elegance. The remainder of this text is devoted to formal exposure to these attributes for diverse applications in the computational engineering sciences. Bon voyage!

Exercises

4.1 Verify the analytical solution (4.4) to the statement (4.1)–(4.3).

4.2 Verify that the Ψ_α and Q_α cited in equation (4.6) for T^N agree with equation (4.5).

4.3 Verify the integrated-by-parts GWS^N algorithm (4.11).

4.4 For the linear FE basis, verify the GWS^h expressions (4.19) and (4.20).

4.5 Verify that the FE contributions (4.23) assemble to form equations (4.25) and (4.26).

4.6 Verify the DOF solution (4.29).

4.7 Confirm the GWS^h algorithm solution for $F3$, hence also $F1$, equation (4.32).

4.8 For the $M = 2$ mesh, compute approximate solutions for the boundary fluxes $F3$ and $F1$ using an appropriate order FD approximation to the definition (4.32).

4.9 Compute the GWS^h algorithm solution for an $M = 3$ nonuniform discretization of domain L for equations (4.1)–(4.3). Note temperature and boundary flux DOF accuracy, hence comment on GWS^h algorithm *error extremization*.

4.10 Regenerate the DOF solution (4.29) assuming $\bar{k}_1 \equiv 2\,\bar{k}_2$. Note that the exact solution (4.4) is no longer pertinent for comparison.

Reference

[1] Oden, J.T. and Demkowicz, L.F. (1996) *Applied Functional Analysis*, CRC Press, Boca Raton, FL.

5

Steady Heat Transfer, $n = 1$: GWS^h implemented with $\{N_k(\zeta_\alpha)\}$, $1 \le k \le 3$, accuracy, convergence, nonlinearity, templates

5.1 Introduction

The weak statement recipe for constructing computable discrete approximate solutions to PDE + BCs systems is completely general in n-dimensions for linear and nonlinear flux BCs on domain closures of arbitrary shape. The geometries of finite elements for $1 \le n \le 3$ dimensions, Figure 5.1, are

- line segments
- triangles and quadrilaterals
- tetrahedra and hexahedra

For each of these element geometries, the approximate solution expansion coefficients, the *degrees of freedom* (DOF), are always defined at least at each geometric singularity, that is, the vertices of each element. Development of higher (than linear) degree FE trial space basis functions for these element geometries add either nonvertex DOF [1], or embed local DOF enrichments (*p*-elements) that are *condensed out* prior to the algebraic solution process [2].

Finite Elements ⇔ Computational Engineering Sciences, First Edition. A. J. Baker.
© 2012 John Wiley & Sons, Ltd. Published 2012 by John Wiley & Sons, Ltd.

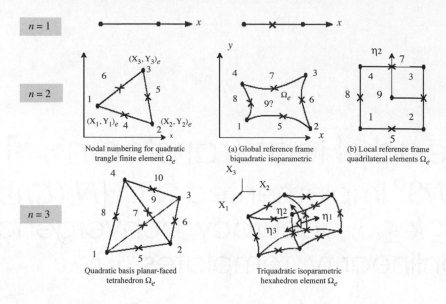

Figure 5.1 Finite element domains Ω_e for $1 \leq n \leq 3$ and $k = 1, 2$

The underlying piecewise-continuous Lagrange interpolation polynomial becomes clearly expressed in the FE *trial space basis* in terms of a pertinent local coordinate system. Specifically, the basis is a column matrix containing polynomials complete to degree $k = 1, 2, 3, \ldots$ which in the literature is usually symbolized as $\{N_k(\cdot)\}$. As detailed in the Chapter 4 tutorial, their union can always be connected to the global trial space $\Psi_\alpha(\mathbf{x})$. Importantly, the need to do so almost never exists.

The dot notation in $\{N_k(\cdot)\}$ is the placeholder for the trial space basis which is always expressed in one of two *local* coordinate system options:

- *natural* system ζ_α, $1 \leq \alpha \leq n + 1$, for lines, triangles, and tetrahedra
- *tensor product* system ηi, $1 \leq i \leq n$, for lines, quadrilaterals, and hexahedra

The linear ($k \equiv 1$) FE basis always defines this local coordinate system, also the *coordinate transformation* $x_j = f(\zeta_\alpha, \eta_i)$ from global to local. The derivation of $k > 1$ FE trial space bases involves relatively elementary algebra, with direct connection to Lagrange and Hermite interpolation polynomials possible.

5.2 Steady Heat Transfer, $n = 1$

To gain familiarity with variable completeness degree k FE trial space bases $\{N_k(\cdot)\}$, a thorough examination of the $n = 1$ heat-transfer problem statement is pursued.

The all-important convective heat transfer (Robin) BC compliments the development, as well as the assumption of variable thermal conductivity $k = k(x, T(x))$.

The DE + BCs statement examined in this chapter is

$$\mathcal{L}(T) = -\frac{d}{dx}\left[k(T, x)\frac{dT}{dx}\right] - s(x) = 0, \text{ on } \Omega \subset \mathfrak{R}^1. \tag{5.1}$$

$$\ell(T) = k\frac{dT}{dx} + h(T - T_r) = 0, \text{ on } \partial\Omega_R. \tag{5.2}$$

$$T(x_b) = T_b, \text{ on } \partial\Omega_D \tag{5.3}$$

In equation (5.1), $\mathfrak{R}^1$ denotes $n = 1$ Euclidean space (the x axis), $s = s(x)$ is the (distributed) source, and the practical situation of *nonlinearity* is introduced by k being temperature-dependent. In efflux BC equation (5.2), h is the convection heat-transfer coefficient with T_r the exchange fluid reference temperature. Statement completion (5.3) is the fixed temperature (Dirichlet) BC.

The WS^N requires definition of the approximate solution and the trial and test spaces. Recalling equation (4.5)

$$T^N(x) = \sum_{\alpha=1}^{N} \Psi_\alpha(x)Q_\alpha \tag{5.4}$$

and selecting the *optimal* Galerkin criterion, (4.9), the GWS^N for equations (5.1)–(5.3) via equation (5.4), for $1 \leq \beta \leq N$ and following integration-by-parts, is

$$\text{GWS}^N = \int_\Omega \Psi_\beta \mathcal{L}(T^N)dx = \int_\Omega \Psi_\beta\left[-\frac{d}{dx}\left(k(T,x)\frac{dT^N}{dx}\right) - s(x)\right]dx$$

$$= \int_\Omega\left[\frac{d\Psi_\beta}{dx}k(T,x)\frac{dT^N}{dx} - \Psi_\beta s(x)\right]dx - \Psi_\beta k\frac{dT^N}{dx}\bigg|_{\partial\Omega} \tag{5.5}$$

As thoroughly detailed in Chapter 4.4, the FE discrete implementation of equation (5.5) amounts to $\{N_k(\zeta_\alpha)\}$ replacing $\Psi_\alpha(x)$, integrals on Ω replaced with integration on the generic element Ω_e, and *assembly* of these local matrices onto the global statement. Equation (5.4) is replaced by

$$T_e(x) = \{N_k(\zeta_\alpha)\}^T\{Q\}_e \tag{5.6}$$

and upon inserting BC (5.2) the GWS^h companion to equation (5.5) is

$$\text{GWS}^h = S_e \left[\int_{\Omega_e} \frac{\mathrm{d}\{N_k\}}{\mathrm{d}x} k_e(T,x) \frac{\mathrm{d}\{N_k\}^T}{\mathrm{d}x} \mathrm{d}x \{Q\}_e - \int_{\Omega_e} \{N_k\} s_e(x) \mathrm{d}x \right.$$

$$\left. + \{N_k\} h \left(\{N_k\}^T \{Q\}_e - T_r \right) \Big|_{\partial \Omega_R} - \{N_k\} k \frac{\mathrm{d}T^h}{\mathrm{d}x} \Big|_{\partial \Omega_D} \right] = \{0\}$$

(5.7)

introducing the *assembly* operator S_e to represent the matrix-addition process fully illustrated in Chapter 4.5. As in Chapter 4, since BC (5.3) fixes the appropriate DOF at T_b the coincident heat efflux becomes the unknown DOF on $\partial \Omega_D$, the last term in equation (5.7).

5.3 FE $k = 1$ Trial Space Basis Matrix Library

The linear basis GWS^h implementation is always the easiest to complete. For $k = 1$, the matrix elements of $\{N_1\}$ are defined in equation (4.14)

$$\{N_1(\zeta_\alpha)\} = \begin{Bmatrix} \zeta_1 \\ \zeta_2 \end{Bmatrix} = \begin{Bmatrix} 1 - \bar{x}/l_e \\ \bar{x}/l_e \end{Bmatrix},$$

(5.8)

where $\bar{x} = (x - X_L)_e$ defines the affine transformation with origin at the left node of Ω_e. The GWS^h thermal conductivity term element matrix, given the temporary label [DIFF]$_e$ (since it is a *diffusive* mechanism), is

$$[\text{DIFF}]_e \equiv \int_{\Omega_e} \frac{\mathrm{d}\{N_1\}}{\mathrm{d}x} k_e(x,T) \frac{\mathrm{d}\{N_1\}^T}{\mathrm{d}x} \mathrm{d}x = \frac{1}{l_e^2} \begin{bmatrix} 1 & -1 \\ -1 & 1 \end{bmatrix} \int_{\Omega_e} k_e(x,T) \mathrm{d}x.$$

(5.9)

Restricting to spatial dependence $k = k(x)$ for now, the integral remaining in equation (5.9) defines the thermal conductivity *average* value on Ω_e times l_e. Labeling it $\bar{k}_e$, this contribution to GWS^h involves element data multiplying a member of the $k = 1$ basis element matrix *library* as

$$[\text{DIFF}]_e \equiv \frac{\bar{k}_e}{l_e} \begin{bmatrix} 1 & -1 \\ -1 & 1 \end{bmatrix}.$$

(5.10)

The distributed source $s_e(x)$, equation (5.1), implemented via interpolation with $\{N_1\}$, generates another member of the $k = 1$ basis element matrix library. Being data, it is entered into the {b} matrix, (4.24), as

$$\{b(s)\}_e \equiv \int_{\Omega_e} \{N_1\} s_e(x) dx = \int_{\Omega_e} \{N_1\}\{N_1\}^T dx \{SRC\}_e$$

$$= \frac{l_e}{6} \begin{bmatrix} 2 & 1 \\ 1 & 2 \end{bmatrix} \{SRC\}_e \tag{5.11}$$

and $\{SRC\}_e$ holds the nodal values of the interpolation of $s(x)$ on Ω_e.

The convection heat transfer BC, equation (5.7), amounts to an endpoint evaluation for $n = 1$. Recalling the Kronecker delta introduced in Chapter 4, the linear basis FE matrix library completion is

$$[BC]_e = h_e \begin{bmatrix} \delta_{e1} & 0 \\ 0 & \delta_{eM} \end{bmatrix}, \quad \{b(h, T_r)\}_e = h_e T_r \begin{Bmatrix} \delta_{e1} \\ \delta_{eM} \end{Bmatrix} \tag{5.12}$$

with [BC] a temporary placeholder and δ_{e1} and δ_{eM} the "on/off" switches, nonzero and equal to unity only when the subscripts are identical.

The $k = 1$ basis GWS^h implementation for *all linear* DE + BCs problem statements on $n = 1$ is specified in equations (5.10)–(5.12). Introducing the placeholder $\{WS\}_e$ for this sum of terms, the GWS^h *computable* matrix statement is

$$GWS^h \equiv S_e(\{WS\}_e) = \{0\}$$

$$\{WS\}_e \equiv ([DIFF]_e + [BC]_e)\{Q\}_e - \{b\}_e. \tag{5.13}$$

Prior to examining programmatic issues a simple example illustrates the operation of equation (5.13). Assume the (given) *data* are $h = 20$, $T_r = 1500$, conductivity distribution $10 \le k_e(x) \le 20$, and $T(x = 2) = 306.85$ in traditional US units. For these data, the available analytical solution confirms the left-end temperature is precisely $1000.000°F$ [3]. Further, since the analytical solution is logarithmic, that is, $T = T(\log x)$, no piecewise continuous polynomial can generate nodally exact discrete solution DOF. This enables precise examination of GWS^h algorithm accuracy and *asymptotic convergence* properties.

Assume a uniform $M = 2$ discretization of the span $1 \le x \le 2$, Figure 5.2, the terminal data specification for a simulation. Hence:

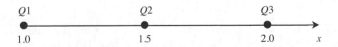

Figure 5.2 $M = 2$ discretization of $\Omega \subset \Re^1$

Example 5.1

The GWS^h DOF array is $\{Q\}^T = \{Q1, Q2, Q3\}$. The GWS^h operation in the second line of equation (5.13) for $1 \le e \le M = 2$ produces

$e = 1$:

$$[\text{DIFF}]_1 = \frac{\bar{k}_e}{l_e}\begin{bmatrix} 1 & -1 \\ -1 & 1 \end{bmatrix} = \frac{(10+15)}{2(0.5)}\begin{bmatrix} 1 & -1 \\ -1 & 1 \end{bmatrix} = \begin{bmatrix} 25 & -25 \\ -25 & 25 \end{bmatrix}$$

$$[\text{BC}]_1 = h_e\begin{bmatrix} \delta_{11} & 0 \\ 0 & 0 \end{bmatrix} = 20\begin{bmatrix} 1 & 0 \\ 0 & 0 \end{bmatrix} = \begin{bmatrix} 20 & 0 \\ 0 & 0 \end{bmatrix}$$

$$\{b\}_1 = h_e T_r\begin{Bmatrix} \delta_{11} \\ 0 \end{Bmatrix} = (20)(1500)\begin{Bmatrix} 1 \\ 0 \end{Bmatrix} = \begin{Bmatrix} 30,000 \\ 0 \end{Bmatrix}$$

$e = 2$:

$$[\text{DIFF}]_2 = \frac{\bar{k}_e}{l_e}\begin{bmatrix} 1 & -1 \\ -1 & 1 \end{bmatrix} = \frac{(15+20)}{2(0.5)}\begin{bmatrix} 1 & -1 \\ -1 & 1 \end{bmatrix} = \begin{bmatrix} 35 & -35 \\ -35 & 35 \end{bmatrix}$$

$$\{b\}_2 = -k\frac{dT}{dx}\begin{Bmatrix} 0 \\ \delta_{22} \end{Bmatrix} = \begin{Bmatrix} 0 \\ F3 \end{Bmatrix}.$$

Then, the operation GWS$^h = S_e\{WS\}_e$ defined in the first line of equation (5.13) generates (a suggested exercise)

$$\text{GWS}^h = S_e\{WS\}_e = \{0\}$$

$$= \begin{bmatrix} 25+20 & -25 & \\ -25 & 25+35 & -35 \\ & -35 & 35 \end{bmatrix}\begin{Bmatrix} Q1 \\ Q2 \\ Q3 \end{Bmatrix} - \begin{Bmatrix} 30,000 \\ 0 \\ -F3 \end{Bmatrix} = \{0\}$$

that cannot be solved until the BC $Q3 = T_b = 306.85$ is enforced. As in Chapter 4, the unknown in the third row of GWS^h is $F3$, and upon Dirichlet BC enforcement the DOF solution is

$$\{Q\} = \{996.86, 594.36, 306.84\}^T$$

$$F3 = 10,062.85.$$

Recalling the exact surface temperature is 1000.0 F, the DOF $Q1$ generated with this $M = 2$ coarsest possible (!) mesh is *not nodally exact*, but is less than 0.5% in error. This attests to the GWS^h intrinsic embedding of the solution-dependent efflux (gradient) BC (5.2).

5.4 Object-Oriented GWS^h Programming

The *element* contributions in $\{WS\}_e$ are always products of element-dependent data with *element-independent* library matrices. This data structure admits algorithm "coding" using an *object-oriented* syntax. The enabling observation is that *every* contribution $\{WS\}_e$ to GWS^h can be expressed as

$$\{WS\}_e = (\text{const})\begin{pmatrix} elem \\ const \end{pmatrix}_e \begin{Bmatrix} distr \\ data \end{Bmatrix}_e^T (metri\ data;\ \det)[Matrix]\begin{Bmatrix} Q\ or \\ data \end{Bmatrix}_e, \tag{5.14}$$

where subscript e denotes element-dependent and the bracket distinctions are

$$(\cdot) \Rightarrow \text{scalars (a number)}$$
$$\{\cdot\} \Rightarrow \text{a column matrix of element DOF length.}$$
$$[\cdot] \Rightarrow \text{a square matrix of element DOF order}$$

In the $\{WS\}_e$ *template* (5.14), the objects are the scalars, arrays, matrices, and so on, which correlate with instruction sequences for handling of data. Thereby, the template corresponds to a theory implementation *pictogram* that communicates algorithm operation essence in a "readable format," analogous to the role of stencils for FD operators.

The key to writing a template is to *name* the *objects* constituting data types and terms forming $\{WS\}_e$. Objects can be named arbitrarily, but a specific FE matrix nomenclature has proven useful to sort details regarding basis degree k, differentiation, and dimension n. Herein, the matrix identification convention is:

$$\text{matrix} \Leftrightarrow [M\ bcccd]\ \text{or}\ \{M\ bcccd\}. \tag{5.15}$$

M = matrix prefix, $M \Rightarrow A, B, C$ for $n = 1, 2, 3$
b = an integer, the number of FE bases $\{N_k\}$ in the integrand forming the $\{WS\}_e$ term
c = a Boolean index $(0, 1)$, repeated b times, to indicate each integrand basis $\{N_k\}$ (no, yes) differentiated
d = a label for any added delineation

In this text, a template enables hands-on computer lab execution in MATLAB®, using the specifically written MATLAB® *toolbox FEmPSE* (Finite Element Problem Solving Environment), a free download from www.wiley.com/go/baker/finite. As this chapter progresses, you will hopefully gain appreciation that equations (5.14) and (5.15) transparently organize algorithm implementation, eventually including *nonlinearity*. As these chapters proceed, the template syntax transparently communicates GWS^h implementations for all $n = 1, 2, 3$ including the $n - 1$-dimensional BCs.

The template statement for the linear basis GWS^h algorithm (5.13), for equations (5.1)–(5.3), is

$$
\begin{aligned}
\{WS\}_e &\equiv ([DIFF]_e + [BC]_e)\{Q\}_e - \{b\}_e \\
&= ( )(COND)\{ \}(0;-1)[A211L]\{Q\} \\
&\quad +(HCON)( )\{ \}(0)[ONE]\{Q\} \\
&\quad -(HCON,\ TR)( )\{ \}(0)[ONE]\{ \} \\
&\quad -( )( )\{ \}(0;1)[A\,200L]\{SRC\}.
\end{aligned}
\tag{5.16}
$$

In equation (5.16) an empty bracket () or { } signifies no data of that type which *FEmPSE* bypasses. In the metric data brackets, the 0 indicates the transformation is affine (no data) for $n = 1$, and the integer following the semicolon is the exponent on element measure l_e. The matrix [ONE] corresponds to δ_{11} in equation (5.12). Finally, in the *matrix convention* renamed $[DIFF]_e$ and $[BC]_e$ matrices, $d \Rightarrow L$ denotes *Linear* FE trial space basis library matrices.

Table 5.1 details the template (5.16) plus data inserted into the MATLAB® .m file for computer lab 5.1. For those not familiar with MATLAB® key syntax definitions for this $M = 4$ .m file are:

```
% signifies a comment
global X1 sets aside an array for the M-dependent node
     coordinates of Ωʰ
semicolon terminates an instruction
superscript prime denotes a row matrix
asjac signifies "assemble jacobian," which for this problem is
          only the line containing [A211L] in (5.16)
the terms with the [ONE] matrix are handled as scalar
     operations
the last row in GWSʰ matrix is modified to enforce Dirichlet
     BC on Q5, i.e., [Matrix]{Q} ⇒ [0 0 0 0 1]Q5 = Tb
backslash denotes matrix solve command
energy norm computation, discussed shortly
```

After insertion of *FEmPSE* into your MATLAB® stream, call the .m file inside a FOR loop on M, the mesh number, to run the regular mesh-refinement study. The results you will generate are summarized in Table 5.2, which confirms DOF $Q1$ tends monotonically towards the analytical solution $T_{exact} = 1000.00$ at $x = 1$ on each successive mesh refinement. Plot the Table 5.2 error in $Q1$ versus mesh measure l_e on log-log scales to confirm the error reduction rate for DOF $Q1$ under regular mesh refinement is quadratic. This observation correlates in Taylor series (TS) terminology, replacing Δx with mesh measure l_e, recall Chapter 3.4, as

$$
error^h \cong O(l_e^2).
\tag{5.17}
$$

Table 5.1 MATLAB® .m file for computer lab 5.1

```
% FEm. PSE template, steady conduction with variable conductivity

global X1;                        % array for node coordinates

% constant data
h = 20;                           % convection coefficient
Tr = 1500;                        % reference temperature
Tb = 306.85282;                   % fixed temperature

% uniform discretization, M = 4
nnodes = 5;
X1 = linspace(0,1,nnodes);

% distributed data at the nodes
COND = X1' * 10 + 10;

% load the FE matrix library
load femlib;

% assemble Matrix for WSe term [DIFF]
DIFF = asjac1D(1,[],COND,-1,A3011L,[]);

% complete Matrix for [HBC] on element 1
DIFFHBC = DIFF;                   % copy DIFF
DIFFHBC(1,1) = DIFFHBC(1,1) + h;  % modify the 1,1 entry

% set up data matrix (b)
b = zeros(nnodes,1);              % an nnodes by 1 matrix of zeros


% modify b for convective BC at node 1
b(1,1) = h * Tr;

% modify b for the Dirichlet boundary condition
b(nnodes,1) = Tb;

% modify DIFFHBC for Dirichlet BC direct solve
DHdiri = DIFFHBC;                 % copy DIFFHBC
DHdiri(nnodes, : ) = zeros(1,nnodes);  % zero out the last row
DHdiri(nnodes,nnodes) = 1;        % i.e., 1 * QM = Tb

% solve the linear system
Q = DHdiri \ b;

% compute Energy Norm using [DIFF + HBC] (without Dirichlet BC)
format long;                      % display Enorm in long format
Enorm = 0.5 * Q' * DIFFHBC * Q;
Format short;
```

Table 5.2 Computer lab 5.1 convergence for DOF Q1

Mesh	M	l_e	Q1	ΔQ1	Error in Q1
Ω^h	1	1	988.6512		11.349
$\Omega^{h/2}$	2	0.5	996.8656	8.2144	3.134
$\Omega^{h/4}$	4	0.25	999.1910	2.3254	0.809
$\Omega^{h/8}$	8	0.125	999.7960	0.6049	0.204
$\Omega^{h/16}$	16	0.0625	999.9489	0.1529	0.051
$\Omega^{h/32}$	32	0.03125	999.9872	0.0383	0.013

5.5 Higher Completeness Degree Trial Space Bases

A GWS^h can be readily implemented using higher completeness degree trial space bases, for example, quadratic ($k=2$) and cubic ($k=3$), recall the approximation definition (5.6). Enriching the trial space basis in this manner invariably yields improved accuracy for solutions driven by *smooth data* on a sufficiently refined mesh. With the smoothness caveat, the result is a significant improvement in the *asymptotic convergence rate* of error reduction, as equation (5.17) generalizes to

$$\text{error}^h \cong O(l_e^{2k}). \tag{5.18}$$

Finite element bases of degree $k>1$ are Lagrange or Hermite piecewise continuous polynomials, recall equation (3.41) for $n=1$. The direct construction for a trial space basis spanning $\Omega_e \subset \Re^1$ starts with the algebraic statement and definition

$$T_e(\bar{x}) = a + b(\bar{x}/l_e) + c(\bar{x}/l_e)^2 + d(\bar{x}/l_e)^3 \dots$$
$$\equiv \{N_k(\zeta_\alpha)\}^T \{Q\}_e \tag{5.19}$$

for the origin of $\bar{x}$ at XL_e and l_e the measure (length) of Ω_e. For the quadratic basis, define the DOF as $T_e(x = XL, XM, XR) \equiv \{QL, QM, QR\}_e$ where "L, M, R" denotes "left, middle, right." Then writing equation (5.19) at these geometric coordinates generates the matrix statement

$$\begin{bmatrix} 1 & 0 & 0 \\ 1 & 1/2 & 1/4 \\ 1 & 1 & 1 \end{bmatrix} \begin{Bmatrix} a \\ b \\ c \end{Bmatrix} = \begin{Bmatrix} QL \\ QM \\ QR \end{Bmatrix}_e. \tag{5.20}$$

Solving equation (5.20) for $\{a,b,c\}^T$ is elementary, an exercise. Extracting common multipliers and recalling $\{\zeta_\alpha\}^T = \{\zeta_1, \zeta_2\} = \{1 - \bar{x}/l_e \ \ \bar{x}/l_e\}$, the natural coordinate $k=2$ trial space basis $\{N_2(\zeta_\alpha)\}$ written in symmetric form is

$$\{N_2(\zeta_\alpha)\} = \begin{Bmatrix} \zeta_1(2\zeta_1 - 1) \\ 4\zeta_1\zeta_2 \\ \zeta_2(2\zeta_2 - 1) \end{Bmatrix}. \tag{5.21}$$

The natural coordinate $k=3$ Lagrange trial space basis is similarly determined (an exercise) as

$$\{N_3(\zeta_\alpha)\} = \frac{9}{2} \begin{Bmatrix} \zeta_1(\zeta_2^2 - \zeta_2 + 2/9) \\ \zeta_1\zeta_2(2 - 3\zeta_2) \\ \zeta_1\zeta_2(3\zeta_2 - 1) \\ \zeta_2(\zeta_2^2 - \zeta_2 + 2/9) \end{Bmatrix}. \tag{5.22}$$

The integral calculus operations forming the matrix library become more complicated when employing $k > 1$ bases. Again using $[DIFF]_e$ as the placeholder, the GWS^h algorithm DE conductivity term replacing equation (5.9) is

$$[DIFF]_e = \int_{W_e} \frac{d\{N_k\}}{dx} k_e(x) \frac{d\{N_k\}^T}{dx} dx. \tag{5.23}$$

The basis derivatives in equation (5.23) now retain the $\{\zeta_\alpha\}$ hence cannot be extracted from the integrand. Further, interpolation using an FE basis is now required for distributed conductivity $k_e(x)$, which need not be of degree identical to that for T_e.
For illustration, this selection is

$$k_e(x) \cong \{COND\}_e^T \{N_1\} \tag{5.24}$$

with $\{COND\}_e$ containing the element vertex node (only) values of thermal conductivity. Substituting equation (5.24) into equation (5.23) generates $[DIFF]_e$ as a *nonstandard* matrix with elements themselves matrices. These weak form theory-generated *hypermatrices* enable precise handling of *nonlinearity* in problem statement definitions, to be thoroughly detailed.
Continuing, the derivative of the $\{N_2\}$ basis is

$$\frac{d\{N_2\}^T}{dx} = \frac{1}{l_e}\{\zeta_2 - 3\zeta_1,\ 4(\zeta_1 - \zeta_2),\ 3\zeta_2 - \zeta_1\} \tag{5.25}$$

and the integral calculus statement essence for forming $[DIFF]_e$ is

$$[DIFF]_e = \frac{\{COND\}_e^T}{l_e^2} \int_{\Omega_e} \begin{Bmatrix} \zeta_1 \\ \zeta_2 \end{Bmatrix} \begin{bmatrix} (\zeta_2 - 3\zeta_1)^2, & \cdot, & \cdot \\ \cdot, & 16(\zeta_1 - \zeta_2)^2, & \cdot \\ \cdot, & \cdot, & (3\zeta_2 - \zeta_1)^2 \end{bmatrix} dx. \tag{5.26}$$

Noting the $\{N_1\}$ matrix multiplier in equation (5.26) is but an array of scalars, the integrand for the (1,1) matrix element is

$$\int_{\Omega_e} \begin{Bmatrix} \zeta_1 \\ \zeta_2 \end{Bmatrix} \begin{bmatrix} (\zeta_2 - 3\zeta_1)^2, & \cdot & \cdot \\ & \cdot & \\ & \cdot & \end{bmatrix} dx \Rightarrow \int_{\Omega_2} \left[\begin{Bmatrix} \zeta_1(\zeta_2^2 - 6\zeta_1\zeta_2 + 9\zeta_1^2) \\ \zeta_2(\zeta_2^2 - 6\zeta_1\zeta_2 + 9\zeta_1^2) \end{Bmatrix}, \cdots \right] dx. \tag{5.27}$$

Evaluating equation (5.27) requires integration of cubic polynomials in the $\{\zeta_\alpha\}$ over Ω_e. For the natural coordinate basis the *closed-form* analytic solution is

$$\int_{\Omega_e} \zeta_1^p \zeta_2^q \, dx = l_e \frac{p!q!}{(1+p+q)!}. \tag{5.28}$$

The (1,1) matrix element in $[DIFF]_e$ is thus readily determined

$$(\text{diff}_{1,1})_e = \frac{l_e}{6}\left\{\begin{array}{c} 11 \\ 3 \end{array}\right\}$$

(5.29)

and an exercise will verify this mixed $\{N_1\} - \{N_2\}$ basis $[DIFF]_e$ matrix is

$$[DIFF]_e = \frac{\{COND\}_e^T}{6l_e}\left[\begin{array}{ccc} \left\{\begin{array}{c}11\\3\end{array}\right\} - \left\{\begin{array}{c}12\\4\end{array}\right\} & \left\{\begin{array}{c}1\\1\end{array}\right\} \\ -\left\{\begin{array}{c}12\\4\end{array}\right\} - \left\{\begin{array}{c}16\\16\end{array}\right\} & -\left\{\begin{array}{c}4\\12\end{array}\right\} \\ \left\{\begin{array}{c}1\\1\end{array}\right\} - \left\{\begin{array}{c}4\\12\end{array}\right\} & \left\{\begin{array}{c}3\\11\end{array}\right\} \end{array}\right] \equiv \frac{\{COND\}_e^T}{l_e}[A3011LQ],$$

(5.30)

which defines the *nonlinear* linear-quadratic library *hypermatrix* $[A3011LQ]_e$. The last expression in equation (5.30) exemplifies the template matrix multiplication operation $\{distributed\ data\}_e\ ^T[hyper\ Matrix]$ inferred in equation (5.14).

The $\{N_2\}$ basis matrix library completion for BCs is readily verified to be,

$$[BC]_e = h_e\begin{bmatrix} \delta_{e1} & 0 & 0 \\ 0 & 0 & 0 \\ 0 & 0 & \delta_{eM} \end{bmatrix}, \{b(h, T_r)\}_e = h_e T_r\left\{\begin{array}{c} \delta_{e1} \\ 0 \\ \delta_{eM} \end{array}\right\}$$

(5.31)

identical to the $k = 1$ basis BC matrices realizing DOF locations!

For a distributed source and interpolating $s_e(x)$ using $\{N_2\}$ on Ω_e, the source data matrix statement (an exercise) is

$$\{b(s)\}_e = \frac{l_e}{30}\begin{bmatrix} 4 & 2 & -1 \\ 2 & 16 & 2 \\ -1 & 2 & 4 \end{bmatrix}\{SRC\}_e \equiv l_e[A200Q]\{SRC\}_e$$

(5.32)

where $\{SRC\}_e$ contains nodal data from the interpolation of $s_e(x)$ at $(XL, XM, XR)_e$, the GWS^h algorithm DOF locations on Ω_e.

Example 5.2

Generate the GWS^h solution for the DE + BCs problem statement using the $\{N_2\}$ basis for an $M = 1$ discretization. The temperature DOF remain $\{Q\}^T = \{Q1, Q2, Q3\}$ where Q2 is now QM in equation (5.20). No assembly is required, and clearing the scalar multiplier on $[DIFF]_{e=1}$, equation (5.30), generates for $GWS^h = S_e(\{WS\}_1) = \{0\}$

$$\begin{bmatrix} 17+20 & -20 & 3 \\ -20 & 48 & -28 \\ 3 & -28 & 25 \end{bmatrix}\left\{\begin{array}{c} Q1 \\ Q2 \\ Q3 \end{array}\right\} - \left\{\begin{array}{c} 18,000 \\ 0 \\ F3 \end{array}\right\} = \{0\}.$$

The $M = 1$, $k = 2$ matrix statement is 3×3, identical to that for $M = 2$, $k = 1$. Reducing it to 2×2 by imposing BC $Q3 = T_b$, then solving for $Q1$ and $Q2$ yields $Q1 = 999.6$ F. This is significantly more accurate than the $k = 1$, $M = 2$ DOF solution $Q1 = 996.86$ F.

The $k = 3$, $M = 1$ GWS^h solution you will discover is even more accurate(!), which illustrates the return on investment of *enriching* the trial space basis. This topic becomes fully quantified upon executing the suggested computer labs which, following added theory developments, define the developing assessment framework.

5.6 Global Theory, Asymptotic Error Estimate

The goal is to identify a global theory for *quantifying* discrete approximation *error* for solutions generated by FE implementations GWS^h. For *smooth data* generating smooth solutions for which a sufficient number of TS derivatives exist, recall equation (3.24), TS truncation error analysis predicts the FE $k = 1,2,3$ trial space basis GWS^h implementations are accordingly 2nd, 4th, and 6th*order accurate* approximations for the DE + BCs statement (5.1)–(5.3). Truncating the TS at order l_e^{2k+1} for C some constant the error expression in FD parlance is

$$e^h \approx O\left(l_e^{2k}\right) = C l_e^{2k}. \tag{5.33}$$

Solution adherence to this theory is verifiable by conducting regular mesh-refinement experiments to generate the *a posteriori* data supporting *error quantification* without knowledge of the constant C. The M refinement process is summarized as

$$\begin{aligned} \text{meshings:} \quad & \Omega^h, \Omega^{h/2}, \Omega^{h/4}, \\ \text{with solutions:} \quad & T^h + e^h = T = T^{h/2} + e^{h/2} = \dots \end{aligned} \tag{5.34}$$

Substituting equation (5.33) for e^h and $e^{h/2}$ into equation (5.34) clears the constant C (an exercise) leading to

$$e^h = 2^{2k} e^{h/2}. \tag{5.35}$$

Substituting equation (5.35) into equation (5.34) yields the deterministic error estimate

$$T^{h/2} - T^h \equiv \Delta\, T^{h/2} = (2^{2k} - 1) e^{h/2} \tag{5.36}$$

the modest rearrangement of which quantifies the error $e^{h/2}$ associated with the finer mesh solution as

$$e^{h/2} = \frac{\Delta T^{h/2}}{2^{2k} - 1}. \tag{5.37}$$

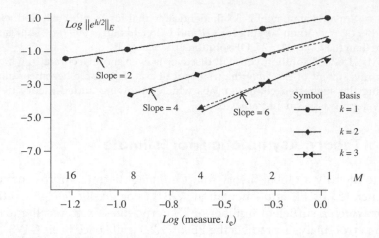

Figure 5.3 Convergence of GWS^h solutions for DOF Q1 under regular mesh refinement

Note that equation (5.37) is accurate *only* when the solution is *smooth* enough such that the truncation error concept underlying equation (5.33) is *uniformly valid* over the mesh sequence. A quick assessment of validity is provided by a *slope* computation of the data. Taking logs of the definition of *slope* of error versus mesh refinement by a factor of 2 yields

$$\text{slope} \equiv \frac{\text{rise}}{\text{run}} = \frac{\log\left(e^{h/M}\right) - \log e^{h/2M}}{\log\left(l_e\right) - \log\left(l_{e/2}\right)} = \frac{\log\left(e^{h/M}/e^{h/2M}\right)}{\log 2}. \tag{5.38}$$

Figure 5.3 summarizes the results you will generate in the suggested computer labs. In the absence of round-off, the FE $k = 1,2,3$ basis GWS^h implementations produce discrete approximate solutions for DOF $Q1$ exhibiting *asymptotic convergence* rates in agreement with the TS prediction (5.33). Note that the $k = 3$ basis algorithm is sufficiently accurate on refined meshes such that computing *round-off* error dominates discrete approximation error, which of course was not considered in theorizing (5.33).

To test the nonlinear formulation leading to equation (5.30), a suggested computer lab variation is to approximate the hypermatrix statement $\{\text{COND}\}_e^T [\text{A3011}LQ]$ with element-average conductivity $\bar{k}_e[\text{A211}Q]$. You will observe an accuracy loss on coarse meshes, which disappears on mesh refinement with no degradation of convergence rate (equation (5.37)).

Most practical engineering problems are driven by *nonsmooth data*, which adversely impacts solution smoothness hence compromises the TS concept of *order of accuracy*. Weak form theoreticians anticipated this *reality* via definition of a *global* integral measure of error called a *norm*. It replaces the highest-order derivative existence requirement of the TS prediction, embedded in equation (5.33), and explicitly introduces the concept of data *roughness* affecting solution convergence.

The *energy norm* is an integral measure intrinsic to boundary value problems of which the DE + BCs statement is an example [4]. The energy norm definition for equations (5.1)–(5.3) is

$$\|T\|_E \equiv \frac{1}{2}\left(\int_\Omega k\frac{\mathrm{d}T}{\mathrm{d}x}\cdot\frac{\mathrm{d}T}{\mathrm{d}x}\,\mathrm{d}x + hT^2\big|_{\partial\Omega}\right). \tag{5.39}$$

Analogous to equation (5.34), a regular mesh-refinement study produces the GWS^h solution sequence

$$\|T^h\|_E + \|e^h\|_E = \|T\|_E = \|T^{h/2}\|_E + \|e^{h/2}\|_E = \cdots \tag{5.40}$$

The mathematicians' *asymptotic error estimate* replacing equation (5.33) is [5]

$$\|e^h\|_E \le Cl_e^{2k}\,\max\left|\frac{\mathrm{d}^{k+1}T}{\mathrm{d}x^{k+1}}\right| \tag{5.41}$$

for C a constant and $\max\left|\mathrm{d}^{k+1}T/\mathrm{d}x^{k+1}\right|$ the extremum $(k+1)^{\mathrm{st}}$ derivative of the exact solution $T(x)$. For FE basis completeness degree k, this TS-analogous multiplier precisely quantifies how *smooth* the solution to equations (5.1)–(5.3) must be for the k-dependent GWS^h algorithm error estimate (5.41) to be valid. (*Note:* this *flaw* in the theory, i.e., assuming knowledge of the exact solution $T(x)$ extremum derivative, will shortly be replaced by a suitable norm of the *data* driving the problem).

Evaluating a GWS^h solution energy norm is direct. The integral (5.39) is generated by direct summing of the element-level computations

$$\|T^h\|_E \equiv \frac{1}{2}\int_\Omega k\frac{\mathrm{d}T^h}{\mathrm{d}x}\frac{\mathrm{d}T^h}{\mathrm{d}x}\,\mathrm{d}x + \frac{1}{2}hT^hT^h\Big|_{\partial\Omega}$$

$$= \frac{1}{2}\sum_{e=1}^M\int_{\Omega_e} k\frac{\mathrm{d}T_e}{\mathrm{d}x}\frac{\mathrm{d}T_e}{\mathrm{d}x}\,\mathrm{d}x + \frac{1}{2}hT_eT_e\Big|_{\partial\Omega_e\cap\partial\Omega}$$

$$= \frac{1}{2}\sum_{e=1}^M\{Q\}_e^T[\mathrm{DIFF}+\mathrm{BC}]_e\{Q\}_e. \tag{5.42}$$

The resultant error prediction for the finer mesh solution during regular mesh refinement which replaces equation (5.37) is

$$\|e^{h/2}\|_E = \frac{\Delta\|T^{h/2}\|_E}{2^{2k}-1}, \tag{5.43}$$

Table 5.3 Error quantization via convergence theory in $\|e^h\|_E$

Mesh	M	l_e	$\|T^{h/2}\|_E 10^7$	$\|e^{h/2}\|_E(\text{est})10^4$	Slope
Ω^h	1	1.00000	1.32607		
$\Omega^{h/2}$	2	0.50000	1.34091	4.947431	
$\Omega^{h/4}$	4	0.25000	1.34511	1.400551	1.8207
$\Omega^{h/8}$	8	0.12500	1.34620	0.364350	1.9426
Ω^{h16}	16	0.06250	1.34648	0.092085	1.9843
$\Omega^{h/32}$	32	0.03125	1.34655	0.023086	1.9966

Mesh	M	l_e	$\|T^{h/2}\|_E 10^7$	$\|e^{h/2}\|_E(\text{est})10^2$	Slope
Ω^h	1	1.00000	1.345937		
$\Omega^{h/2}$	2	0.50000	1.346519	3.884499	
$\Omega^{h/4}$	4	0.25000	1.346569	0.331436	3.5509
$\Omega^{h/8}$	8	0.12500	1.346572	0.023085	3.8437
$\Omega^{h/16}$	16	0.06250	1.346573	0.001488	3.9553
$\Omega^{h/32}$	32	0.03125	1.346573	0.000094	3.9883

Mesh	M	l_e	$\|T^{h/2}\|_E 10^7$	$\|e^{h/2}\|_E(\text{est})$	Slope
Ω^h	1	1.00000	1.346554		
$\Omega^{h/2}$	2	0.50000	1.346572	2.981042	
$\Omega^{h/4}$	4	0.25000	1.346573	0.080742	5.2064
$\Omega^{h/8}$	8	0.12500	1.346573	0.001657	5.6064
$\Omega^{h/16}$	16	0.06250	1.346573	0.000439	1.9154
$\Omega^{h/32}$	32	0.03125	1.346573	0.001654	−1.9128

where $\Delta\|T^{h/2}\|_E \equiv \|T^{h/2}\|_E - \|T^h\|_E$. Solution adherence to asymptotic error estimate (5.41) for the finer mesh solution remains solidly verifiable via

$$\text{slope} = \frac{\log\|e^{h/M}\|_E / \|e^{h/2M}\|_E}{\log 2}. \tag{5.44}$$

With these developments, the goal is to firmly grasp asymptotic convergence estimation for GWS^h solution error quantization for FE basis completeness degrees $1 \le k \le 3$. Computer lab 5.2 enables this operation and never requires knowledge of the exact solution nor C expressed in equation (5.41).

Table 5.3 summarizes the data you will generate in computing $\|T^h\|_E$, (5.42), then using equations (5.43) and (5.44) to quantify error. The coarsest M-error estimate $\|e^{h/2}\|_E$ is essentially independent of basis degree k, but thereafter the rate of estimated error decrease is strongly k-dependent. The $k = 1, 2$ basis solutions agree well with the theoretical slope of two and four on all but the coarsest mesh M.

The modest $k = 1, 2, 3$ data departures from predicted slope for the coarsest M solution illustrate the *coarse mesh accuracy* often ascribed to an FE algorithm, that is, the error is actually *less* than that predicted by the theory (equation (5.41)). This is not a theory flaw as its derivation [5], assumes the mesh M sufficiently refined such that C in equation (5.41) is indeed a constant, since on very coarse meshes C depends weakly on l_e.

The $k = 3$ data also evidences departure from theory for the finest meshes M. Again, the GWS^h theory has not flawed; instead these meshes are sufficiently refined such that the $k = 3$ FE basis solutions are identical with the cubic Lagrange interpolation of the exact (logarithmic) solution to within the significance of computer *finite arithmetic*.

5.7 Nonsmooth Data, Theory Generalization

GWS^h implementation via $1 \leq k \leq 3$ trial space bases generates attributes including basis completeness degree *matrix library* and energy norm and TS theory prediction of smooth data asymptotic convergence under regular mesh refinement. The template converts theory to compute practice leading to approximation error *quantification*.

Continuing weak form theory exposition, the $n = 1$ DE + BCs statement moves to *nonsmooth data* specification prompting theory modification. Specifically, the $\max \left| d^{k+1} T(x) / dx^{k+1} \right|$ error bound in equation (5.41) is replaced with the square of the L2 (mean square) norm of the domain data definition, [6]. Thereby, equation (5.41) loses its explicit TS appearance in becoming replaced with

$$\left\| e^h \right\|_E \leq C l_e^{2k} \| \text{data} \|_{\Omega, L2}^2 . \tag{5.45}$$

Validation of equation (5.45) is enabled by the specification of a variety of source terms $s(x)$ in equation (5.1). For the domain source-driven DE statement the L2 norm definition is

$$\| \text{data} \|_{\Omega, L2} \equiv \left[\int_\Omega (s^h)^2 dx \right]^{1/2} . \tag{5.46}$$

The source distributions for computer lab 5.3 are of the form $A \sin(\lambda x)$, hence are square integrable and equation (5.46) exists. The lab inputs source distributions by interpolation at the mesh geometric nodes, also at the DOF coordinates for $k > 1$ basis implementations. Hence, the *data* bounding the *size* of the error (5.45) for any M is C times

$$\| \text{data} \|_{L2}^2 \equiv \int_\Omega (s^h)^2 dx$$

$$= \sum_{e=1}^M \int_{\Omega_e} \{SRC\}_e^T \{N_k\} \{N_k\}^T \{SRC\}_e dx \tag{5.47}$$

$$= \sum_{e=1}^M l_e \{SRC\}_e^T [A \, 200d] \{SRC\}_e .$$

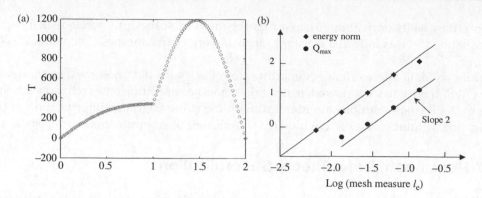

Figure 5.4 Computer lab 5.3 GWS^h nonsmooth data performance characterization; (a) two material DOF solution, $k = 1, 2$, (b) asymptotic convergence in energy norm and Q_{max}, $k = 1$

The GWS^h $k = 1, 2$ basis source library matrices are defined, equations (5.11) and (5.32). Only modest edits to the computer lab 5.2 .m file are required to specify the suggested regular mesh-refinement sequence. The DE + BCs statement remains (5.1)–(5.3) with source defined first as a single half-sine wave, then double half-sine waves with distinct amplitudes. In US customary units, these *data* definitions are $T_b(x) = 0$ at x_L and x_R and: (a) $k = 0.1$, $s = A \sin(\pi x/l)$, $A = 100$, and (b) $k = 0.1, 0.001$, $s = A \sin(\pi x/l/2)$, $A = 100, 10$.

The $k = 1$ basis $M = 4, 8$ interpolations of $s = A\sin(\lambda x)$ are quite inaccurate. This leads to underprediction of the *magnitude* of equation (5.47) which in turn *pollutes* $\|data\|^2_{\Omega,L2}$ in equation (5.45). This error disappears for $k = 1$ basis $M \geq 16$ solutions as data-interpolation error vanishes as $O\left(l_e^4\right)$, twice the $O\left(l_e^2\right)$ rate of approximation error [5].

The $k = 2$ basis source interpolation error is vanishingly small for all meshes $M \geq 4$. Your lab convergence graphs will clearly illustrate these issues, ultimately predicting the M necessary for $k = 1,2$ basis solutions to agree with the theoretical estimate (5.45), hence "get onto" the theoretical slope, equation (5.44).

The computer lab 5.3 continuation for dual conductivity and source data generates GWS^h $k = 1,2,3$ basis solutions that are now only *piecewise-continuous* (Figure 5.4a). A TS theory truncation error concept is thereby problematic as the necessary derivatives do not exist everywhere. The consequence is the maximum DOF (Q_{max}) for larger M departs from the TS predicted convergence rate of two for the $k = 1$ basis (Figure 5.4b).

These data impugn the TS theory prediction of $2k$-order accurate, (5.33). In distinction, the energy norm theory intrinsically compensates for data nonsmoothness in an elegant global manner. The *asymptotic error estimate* (5.45) requires only that the data be smooth enough to be square-integrable, (5.46), a property of the lab 5.3 continuation data. The $k = 1$ basis convergence in energy, Figure 5.4b, adheres to the theory slope of two for all but the coarsest M solution! The $k = 2$ basis computer lab completion expands on these observations.

5.8 Temperature-Dependent Conductivity, Nonlinearity

Thermal conductivity has been a specified function of location. In practice, conductivity variability depends on temperature as well. Taking this step towards reality renders the DE + BCs statement (5.1)–(5.3) explicitly *nonlinear*. The discretely implemented GWS^h algorithm response is to remain absolutely unchanged, that is,

$$GWS^h \equiv S_e(\{WS\}_e) = \{0\}$$

$$\{WS\}_e \equiv ([DIFF]_e + [BC]_e)\{Q\}_e - \{b\}_e. \tag{5.13}$$

However the $[DIFF]_e$ matrix will now be explicitly nonlinear requiring library hypermatrix generation for any basis completeness degrees k as

$$[DIFF]_e = l_e^{-1}\{COND(Q)\}_e^T[A3011d]. \tag{5.48}$$

Conductivity temperature-dependence is typically input via interpolation of given data. For computer lab 5.4 the specification is $k(T) = a + bT + cT^2$ and the resultant template for $[DIFF]_e$ contributions to $\{WS\}_e$ for any basis degree k is

$$
\begin{aligned}
\{WS\}_e = & (A) \, () \, \{ \, \} \, (0:-1) \, [A211d] \, \{Q\} \\
& + (B) \, () \, \{Q\} \, (0:-1) \, [A3011d]\{Q\} \\
& + (C) \, () \, \{Q,Q\} \, (0:-1) \, [A3011d]\{Q\}.
\end{aligned}
\tag{5.49}
$$

In equation (5.49) the cT^2 term distribution on Ω_e is interpolated using the DOF product $\{Q,Q\}$ rather than hypermatrix elevation to $[A40011d]$. This amounts to introduction of interpolation error that vanishes more rapidly than does approximation error on mesh refinement [5].

As GWS^h formed with equation (5.49) is explicitly nonlinear an iteration procedure is required to solve the global matrix statement (5.13). A Newton iteration algorithm, which exhibits a quadratic rate of iterative convergence to the solution, is selected. The generic Newton algorithm statement is

$$[JAC]^p\{\delta Q\}^{p+1} = -\{FQ\}^p, \tag{5.50}$$

where $p = 0, 1, 2, \ldots$ is the iteration index.

The term definitions for equation (5.50) are the iterative modifications to equations (5.13)

$$GWS^h \equiv \{FQ\}^p = S_e\{WS\}_e^p = \{0\}$$

$$\{WS\}_e^p = ([DIFF(\{Q^p\})]_e + [BC]_e)\{Q\}_e^p - \{b\}_e \tag{5.51}$$

the *jacobian* definition is

$$[JAC] \equiv \partial\{FQ\}/\partial\{Q\} \tag{5.52}$$

and the update process for the DOF column matrix is

$$\{Q\}^{p+1} = \{Q\}^p + \{\delta Q\}^{p+1}. \tag{5.53}$$

Iterative convergence is defined to occur when $\max\left|\{\delta Q^{p+1}\}\right| \leq \varepsilon$ with a representative convergence criterion being $\varepsilon \leq O(10^{-4})$.

The step necessary for algorithm implementation is derivation of the Newton jacobian (5.52). As with all GWS^h algorithms it is formed via assembly

$$[JAC] = S_e([JAC]_e), \quad [JAC]_e \equiv \frac{\partial\{FQ\}_e}{\partial\{Q\}_e} \tag{5.54}$$

and matrix differentiation of $\{FQ\}_e$ by $\{Q\}_e$ is an *analytical* operation involving the chain rule. The suggested exercise requests verification that for any basis completeness degree k

$$
\begin{aligned}
[JAC]_e &\equiv \partial\{WS\}_e/\partial\{Q\}_e \\
&= (A)(\,)\{\,\}(0;-1)[A211d]\,[\,] \\
&\quad +(B)(\,)\{Q\}(0;-1)[A3011d]\,[\,] \\
&\quad +(B)(\,)\{Q\}(0;-1)[A3110d]\,[\,] \\
&\quad +(C)(\,)\{Q,Q\}(0;-1)[A3011d]\,[\,] \\
&\quad +(2C)(\,)\{Q\}(0;-1)[A3110d]\,[Q] \\
&\quad +(HCON)(\,)\{\,\}(0)[ONE]\,[\,].
\end{aligned}
\tag{5.55}
$$

In equation (5.55), the template sixth bracket when empty contains the *identity matrix*, a diagonal matrix of ones, as generated by DOF column matrix analytical differentiation

$$\frac{\partial\{Q\}_e}{\partial\{Q\}_e} = \frac{\partial\{Q1,Q2\}_e^T}{\partial\{Q1,Q2\}_e^T} = \begin{bmatrix} 1 & 0 \\ 0 & 1 \end{bmatrix}. \tag{5.56}$$

The next to last line in equation (5.55) is for the quadratic nonlinearity. When differentiating the hypermatrix nonlinearity the DOF arrays $\{Q\}$ in the template third and sixth brackets must be interchanged in using the chain rule. This results in [A3110d] replacing [A3011d], an exercise. The sixth bracket remains a diagonal matrix containing the DOF entries

$$\frac{\partial\{Q,Q\}_e}{\partial\{Q\}_e} \Rightarrow 2\begin{bmatrix} Q1 & 0 \\ 0 & Q2 \end{bmatrix}_e. \tag{5.57}$$

Table 5.4 Newton algorithm template .m file snippet

```
%        form      residual

resmat = asres1d (A, [ ], [ ], -1, A3011L. [Q]);
resmat = resmat + asres1d (B, [ ], Q', -1, A3011L. [Q]);
resmat = resmat + asres1d (B, [ ], [ ], Q^2, -1, A3011L. [Q]);
       res = resmat

%        form      jacobian

jac = asjac1d (A, [ ], [ ], -1, A3011L. [ ]);
jac = jac + asjac1d (B, [ ], Q', -1, A3011L. [ ]);
jac = jac + asjac1d (B, [ ], Q', -1, A3110L. [ ]);
jac = jac + asjac1d (C, [ ], Q', Q^2, A3011L. [ ]);
jac = jac + asjac1d (2*C, [ ], Q', -1, A3110L. [Q]);

%           modify the jacobian to enforce Dirichlet BCs at both ends
jac (1, 1)                =        1E10;
jac (nnodes, nnodes)      =        1E10;
dQ = jac\(-1*(res));          %      linear solve for dQ
Q = Q + dQ;                   %         update Q
```

This nonlinear DE + BCs statement well illustrates template hypermatrix syntax directly facilitating precise GWS^h algorithm conversion to computable form. The associated $k = 1$ basis MATLAB® .m file snippet for computer lab 5.4 is listed in Table 5.4. The complete template augments these data operations for [BC] and source terms, as previously developed.

The Newton algorithm must enforce that select DOF in $\{Q\}$ are constrained by a Dirichlet BC, hence not variables. This is easily accomplished by multiplying the matrix diagonal entry in [JAC] for each such DOF by a large number, 1E10 in Table 5.4. The algebraic solution process is thereby *informed* that such constrained δQ will be computed essentially zero. During the following DOF update process (5.53), DOF in $\{Q\}$ constrained by T_b are *not* updated, hence no error enters $\{Q\}^{p+1}$.

Computer lab 5.4 suggests the pertinent accuracy/convergence study. Thus generated *a posteriori* data verifies $k = 1, 2, 3$ basis GWS^h solution adherence to the asymptotic theory (5.45) for all conductivity nonlinear definitions. Interestingly, these GWS^h solutions generate DOF that are *nodally exact* on any mesh M(!) for the conductivity linear-dependence specification. This validates weak form foundation robustness of the nonlinear GWS^h algorithm.

Continuing, these data will also confirm Newton algorithm quadratic convergence for all degree k bases, once computed $\max|\{\delta Q\}|$ decreases below unity. Remove the template nonlinear jacobian contributions by inserting % on .m file jacobian lines with matrix [A3110d], then observe this *quasi*-Newton algorithm convergence rate degrades to about linear.

5.9 Static Condensation, *p*-Elements

For $n=1$ DE + BCs statements the developed theory predicts, and chapter *a posteriori* data confirm, significantly improved accuracy for Lagrange $k>1$ trial space basis GWS^h implementations. However, for real-world $n>1$ dimensional PDE systems, the rapid size escalation of the global [Matrix] created by use of $k>1$ bases leads to significantly more compute-intense algebraic processes.

One approach to moderation of this detraction is the development of *p*-elements wherein all nonvertex DOF are removed via *static condensation* prior to entering the algebraic process. The result is the containment of $k>1$ [Matrix] order escalation while benefiting from utilization of the better performing FE $k>1$ bases.

Static condensation is readily illustrated for linear DE + BCs without source, which enables an element level illustration. Selecting the $k=2$ basis the nonvertex DOF QM is to be removed. The parent $\{WS\}_e$ term is

$$\{WS\}_e = [\text{DIFF}]_{e3\times 3}\{Q\}_{e3\times 1} - \{b\}_{e3\times 1}, \text{ and } \{Q\}_e = \left\{\begin{array}{c} \text{QL} \\ \text{QM} \\ \text{QR} \end{array}\right\}_e \tag{5.58}$$

and subscripts in equation (5.58) denote matrix order. Shuffle matrix rows and columns to move QM to the bottom location in $\{Q\}_e$, then partition equation (5.58) into the form

$$\{WS\}_e = \begin{bmatrix} \text{DIFF}_{\alpha\alpha} & \text{DIFF}_{\alpha\beta} \\ \text{DIFF}_{\beta\alpha} & \text{DIFF}_{\beta\beta} \end{bmatrix}_e \left\{\begin{array}{c} Q_\alpha \\ Q_\beta \end{array}\right\}_e - \left\{\begin{array}{c} b_\alpha \\ b_\beta \end{array}\right\}_e, \left\{\begin{array}{c} Q_\alpha \\ Q_\beta \end{array}\right\}_e = \left\{\begin{array}{c} \text{QL} \\ \text{QR} \\ \text{QM} \end{array}\right\}_e. \tag{5.59}$$

The lower matrix partition in equation (5.59) is solved for DOF Q_β, which for $k=2$ is only QM, but for $k=3$ would contain the two nonvertex DOF (an exercise). The result is

$$\{Q_\beta\}_e = [\text{DIFF}_{\beta\beta}]_e^{-1}\left((b_\beta)_e - [\text{DIFF}_{\beta\alpha}]_e\{Q_\alpha\}_e\right). \tag{5.60}$$

Figure 5.5 Uniform $M = 2$, $k = 2$ basis mesh

Substituting equation (5.60) into equation (5.59) produces the *statically-condensed* 2×2 element level matrix statement

$$\{WS\}_e = \left[DIFF_{\alpha\alpha} - DIFF_{\alpha\beta}DIFF_{\beta\beta}^{-1}DIFF_{\beta\alpha}\right]_e \{Q_\alpha\}_e$$
$$- \{b_\alpha\}_e + \left[DIFF_{\alpha\beta}DIFF_{\beta\beta}^{-1}\right]_e \{b_\alpha\}_e. \tag{5.61}$$

This reduced order (superscript R) matrix expression for $\{WS\}_e$ is

$$\{WS\}_e = [DIFF]_e^R \{Q\}_e^R - \{b\}_e^R \tag{5.62}$$

for the definitions

$$[DIFF]_e^R = \left[DIFF_{\alpha\alpha} - DIFF_{\alpha\beta}DIFF_{\beta\beta}^{-1}DIFF_{\beta\alpha}\right]_{2\times2}$$
$$\{b\}_e^R = \{b\}_e - \left[DIFF_{\alpha\beta}DIFF_{\beta\beta}^{-1}\right]_e \{b_\beta\}_e \tag{5.63}$$
$$\{Q\}_e^R = \{QL, QR\}_e^T.$$

Example 5.3

Resolve Example 5.2 using the statically condensed $k = 2$ formulation on an $M = 2$ uniform mesh. The solution domain with DOF notation is given in Figure 5.5. The data arrays are

$$\{COND\}_{e=1}^T = \{10., 15.\}, \; l_e = 0.5, \; h = 20., \; T_r = 1500$$

$$\{COND\}_{e=2}^T = \{15., 20.\}, \; l_e = 0.5.$$

On $\Omega_{e=1}$ the operation (5.63) generates

$$[DIFF]_e\{Q\}_e = \frac{\{10,15\}}{6*0.5}\left[\begin{array}{ccc}\begin{Bmatrix}11\\3\end{Bmatrix} & -\begin{Bmatrix}12\\4\end{Bmatrix} & \begin{Bmatrix}1\\1\end{Bmatrix}\\ -\begin{Bmatrix}12\\4\end{Bmatrix} & \begin{Bmatrix}16\\16\end{Bmatrix} & -\begin{Bmatrix}4\\12\end{Bmatrix}\\ \begin{Bmatrix}1\\1\end{Bmatrix} & -\begin{Bmatrix}4\\12\end{Bmatrix} & \begin{Bmatrix}3\\11\end{Bmatrix}\end{array}\right]\begin{Bmatrix}QL\\QM\\QR\end{Bmatrix}_e$$

$$= \frac{10}{3}\begin{bmatrix}12.5 & 2.5 & -18\\2.5 & 19.5 & -22\\-18 & -22 & 40\end{bmatrix}\begin{Bmatrix}QL\\QM\\QR\end{Bmatrix}_e$$

hence

$$[\text{DIFF}]_{e=1}^{R} = \frac{10}{3}\begin{bmatrix} 15.5 & 2.5 \\ 2.5 & 19.5 \end{bmatrix} - \begin{Bmatrix} -18 \\ -22 \end{Bmatrix}\frac{1}{40}\{-18 \quad -22\}$$

$$= \frac{10}{3}\begin{bmatrix} 7.4 & -7.4 \\ -7.4 & 7.4 \end{bmatrix}.$$

Similarly, on $\Omega_{e=1}$

$$[\text{DIFF}]_{e=2}^{R} = \frac{10}{6}\begin{bmatrix} 22.5 & 3.5 \\ 3.5 & 26.5 \end{bmatrix} - \begin{Bmatrix} -26 \\ -30 \end{Bmatrix}\frac{1}{56}\{-26 \quad -30\}$$

$$= \frac{10}{3}\begin{bmatrix} 10.43 & -10.43 \\ -10.43 & 10.43 \end{bmatrix}.$$

The Robin BC application is unaffected and $\{b_\beta\}_e = \{0\}$. Hence

$$\text{GWS}^h = S_e\{\text{WS}\}_e$$

$$= \frac{10}{3}\begin{bmatrix} 7.4+6, & -7.4, & 0 \\ -7.4, & 7.4+10.43, & -10.43 \\ 0, & 0, & 3/10 \end{bmatrix}\begin{Bmatrix} Q_1 \\ Q_{1.5} \\ Q_2 \end{Bmatrix} - \begin{Bmatrix} 30,000. \\ 0. \\ 306.85 \end{Bmatrix} = 0\}. \tag{5.64}$$

The solution of equation (5.64) for the left DOF is $Q_1 = 999.97042$ F, which is identical to within round-off with the 999.97043 F obtained via the standard $M=2$, $k=2$ GWS^h algorithm, an extension on Example 5.2.

5.10 Chapter Summary

The goal of the chapter is the introduction of basic operational features of FE trial space basis discrete implementations GWS^h for the well understood $n=1$ DE + BCs statement. Chapter content illustrates "everything of importance" about the theory via hand examples and computer labs detailing $1 \leq k \leq 3$ basis constructions, asymptotic error estimation, nonsmooth data, *nonlinearity*, and static condensation. Importantly, each issue is clearly organized for computation via object-oriented *templates*, MATLAB®-enabled using the *FEmPSE* toolbox.

Hopefully your conclusion is that the process GWS^N ⇒ GWS^h precisely converts solution of a conservation principle differential equation with BCs into an algebraic matrix statement amenable to computing. In context, the underlying weak form theory embodies the classical SOV tools of trial space (including enhancement), function orthogonality, and integral/differential calculus in providing a precise *recipe* for addressing all mathematical issues, complete with the ability to quantify *solution quality*.

In summary, for any linear $n = 1$ two-point boundary value problem with applicable boundary conditions and square-integrable data, the FE implementation of GWS^N on a mesh Ω^h produces GWS^h, the solvable algebraic form of which is

$$\mathrm{GWS}^h = S_e\{\mathrm{WS}\}_e = \{0\}$$
$$\{\mathrm{WS}\}_e = [\mathrm{DIFF}]_e + [\mathrm{BC}]_e\{Q\}_e - \{b(s, T_r)\}_e. \tag{5.13}$$

Alternatively, for the two-point boundary value problem containing nonlinearity equation (5.13) is replaced as

$$\mathrm{GWS}^h \equiv \{FQ\}^p = S_e\{\mathrm{WS}\}_e^p = \{0\}$$
$$\{\mathrm{WS}\}_e^p = ([\mathrm{DIFF}(\{Q^p\})]_e + [\mathrm{BC}]_e)\{Q\}_e^p - \{b(\{Q^p\})\}_e \tag{5.51}$$

with iterative algebraic solution via the Newton algorithm

$$[\mathrm{JAC}]^p\{\delta Q\}^{p+1} = -\{FQ\}^p \tag{5.50}$$

with analytical jacobian formation using calculus and chain rule

$$[\mathrm{JAC}] = S_e([\mathrm{JAC}]_e), \quad [\mathrm{JAC}]_e \equiv \frac{\partial\{FQ\}_e}{\partial\{Q\}_e}. \tag{5.54}$$

In all instances, upon recognizing the *objects* in statements (5.13), (5.50)–(5.54) are universally classifiable by type, all contributions to a GWS^h algorithm are expressible in the *template* form

$$\{\mathrm{WS}\}_e = (\mathrm{const})\begin{pmatrix} elem \\ const \end{pmatrix}_e \begin{Bmatrix} distr \\ data \end{Bmatrix}_e^T (metric\ data;\ \det)\ [Matrix]\begin{Bmatrix} Q\ \mathrm{or} \\ data \end{Bmatrix}_e. \tag{5.14}$$

The *accuracy* of a GWS^h solution is precisely quantified using *a posteriori* data generated during regular mesh refinement. For arbitrary *nonsmooth data* driving the problem, provided it is *square integrable*, and realizing data exists as well on the domain boundary $\partial\Omega$, the rate at which the GWS^h solution error approaches zero is

$$\|e^h\|_E \leq C l_e^{2k}\left(\|\mathrm{data}\|_{\Omega,L2}^2 + \|\mathrm{data}\|_{\partial\Omega,L2}^2\right) \tag{5.65}$$

as measured in the energy norm pertinent to the conservation principle and its BCs. The energy norm pertinent to $n = 1$ DE + BCs statements is the computable form

$$\|T^h\|_E \equiv \frac{1}{2}\int_\Omega k\frac{dT^h}{dx}\frac{dT^h}{dx}\,dx + \frac{1}{2}hT^hT^h\bigg|_{\partial\Omega} = \frac{1}{2}\sum_{e=1}^M \{Q\}_e^T[\mathrm{DIFF}+\mathrm{BC}]_e\{Q\}_e \qquad (5.66)$$

in concert with equation (5.65) leading to the *quantitative* error estimate

$$\|e^{h/2}\|_E = \frac{\Delta\|T^{h/2}\|_E}{2^{2k}-1}. \qquad (5.43)$$

Exercises

5.1 For the DE + BCs statement (5.1)–(5.3), proceed through the GWS^N steps leading to GWS^h, (5.7).

5.2 Verify the FE linear basis matrix library (5.10)–(5.12).

5.3 Verify the $k = 1$ basis M = 2 solution, Example 5.1.

5.4 Confirm the basis entries for $T_e = \{N_k\}^T\{Q\}_e$ for $k = 2$, (5.21), and confirm they possess required (0,1) consistency for each DOF.

5.5 Derive the basis entries for $T_e = \{N_k\}^T\{Q\}_e$ for $k = 3$, (5.22), and confirm they possess required (0,1) consistency for each DOF.

5.6 Verify the hypermatrix $[\mathrm{DIFF}]_e$ formed using $\{N_2\}$, (5.30).

5.7 Confirm the truncation error estimate $e^{h/2}$, (5.37).

5.8 Verify the convergence slope equation, (5.38).

5.9 For temperature-dependent conductivity, proceed through the GWS^h algorithm steps (5.49)–(5.54), hence thoroughly verify the Newton jacobian template (5.55)–(5.57).

5.10 Verify the statically condensed $[\mathrm{DIFF}]_e^R$ matrices in Example 5.3.

5.11 Repeat the Example 5.3 static condensation exercise for the $k = 3$ basis implementation.

Computer Labs

The *FEmPSE* toolbox and all MATLAB® .m files required for conducting the computer labs are available for download at www.wiley.com/go/baker/finite. The Chapter 5 computer labs are:

5.1 Complete the DE + BCs $n = 1$ GWS^h linear basis convergence study to generate the data in Table 5.2. Examine the MATLAB® .m file instructions, Table 5.1, for correctness in forming the $[DIFF]_e$ matrix, also input for $[BC]_e$ and enforcement of the Dirichlet BC. Recalling $T_{exact} = 1000.00F$, verify the error relationship (5.18).

5.2 Augment the lab 5.1 .m file to enable the $1 \leq k \leq 3$ FE bases regular mesh-refinement study for uniform discretizations Ω^h. For generated *a posteriori* data verify the asymptotic error estimates in T_{max}, (5.37), and in the energy norm (5.42), via the slope formulations (5.38) and (5.44). Summarize your data presentation as in Table 5.3.

5.3 Execute a regular mesh-refinement study for the discontinuous conduction with distributed source problem statement, first for a single material and source, then for dissimilar materials with double sine source, for $1 \leq k \leq 2$ bases. Add code to compute the boundary efflux DOF at each end of the domain using GWS^h *a posteriori* data. Document convergence in $\left\| T^h \right\|_E$ and T_{max} and compare to theory (5.43), (5.37).

5.4 Edit the linear $DE + BCs$ GWS^h algorithm template, Table 5.1, to insert the Newton algorithm template snippet, Table 5.4, for the nonlinear $DE + BCs$ statement with temperature-dependent conductivity. For the linear dependence case, compare iteration convergence rates with and without the nonlinear jacobian contribution. Execute a regular mesh-refinement study to generate error estimates in T_{mid} and $\left\| T^h \right\|_E$ for $k = 1$, 2 basis algorithms. Edit the .m file data specification to quadratic temperature dependence, then repeat the Newton regular mesh-refinement experiment; also evaluate the quasi-Newton jacobian option for a select M specification.

References

[1] Zienkiewicz, O.C. and Taylor, R.L. (1989) *The Finite Element Method*, McGraw-Hill, New York, NY.

[2] Oden, J.T. (1994) Optimal *h–p* finite element methods. *Comput. Method. Appl. M.*, **112**, 309–331.

[3] Carslaw, H.S. and Jaeger, J.C. (1959) *Conduction of Heat in Solids*, Clarendon Press, UK.

[4] Baker, A.J. (2013) *Optimal Modified Continuous Galerkin CFD*, Wiley, New York, NY.

[5] Strang, G. and Fix, G.J. (1973) *An Analysis of the Finite Element Method*, Prentice-Hall, Englewood Cliffs, NJ.

[6] Oden, J.T. and Reddy, J.N. (1976) *An Introduction to the Mathematical Theory of Finite Elements*, Wiley-Interscience, New York, NY.

6

Engineering Sciences, $n = 1$: GWS^h $\{N_k(\zeta_\alpha)\}$ implementations in the computational engineering sciences

6.1 Introduction

To this point, the engineering science discipline supporting exposition of WS$^N \Rightarrow$ GWS$^N \Rightarrow$ GWS^h is heat transfer, recognizing that almost every technical individual possesses a certain familiarity. The illustrated GWS^N construction, discretely implemented via finite element (FE) trial space bases, is applicable to a wide range of problem statements in the *computational engineering sciences*, for example

heat transfer
structural mechanics
mechanical vibrations
heat/mass transport
electromagnetics
fluid mechanics

This chapter addresses this range of engineering science disciplines. The content is restricted to $n = 1$ to support transparent differential equation system conversion to GWS^h statement templates supporting hands-on computing for a range of completeness

Finite Elements ⇔ Computational Engineering Sciences, First Edition. A. J. Baker.
© 2012 John Wiley & Sons, Ltd. Published 2012 by John Wiley & Sons, Ltd.

Figure 6.1 Computational engineering sciences enlightenment

degree FE trial space bases. The focus remains *error quantification* via the available theory, with attention to the caveats of *nonsmooth* data.

Thus, in proceeding through this chapter the reader is exposed to a multidisciplinary *problem-solving environment* supported by MATLAB®. The mission is generating a knowledge base providing detailed insight into requirements for cogent utilization of weak form based computational methodology (Figure 6.1).

6.2 The Euler–Bernoulli Beam Equation

Structural mechanics is the engineering discipline where FE methods were first reduced to computing practice [1]. The elementary structures problem statement is a beam subjected to a transverse load, as occurs in myriad real-world applications. Figure 6.2 illustrates the classic problem definition.

The basic formulation involves stationary momentum conservation ($DP = 0$) combined with a linear elastic constitutive closure model. The other two conservation principles are intrinsically satisfied, $DM = 0 = DE$. Assuming a linear homogeneous isotropic medium, recall Chapter 2.4, and assuming transverse plane sections remain plane results in the *Euler–Bernoulli* (E-B) beam theory differential equation

$$DP \cdot j: \ \mathcal{L}(y) = \frac{d^2}{dx^2}\left(E\,I_{zz}(x)\frac{d^2y}{dx^2}\right) + p(x) = 0. \qquad (6.1)$$

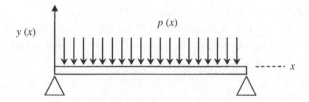

Figure 6.2 Structural beam with transverse loading $p(x)$

In equation (6.1), $y(x)$ is the distribution of vertical displacement from the horizontal, $p(x)$ is the distributed load, E the Youngs modulus, and $I_{zz}(x) = $ the z-axis principle moment of inertia *distribution* of the beam. The practical BCs include fixed displacement and/or slope (dy/dx), plus an imposed moment (M) or shear force (V), at each end of the beam.

The GWS^N construction for equation (6.1) requires dual integration-by-parts to symmetrize the test and trial function derivative support requirement. Completing the suggested exercise will confirm that this operation produces, for $1 \le \beta \le N$

$$
\begin{aligned}
\text{GWS}^N &\equiv \int_\Omega \Psi_\beta(x)\mathcal{L}\left(y^N\right)dx \equiv 0 = \int_\Omega \Psi_\beta \left[\frac{d^2}{dx^2}EI_{zz}\frac{d^2 y^N}{dx^2} + p\right]dx \\
&= \int_\Omega \left[-\frac{d}{dx}\Psi_\beta \frac{d}{dx}EI_{zz}\frac{d^2 y^N}{dx^2} + \Psi_\beta p\right]dx + \Psi_\beta \frac{d}{dx}EI_{zz}\frac{d^2 y^N}{dx^2}\Bigg|_L^R \\
&= \int_\Omega \left[\frac{d^2}{dx^2}\Psi_\beta EI_{zz}\frac{d^2 y^N}{dx^2} + \Psi_\beta p\right]dx - \frac{d\Psi_\beta}{dx}M\Bigg|_L^R + \Psi_\beta V\Bigg|_L^R = 0.
\end{aligned}
$$ (6.2)

Importantly, the two integration-by-parts operations *directly* introduce the eligible moment and shear BCs at the beam ends (x_R, x_L) with definitions

$$
M \equiv EI\frac{d^2 y}{dx^2}, \quad V \equiv \frac{d}{dx}\left(EI\frac{d^2 y}{dx^2}\right).
$$ (6.3)

The FE discrete implementation GWS^h of equation (6.2) leads to

$$
\text{GWS}^h \equiv S_e\{\text{WS}\}_e \equiv \{0\}
$$

$$
\{\text{WS}\}_e \equiv \int_{\Omega_e} \frac{d^2\{N_k\}}{dx^2}EI_{zz}(x)\frac{d^2\{N_k\}^T}{dx^2}dx\{Y\}_e
$$ (6.4)

$$
+ \int_{\Omega_e}\{N_k\}p(x)dx - \frac{d\{N_k\}}{dx}M\Big|_{\partial\Omega_e} + \{N_k\}V\Big|_{\partial\Omega_e}
$$

Recalling template syntax

$$
\{\text{WS}\}_e = (\text{const})\begin{pmatrix}\text{elem}\\\text{const}\end{pmatrix}_e \begin{Bmatrix}\text{distr}\\\text{data}\end{Bmatrix}_e^T (\text{metric data; det})\,[\text{Matrix}]\begin{Bmatrix}Q \text{ or}\\\text{data}\end{Bmatrix}_e
$$ (6.5)

the template for the E-B beam theory GWS^h algorithm (6.4) is

$$
\{\text{WS}\}_e = (E)(\ )\{IZZ\}(-3)[A3022d]\{Y\} + (\ )(\ )\{\ \}(1)[A200d]\{P\} + \{BCs\}.
$$ (6.6)

The matrix Boolean indices for the diffusion term in equation (6.1) (with four spatial derivatives) are now 2 resulting from dual integration by parts. These clearly indicate the differentiation order the selected FE trial space basis must support. The exponent -3 on l_e is generated by $2 \times 2 = 4$ differentiations followed by one integration.

As the FE trial space basis must possess square integrable second-derivative products, the requirement for $\{N_k(\zeta_\alpha)\}$ is $k \geq 2$. Implementation via the developed Lagrange quadratic or cubic bases is admissible, but implementing the E-B beam theory *intrinsic* moment and shear BCs (6.2) is problematic.

The alternative is the cubic Hermite basis that is both natural and highly informative. The starting point is the basic cubic polynomial

$$y_e(\bar{x}) = a + b(\bar{x}/l_e) + c(\bar{x}/l_e)^2 + d(\bar{x}/l_e)^3. \tag{6.7}$$

The coefficient set (a,b,c,d) is then replaced in terms of the degrees of freedom (DOF) now defined as slope and displacement at each end of an element, Figure 6.3. The resultant solution (an exercise) that determines the cubic Hermite basis for discrete implementation of equation (6.4) is

$$y_e(\bar{x}) \equiv \{N_3^H\}^T \{Q\}_e, \{N_3^H\}_e = \left\{ \begin{array}{c} 1 - (\zeta_2)^2(1 + 2\zeta_1) \\ \zeta_2(\zeta_1)^2/l_e \\ (\zeta_2)^2(1 + 2\zeta_1) \\ -\zeta_1(\zeta_2)^2/l_e \end{array} \right\}, \{Q\}_e = \left\{ \begin{array}{c} YL \\ DYL \\ YR \\ DYR \end{array} \right\}_e. \tag{6.8}$$

In equation (6.8), the DOF YL denotes left-node displacement, DYL the left-node slope (Derivative), with corresponding interpretations for YR and DYR. Another exercise asks that you verify that each element of the Hermite basis $\{N_3^H\}_e$ satisfies all $(0, 1)$ Kronecker delta requirements at both geometric nodes of Ω_e. Finally, note the Hermite basis is element-dependent as it contains l_e, which did not occur for the Lagrange trial space basis constructions.

In structural mechanics, the GWS^h theory-generated lead term in equation (6.4) is mathematically analogous to the diffusion matrix $[DIFF]_e$ of the previous chapter heat-transfer development. In the structures FE community it is called the *stiffness matrix*, and the formation of $[STIFF]_e$ requires integrals of polynomial products in the ζ_α coordinates. For variable moment of inertia $I_{zz}(x)$, the practical design choice, it is a

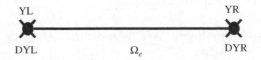

Figure 6.3 Cubic Hermite basis nodes and DOF

hypermatrix. The required Hermite basis second derivatives are readily formed from equation (6.8) yielding

$$[\text{STIFF}]_e \equiv \int_{\Omega_e} \frac{\mathrm{d}^2\{N_3^{\text{H}}\}}{\mathrm{d}x^2} E\, I_{zz}(x) \frac{\mathrm{d}^2\{N_3^{\text{H}}\}^T}{\mathrm{d}x^2}\,\mathrm{d}x$$

$$= E\{IZZ\}_e^T \int_{\Omega_e} \left\{ \begin{matrix} \zeta_1 \\ \zeta_2 \end{matrix} \right\} \frac{2}{l_e^2} \left\{ \begin{matrix} -3 + 6\zeta_2 \\ (-2 + 3\zeta_2)l_e \\ 3 - 6\zeta_2 \\ (-1 + 3\zeta_2)l_e \end{matrix} \right\} \frac{2}{l_e^2} \{\cdot\}^T \mathrm{d}x \tag{6.9}$$

with the logical design option to interpolate $I_{zz}(x)$ via the $k = 1$ Lagrange basis. One certainly would not elect to use the Hermite basis, as the moment of inertia data varies only slowly or discontinuously, for example, tapered web or a welded flange plate.

For the same reasons, the distributed load $p(x)$ would not employ the Hermite basis for interpolation. Selecting $\{N_1\}$ for all data, and making ample use of analytical integration (5.12) produces the E-B beam theory GWS^h algorithm matrix library

$$[\text{STIFF}]_e = \frac{E\{IZZ\}_e^T}{l_e^3} \begin{bmatrix} \left\{\begin{matrix}6\\6\end{matrix}\right\} & \left\{\begin{matrix}4\\2\end{matrix}\right\}l_e & \left\{\begin{matrix}-6\\-6\end{matrix}\right\} & \left\{\begin{matrix}2\\4\end{matrix}\right\}l_e \\ & \left\{\begin{matrix}3\\1\end{matrix}\right\}l_e^2 & \left\{\begin{matrix}-4\\-2\end{matrix}\right\}l_e & \left\{\begin{matrix}1\\1\end{matrix}\right\}l_e^2 \\ (\text{sym}) & & \left\{\begin{matrix}6\\6\end{matrix}\right\} & \left\{\begin{matrix}-2\\-4\end{matrix}\right\}l_e \\ & & & \left\{\begin{matrix}1\\3\end{matrix}\right\}l_e^2 \end{bmatrix} \tag{6.10}$$

$$\equiv E\{IZZ\}_e^T[A3022LH]_e$$

$$\{b\}_e = \frac{l_e}{60} \begin{bmatrix} 21 & 9 \\ 3l_e & 2l_e \\ 9 & 21 \\ -2l_e & -3l_e \end{bmatrix} \left\{ \begin{matrix} PL \\ PR \end{matrix} \right\}_e \equiv [A200HL]_e\{P\}_e \tag{6.11}$$

wherein matrix suffixes $d \Rightarrow LH$ and HL clearly define the selected *data*/solution trial space basis mixed interpolations.

In distinction to the heat-transfer algorithm, the library matrices $[A3022LH]_e$ and $[A200HL]_e$ are element-dependent via l_e embeddings. Another exercise will confirm the moment and shear BC matrix contributions to equation (6.4) are

$$-\frac{d\{N_3^H\}}{dx}M\Big|_{\partial\Omega_e} \Rightarrow -MOM\begin{Bmatrix} 0 \\ \delta_{e1} \\ 0 \\ \delta_{eM} \end{Bmatrix}, \{N_3^H\}V\Big|_{\partial\Omega_e} \Rightarrow VSH\begin{Bmatrix} \delta_{e1} \\ 0 \\ \delta_{eM} \\ 0 \end{Bmatrix}. \qquad (6.12)$$

The GWS^h theory-generated Kronecker delta switches *automatically* apply these data to the appropriate DOF! The resulting completion of template (6.6) is

$$\begin{aligned} \{WS\}_e = (E)(\ )\{IZZ\}(-3)[A3022LH]\{Q\} \\ -(\ )(\ )\{\ \}(1)[A200HL]\{P\} \\ -(MOM)(\ )\{\ \}(\ )[ONEM]\{\ \} \\ +(VSH)(\ )\{\ \}(\ )[ONEV]\{\ \}, \end{aligned} \qquad (6.13)$$

where [ONEM] and [ONEV] are template names for the Kronecker delta matrices defined in equation (6.12).

The textbook verification for the E-B beam theory GWS^h formulation specifies a uniformly distributed load on a uniform cross-section simply supported beam. For these data, the GWS^h algorithm $\{N_3^H\}$ Hermite basis implementation produces displacement DOF in *exact* agreement with the analytical solution for any mesh M!

Notwithstanding, $y^h(x)$ does possess *error* $e^h(x)$ and convergence in the energy norm $\|e^h\|_E$ can indeed be verified. Computer lab 6.1 will produce *a posteriori* data that is surprising regarding confirmed convergence rate. The theory developed in Chapter 5, equation (5.45), accurately predicting convergence rates of 2k is well verified. However, equation (6.1) is a *bi-harmonic* differential equation that requires its solution possess increased differentiability. The required theory refinement is, [2]

$$\left\|e^h\right\|_E \le Cl_e^{2\gamma}\left\|data\right\|_{L2}^2, \quad \gamma = (k+1-m), \qquad (6.14)$$

where k remains FE basis completeness degree. For the 4th order (biharmonic) derivative in equation (6.1) the coefficient is $m \equiv 2$ in equation (6.14) while for the 2nd order derivative in heat conduction, equation (5.1), $m \equiv 1$. The data the lab generates is summarized in Table 6.1 which confirms the theory accurately

Table 6.1 E-B beam theory GWS^h $\{N_3^H\}$ basis solution convergence

Mesh	$\|y^h\|_E$	$\|y^h\|_{exact}$ (est.)	Slope
Ω^h	11.39999270400065		
$\Omega^{h/2}$	11.51249263200054	11.51999262720054	
$\Omega^{h/4}$	11.51952387749862	11.51999262719849	4.0000
$\Omega^{h/8}$	11.51996333031451	11.51999262716890	4.0000
$\Omega^{h/16}$	11.51999079587107	11.51999262690817	4.0000
$\Omega^{h/32}$	11.51999251282745	11.51999262729121	3.9997
$\Omega^{h/64}$	11.51999261495366	11.51999262176208	4.0714

predicts asymptotic convergence in displacement energy norm, equation (6.14), as $2\gamma = 2(3 + 1 - 2) = 2(2) = 4$.

6.3 Euler–Bernoulli Beam Theory GWS^h Reformulation

Implementing the cubic Hermite basis GWS^h algorithm required significant programming effort due to embedded l_e in the library matrix, equations (6.10)–(6.11). This detail is hidden from view in the *FEmPSE* toolbox, also in a commercial code. Factually, equation (6.1) results from direct substitution of definition (6.3) for moment M into $DP \cdot j$. The alternative is to formulate GWS^N for a pair of differential equations considering the moment $M \equiv EI(x)d^2y/dx^2$ as displacement distribution differential equation.

Simplifying to a uniform moment of inertia and generalizing for application of a *point load P*, the resultant ordinary differential equation (ODE) pair description of E-B beam theory is

$$\mathcal{L}(M) = -\frac{d^2 M}{dx^2} - p(x) - P = 0$$

$$\mathcal{L}(y) = -EI\frac{d^2 y}{dx^2} + M(x) = 0$$

$$(6.15)$$

The sequence GWS$^N \Rightarrow$ GWS^h is implemented via *any* completeness degree k basis to alter the term contributions in equation (6.4) to

$$\{WS(M)\}_e = \int_{\Omega_e} \left(\frac{d\{N\}}{dx}\frac{d\{N\}^T}{dx}\{M\}_e - \{N\}\{N\}^T\{P\}_e \right) dx - \{N\}\frac{dM}{dx}\Big|_L^R - P\{\delta\} \qquad (6.16)$$

$$\{WS(Y)\}_e = \int_{\Omega_e} \left(EI\frac{d\{N\}}{dx}\frac{d\{N\}^T}{dx}\{Y\}_e + \{N\}\{N\}^T\{M\}_e \right) dx - \{N\}EI\frac{dy}{dx}\Big|_L^R . \qquad (6.17)$$

In equation (6.16) the matrix $\{\delta\}$ is again the Kronecker delta that applies point load P at a geometric node (*only!*). These GWS^h algorithm statements are of standard form and the resultant *two-group template* including BCs for the dual ODE E-B beam formulation are

$$\{WS(M)\}_e = (\,)(\,)\{\,\}(-1)[A211d]\{M\} - (\,)(\,)\{\,\}(1)[A200d]\{P\}$$
$$-(SHR)(\,)\{\,\}(\,)[ONE]\{\,\} - P\{\delta\}$$

$$(6.18)$$

$$\{WS(Y)\}_e = (EI)(\,)\{\,\}(-1)[A211d]\{Y\} + (\,)(\,)\{\,\}(1)[A200d]\{M\}$$
$$-(SLP)(\,)\{\,\}(\,)[ONE]\{\,\} .$$

$$(6.19)$$

Table 6.2 Euler–Bernoulli beam, uniform distributed load, $k = 1$

M	l_e	$\|\|M\|\|_E$	$\|\|Y\|\|_E$	Error$\|\|Y\|\|_E$	Slope (Y)	Error$\|\|M\|\|_E$	Slope (M)
8	0.125	41015625	1460.00				
16	0.062	41503906	1503.14	14.3787		162.760	
32	0.031	41625976	1514.07	3.64495	1.9799	40.6901	2
64	0.015	41656494	1516.82	0.91439	1.9950	10.1725	2
128	0.007	41664123	1517.50	0.22879	1.9987	2.54313	2

Computer lab 6.2 suggests Lagrange $k = 1,2$ FE trial space basis implementations confirming validity of the asymptotic error estimate (6.14) for $m \equiv 1$, for both a uniformly distributed load and a single point load P at the beam center. Tables 6.2 and 6.3 tabulate the $k = 1$ basis *a posteriori* data and Figures 6.4 and 6.5 graph the energy norm prediction comparison to theory. For the point load case, the GWS^h moment solution is mesh independent, which confirms these solutions are the *exact* interpolant of the *only* piecewise-continuous moment distribution.

For FE $k = 1, 2$ basis GWS^h implementations, Figure 6.6 graphs the deflection state variable member energy norm data lab 6.2 will produce for uniform distributed and central point loadings. Asymptotic convergence is accurately predicted by theory, equation (6.14), for both constructions. The *error* level in the $k = 2$ basis algorithm solution is a very substantial *order* two to four decile improvement over the $k = 1$ solution on the refined meshes.

In practice, beams are subjected to a mixture of distributed and point loads. Computer lab 6.2 suggests specifying a mixed loading of your choice and repeating the convergence study. For a uniform load on the beam left half and a point load at the beam three-quarter span, Table 6.4 presents the lab regular mesh refinement *a posteriori* data for the $k = 1$ basis convergence rate. Figure 6.7 confirms existence of the *suboptimal* convergence rate approaching *linear*, not quadratic, the rate *all* $k > 1$ basis solutions will exhibit!

The *suboptimal* asymptotic convergence for *any* $k \geq 1$ basis points to another needed theory refinement. To this point in GWS^h development, that is, transition from equation (5.41) to (5.45) to (6.14), *data nonsmoothness* has not affected asymptotic

Table 6.3 Euler–Bernoulli beam, point load at the center, $k = 1$

M	le	$\|\|M\|\|_E$	$\|\|Y\|\|_E$	Error$\|\|Y\|\|_E$	Slope (Y)	Error$\|\|M\|\|_E$	Slope (M)
8	0.125	1250000	37.0148				
16	0.0625	1250000	37.3753	0.120153		0	
32	0.03125	1250000	37.4665	0.030395	1.98292	0	—
64	0.01562	1250000	37.4893	0.007621	1.99576	1.3349E-11	—
128	0.00781	1250000	37.4950	0.001906	1.99894	−3.5002E-11	—

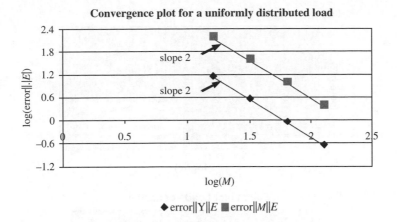

Figure 6.4 for convergence plot for a uniformly distributed load

Figure 6.4 Convergence in energy norm, uniform load, $k=1$ basis

convergence of generated *a posteriori* data. The data nonsmoothness issue is fundamental and the mathematician's response is replacement of equation (6.14) as [2]

$$\|e^h\|_E \le Cl_e{}^{2\gamma}\|data\|_{L2}^2, \quad \gamma \equiv \min\,(k+1-m,\, r-m) \tag{6.20}$$

with term definitions:

$k=$ FE trial space basis completeness degree
$m=$ integer order (1,2) of the underlying variational principle
$r=$ for specified *data*, solution differentiability order.

The GWS^h $k=1$ basis asymptotic convergence graphed in Figure 6.7, as well as for any $k>1$ algorithm implementation, clearly confirms that $(r-1)<k$ is the controlling

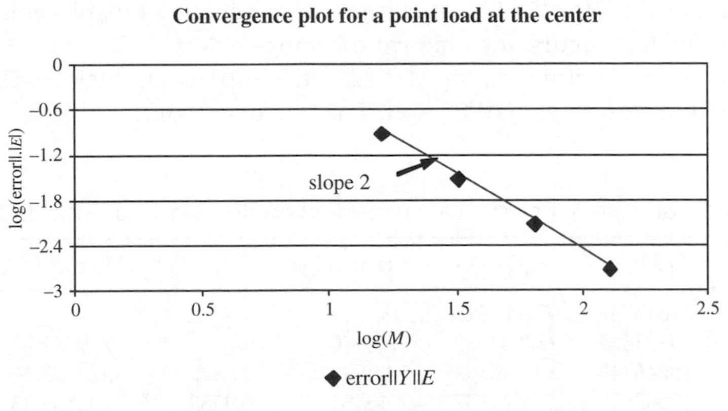

Convergence plot for a point load at the center

Figure 6.5 Convergence in energy norm, central point load, $k=1$ basis

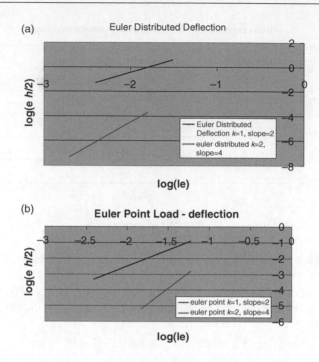

Figure 6.6 Convergence in displacement energy norm, E-B beam GWS^h algorithm, $k = 1$, 2 FE bases, (a) uniform load, (b) central point load

factor in equation (6.20). The forcing is *data nonsmoothness*, and that this is the character of this *truly* simple mixed load specification the author finds surprising. Notwithstanding, the fact that the E-B beam theory GWS^h solution process is convergent, even though suboptimal, does enable *quantitative* determination of the mesh M required to produce an *engineering accurate* solution in, say, displacement DOF extremum.

Table 6.5 summarizes the $k=1$ basis extremum moment and displacement DOF, also the node at which this occurs, for uniform meshings $8 \leq M \leq 128$. The $k=1$ solution is thus *verified accurate* to within 1% for $M = 128$. In comparison, the lab will confirm the $k=2$, $M=16$ uniform mesh GWS^h solution achieves comparable accuracy using

Table 6.4 Euler–Bernoulli beam with mixed distributed and point loads, $k=1$ basis

M	le	$\|M\|_E$	$\|Y\|_E$	Error$\|Y\|_E$	Slope (Y)	Error$\|M\|_E$	Slope (M)
8	0.125	21615532	736.78515				
16	0.062	19395675	658.04988	26.2450		739952.37	
32	0.031	18248243	614.60193	14.4826	0.8577	382477.26	0.9520
64	0.0156	17667638	592.24137	7.45351	0.9583	193535.23	0.9827
128	0.0078	17375907	580.94942	3.76398	0.9856	97243.61	0.9929

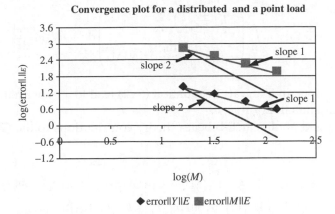

Figure 6.7 E-B beam GWS^h convergence in energy, mixed load, $k = 1$

one-fourth the DOF of the $k = 1$ algorithm, the consequence of superior attributes of enriched trial space bases.

To round out E-B beam theory development, hypothesize the limiting situation where EI becomes dependent on displacement. The associated physical phenomenon is inelastic creep leading to (prediction of) failure. The resultant issue is the elastic stress–strain correlation becoming nonlinear, thus $EI(x) \Rightarrow E(y)I(y(x))$ assuming the moment of inertia is also affected.

This E-B beam theory GWS^h algorithm must be formulated iterative, as introduced in Chapter 5.8. The moment principle is unchanged hence equation (6.16) remains. The displacement formulation inherits complexity from the integration-by-parts operation. Via calculus the result is

$$GWS^N = \int_\Omega \left(\Psi_\beta (-E(y)I_e(y(x)) \frac{d^2y}{dx^2} + M \right) dx \equiv \{0\} \tag{6.21}$$

$$\Rightarrow GWS^h \equiv S_e\{WS(Y)\}_e$$

Table 6.5 E-B beam GWS^h solution DOF accuracy, half distributed plus a point load, $k = 1$

M	Le	MMAX	MMAX (node)	YMAX	YMAX (node)	% change (MMAX)	% change (YMAX)
8	0.125	9.04E+03	4	0.032851	5		
16	0.0625	8.58E+03	8	0.030844	9	−5.066594	−6.109738426
32	0.0312	8.31E+03	14	0.029805	16	−3.214759	−3.368440896
64	0.0156	8.16E+03	27	0.029240	32	−1.850230	−1.894233557
128	0.0078	8.08E+03	52	0.028961	62	−0.953917	−0.95623456

$$\{WS(Y)\}_e = \int_{\Omega_e} \left(\begin{array}{c} EI_e \left(\dfrac{\mathrm{d}\{N\}}{\mathrm{d}x} \right) \dfrac{\mathrm{d}y_e}{\mathrm{d}x} \\[2mm] + \{N\} \dfrac{\mathrm{d}EI_e}{\mathrm{d}x} \dfrac{\mathrm{d}y_e}{\mathrm{d}x} + \{N\} M_e \end{array} \right) \mathrm{d}x - \{N\} EI_e \dfrac{\mathrm{d}y}{\mathrm{d}x} \Big|_L^R \tag{6.22}$$

and the first two domain integrals in equation (6.22) lead to a hypermatrix formulation to embed the variation in $E(y)I(y(x))$.

This hypothesized nonlinear E-B beam theory contributions to the GWS^h algorithm definition (6.4) are

$$\{WS(M)\}_e = l_e^{-1}[A\,211d]\{M\}_e - l_e\,[A\,200d]\{P\}_e - \mathrm{SHR}\big|_L^R - P\{\delta\}$$

$$\{WS(Y)\}_e = l_e^{-1}\{EI\}^T[A\,3011d]\{Y\}_e + l_e^{-1}\{EI\}_e^T[A\,3101d]\{Y\}_e$$
$$+ \ell_e[A\,200d]\{M\}_e - EI * \mathrm{SLP}\big|_L^R. \tag{6.23}$$

Recalling equation (5.50), the Newton algorithm is

$$[\mathrm{JAC}]^p\{\delta Q\}^{p+1} = -\{FQ\}^p \tag{6.24}$$

for $p = 0, 1, 2, \ldots$ the iteration index and $\{FQ\}^p$ the placeholder for the assembly of equation (6.23). Similarly, $[\mathrm{JAC}]^p$ is the assembly of the element contributions

$$[\mathrm{JAC}]_e \equiv \frac{\partial\{FQ\}_e}{\partial\{Q\}_e} \tag{6.25}$$

with sole nonlinear self-coupling contribution (a suggested exercise)

$$[\mathrm{JAC}(Y,Y)]_e \equiv \frac{\partial\{FY\}_e}{\partial\{Y\}_e} = \frac{\partial\{WS(Y)\}_e}{\partial\{Y\}_e}$$

$$= \frac{1}{\ell_e}\{EI\}_e^T[A3011] + \frac{1}{\ell_e}\{Y\}_e^T[A3110]\frac{\partial\{EI\}_e}{\partial\{Y\}_e} \tag{6.26}$$

$$+ \frac{1}{\ell_e}\{EI\}_e^T[A3101] + \frac{1}{\ell_e}\{Y\}_e^T[A3101]\frac{\partial\{EI\}_e}{\partial\{Y\}_e}.$$

Evaluating the specifics of $\partial\{EI\}_e/\partial\{Y\}_e$, typically from experimental data, will generate a diagonal matrix embedding of the nonlinearity $E(y)I(y(x))$. Recalling an empty sixth bracket is a diagonal matrix of ones, the resultant nonlinear dual equation template statement for equations (6.23)–(6.26) is

$$\{WS(M)\}_e = ()(){ }(-1)[A211d]\{M\}$$
$$-()(){ }(1)[A200d]\{P\} - P\{\delta\}$$
$$(SHR)(){ }()[ONE]{ }$$
$$\{WS(Y)\}_e = ()(){EI}(-1)[A3011d]\{Y\} \tag{6.27}$$
$$+()(){EI}(-1)[A3101d]\{Y\}$$
$$+()(){ }(1)[A200d]\{M\}$$
$$-(SLP)(){ }()[ONE]\{EI\}$$

$$[JAC(M,M)]_e = ()(){ }(-1)[A211d][]$$
$$[JAC(Y,Y)]_e = ()(){EI}(-1)[A3011d][]$$
$$= ()(){EI}(-1)[A3110d][\partial EI/\partial Y]$$
$$+()(){EI}(-1)[A3101d][] \tag{6.28}$$
$$+()(){Y}(-1)[A3101d][\partial EI/\partial Y]$$
$$[JAC(Y,M)]_e = -()(){ }(1)[A200d][]$$
$$[JAC(M,Y)]_e = [0].$$

6.4 Timoshenko Beam Theory

Direct GWS^h formulation for the E-B beam theory 4th order biharmonic ($m = 2$) equation (6.1), afforded introduction of the Hermite cubic trial space basis. Recognizing equation (6.1) combines two independent conservation principles that led to the alternative GWS^h construction for the pair of 2nd order ODEs.

Timoshenko beam theory is directly expressed by the state variable pair of transverse plane cross-section rotation (r) and vertical displacement (w). The theory coupled two-point ($m = 1$) boundary value problem statement are

$$\mathcal{L}(r) = -EI\frac{d^2 r}{dx^2} - kGA\left(\frac{dw}{dx} - r\right) = 0 \tag{6.29}$$

$$\mathcal{L}(w) = -kGA\left(\frac{d^2 w}{dx^2} - \frac{dr}{dx}\right) + p(x) + P = 0. \tag{6.30}$$

The additional *data* include the beam cross-section area $A(x)$, the beam shear modulus G, and the theory model constant $k \approx O(1)$.

The significant detraction of the FE trial space basis implementation GWS^h for equations (6.29) and (6.30) is termed *excessive stiffness*. This attribute is assigned to structural mechanics algorithms that underpredict displacements computed using relatively coarse meshes, that is, the implementation is *too* stiff. The traditional *correction* for this behavior is to *underintegrate* the FE matrix generated by the offending term.

For Timoshenko beam theory, the offender is the $kGAr$ term in equation (6.29), and underintegration alters the [A200d] matrix. Full discussion of numerical integration of FE matrices occurs in the following chapter. For now, simply state the $k = 1$ basis

altered matrix $[A200LU]$ is

$$[A200LU] \equiv \frac{l_e}{4}\begin{bmatrix} 1 & 1 \\ 1 & 1 \end{bmatrix} = [A200L] - [A2??L]. \tag{6.31}$$

The second equality identifies the analysis requirement, that is, precisely what is $[A2??L]$? Recalling the $[A200L]$ definition, equation (5.10), the completed exercise determines that

$$[A2??L] = \frac{l_e}{12}\begin{bmatrix} 1 & -1 \\ -1 & 1 \end{bmatrix} = \frac{l_e}{12}[A211L] \Rightarrow \frac{l_e^2}{12}[\text{DIFF}]. \tag{6.32}$$

Thereby, the FE structural mechanics practice of *underintegration* corresponds one-to-one with introduction of an *artificial diffusion* mechanism. Recalling the genuine diffusion matrix $[A211L]$ multiplier $1/l_e$, the artificial diffusion coefficient in equation (6.32) is instead proportional to the square of element measure l_e. Thereby, this term impact rapidly diminishes under mesh refinement, hence is operational only with coarse meshings.

The Timoshenko beam theory GWS^h algorithm is a direct adaptation on the E-B dual-variable script, amenable to nonuniform cross-section area $A(x)$, hence $I(x)$, non-constant loadings, and so on. Written as a Newton algorithm, which for the linear problem will converge in one iteration, the coupled Timoshenko beam template is

$$
\begin{aligned}
\{WS(R)\}_e &= (E)(\ )\{IZZ\}(-1)[A3011d]\{R\} \\
&\quad +(E)(\ )\{IZZ\}(-1)[A3101d]\{R\} \\
&\quad +(GK)(\ )\{A\}(1)[A3000d]\{R\} \\
&\quad \% + (GK,1/12)(\ )\{A\}(1)[A3011d]\{R\} \\
&\quad -(GK)(\ )\{A\}(0)[A3001d]\{W\} \\
\{WS(W)\}_e &= (GK)(\ )\{A\}(0)[A3001d]\{R\} \\
&\quad +(GK)(\ )\{A\}(-1)[A3011d]\{W\} \\
&\quad +(GK)(\ )\{A\}(-1)[A3101d]\{W\} \\
&\quad -(\ )(\ )\{\ \}(1)[A200d]\{P\} - P\{\delta\} \\
&\quad -(GK,A,DW)(\ )\{\ \}(\ )[\text{ONE}]\{\ \} \\
[JAC(R,R)]_e &= (E)(\ )\{IZZ\}(-1)[A3011d][\ ] \\
&\quad +(E)(\ )\{IZZ\}(-1)[A3101d][\ ] \\
&\quad +(GK)(\ )\{A\}(1)[A3000d][\ ] \\
&\quad \% + (GK,1/12)(\ )\{A\}(-1)[A3011d][\ ] \\
[JAC(R,W)]_e &= (-GK)(\ )\{A\}(0)[A3001d][\ ] \\
[JAC(W,W)]_e &= (GK)(\ )\{A\}(-1)[A3011d][\ ] \\
&\quad +(GK)(\ )\{A\}(-1)[A3101d][\ ] \\
&= (GK)(\ )\{A\}(0)[A3001d][\ ].
\end{aligned}
\tag{6.33}
$$

In equation (6.33) the underintegration option is enabled by appropriate exchange of the MATLAB® comment sign (%).

The performance of Timoshenko beam theory GWS^h $k=1$ basis implementation for uniform loading emulates that of the E-B beam theory solution. Table 6.6 summarizes

Table 6.6 Timoshenko beam, uniformly distributed load

M	Le	$\|R\|_E$	$\|W\|_E$	Error$\|W\|_E$	Slope (W)	Error$\|R\|_E$	Slope (R)
8	0.125	3639.530	4344.210				
16	0.0625	4194.528	4952.081	202.6238		184.9995	
32	0.03125	4358.273	5130.006	59.3083	1.772	54.5816	1.761
64	0.015625	4395.758	5171.091	13.6951	2.115	12.4952	2.127
128	0.007812	4405.180	5181.417	3.4420	1.992	3.1407	1.992
256	0.00390	4407.539	5184.002	0.8616	1.998	0.7862	1.998
512	0.001953	4408.129	5184.648	0.2155	2.000	0.1966	2.000

computer lab 6.2 altered theory data, and Figure 6.8 graphs convergence confirming $2(k + 1 - m) = 2$ dominates in equation (6.20). For the central point load case the Timoshenko GWS^h algorithm performance is unique in generating a local error extremum for mesh $M = 64$, Table 6.7. However, as regular mesh refinement continues adherence to $2(k + 1 - m) = 2k = 2$ convergence is recorded, Figure 6.9.

The Timoshenko GWS^h $k = 2$ basis algorithm performance exhibits the several decile error reduction observed for this E-B beam algorithm for uniformly distributed loading. However, the point load $k = 2$ basis asymptotic convergence rate is identical to that for $k = 1$. Figure 6.10 summarizes the point load convergence and the lack of adherence of the $k = 2$ basis solution performance to theory is perplexing. This may be connected to the first spatial derivatives existing in equations (6.29) and (6.30), the presence of which (theoretically) eliminates a companion variational principle which the theory assumes.

Recalling the revised theory statement (6.20), convergence rate dominance of $(r - 1) < k = 1,2$ recorded for the E-B beam mixed-load case also occurs for Timoshenko beam theory GWS^h solutions. This occurs both with and without underintegration and for

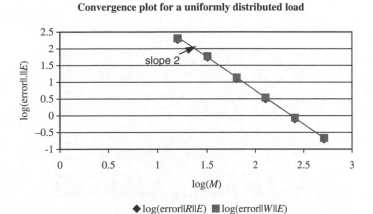

Convergence plot for a uniformly distributed load

slope 2

log(error$\|R\|E$)   log(M)

◆ log(error$\|R\|E$)   ■ log(error$\|W\|E$)

Figure 6.8 Timoshenko theory GWS^h convergence, $k = 1$, uniform load

Table 6.7 Timoshenko beam, point load at the center

M	Le	$\|R\|_E$	$\|W\|_E$	Error$\|W\|_E$	Slope (W)	Error$\|R\|_E$	Slope (R)
8	0.125	1.393	2.068				
16	0.0625	1.903	2.658	0.197		0.170	
32	0.03125	1.917	2.682	0.008	4.581	0.005	5.183
64	0.015625	1.827	2.587	0.032	−1.954	0.030	−2.676
128	0.007812	1.804	2.562	0.008	1.943	0.008	1.951
256	0.003906	1.798	2.556	0.002	1.986	0.002	1.988
512	0.0019531	1.797	2.554	0.001	1.997	0.000	—

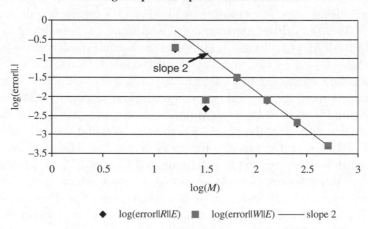

Figure 6.9 Timoshenko theory GWS^h convergence, $k=1$, point load

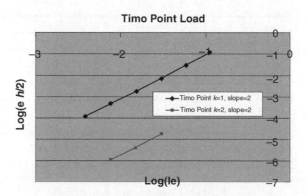

Figure 6.10 Timoshenko theory GWS^h algorithm solution asymptotic convergence, $k=1$, 2 bases, central point load

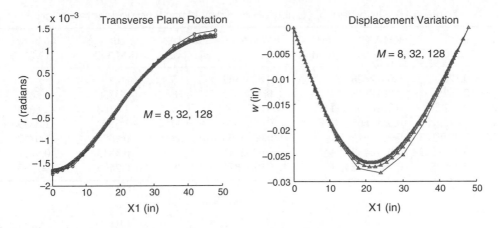

Figure 6.11 Timoshenko beam theory rotation and displacement DOF distributions, distributed and point loads, $k=1$ basis, $M=8$, 32, 128

$k=1,2$ basis implementations. Figure 6.11 graphs the $k=1$ basis rotation and displacement DOF distributions for uniform $M=8$, 32, 128 mesh solutions. Tables 6.8 and 6.9 summarize the *energy norm* data documenting the generated *suboptimal* linear convergence rate.

Table 6.8 Timoshenko distributed + point loads, no underintegration

M	le	$||R||$	$||W||$	Error$||W||$	Slope (W)	Error$||R||$	Slope (R)
8	0.125	1.40E + 03	1.68E + 03				
16	0.062	1.37E + 03	1.63E + 03	16.4847		9.691E + 00	
32	0.031	1.30E + 03	1.55E + 03	26.5546	−0.6878	2.137E + 01	−1.1410
64	0.015	1.25E + 03	1.49E + 03	18.9059	0.4901	1.593E + 01	0.4238
128	0.007	1.23E + 03	1.46E + 03	10.3156	0.8739	8.747E + 00	0.8651
256	0.003	1.21E + 03	1.44E + 03	5.35360	0.9462	4.550E + 00	0.9427
512	0.001	1.21E + 03	1.44E + 03	2.72319	0.9752	2.317E + 00	0.9736

Table 6.9 Timoshenko, distributed + point loads, underintegrated

M	le	$||R||$	$||W||$	Error$||W||$	Slope (W)	Error$||R||$	Slope (R)
4	0.25						
8	0.125	1.23E + 03	1.49E + 03				
16	0.0625	1.33E + 03	1.59E + 03	−30.4017533		−3.341E + 01	
32	0.03125	1.29E + 03	1.54E + 03	15.8178526	—	1.163E + 01	—
64	0.01562	1.25E + 03	1.49E + 03	16.2698758	−0.040	1.354E + 01	−0.2188
128	0.00781	1.23E + 03	1.46E + 03	9.6851813	0.7483	8.174E + 00	0.7277
256	0.00390	1.21E + 03	1.44E + 03	5.2000386	0.8972	4.411E + 00	0.8900
512	0.00195	1.21E + 03	1.44E + 03	2.6853378	0.9534	2.283E + 00	0.9503

Table 6.10 Timoshenko, distributed and point loads, no underintegration

M	Le	RMAX	RMAX (node)	WMAX	WMAX (node)	% change (RMAX)	% change (WMAX)
8	0.125	0.0017328	1	0.0283864	5		
16	0.0625	0.0017318	1	0.0279602	8	−0.3730	−1.5015
32	0.03125	0.0016932	1	0.0272158	15	−1.9168	−2.6623
64	0.015625	0.0016665	1	0.0267028	30	−1.5802	−1.8849
128	0.0078125	0.0016513	1	0.0264155	58	−0.9087	−1.0759
256	0.0039063	0.0016433	1	0.0262642	115	−0.4843	−0.5727
512	0.0019531	0.0016392	1	0.0261871	228	−0.2496	−0.2937

Table 6.11 Timoshenko, distributed and point loads, underintegrated

M	le	RMAX	RMAX (node)	WMAX	WMAX (node)	% change (RMAX)	% change (WMAX)
8	0.125	0.001616	1	0.026779	5		
16	0.0625	0.001698	1	0.027591	8	5.1128	3.0320
32	0.0312	0.001686	1	0.027124	15	−0.7391	−1.694
64	0.0156	0.001664	1	0.026680	30	−1.2795	−1.6360
128	0.0078	0.001650	1	0.026410	58	−0.8333	−1.0133
256	0.00390	0.001643	1	0.026263	115	−0.4654	−0.5571
512	0.00195	0.001639	1	0.026180	229	−0.2449	−0.2898

Due to suboptimal convergence rate, fairly dense meshes M are required to generate extremum solution DOF meeting a specified engineering accuracy. Tables 6.10 and 6.11 summarize the $k = 1$ basis data confirming the (uniform) mesh required for maximum deflection and rotation DOF to be accurate to within 1% is $M = 128$.

The impact of underintegration on Timoshenko theory GWS^h displacement DOF error reduction for coarser mesh $k = 1$ basis solutions is clearly illustrated in Figure 6.12. The direct comparison of normalized displacement DOF error for E-B versus Timoshenko beam theory GWS^h $k = 1$ basis solutions as a function of M is graphed in Figure 6.13. The underintegrated Timoshenko theory error is notably smaller on the coarsest mesh; thereafter, the theory implementation results are identical.

6.5 Mechanical Vibrations of a Beam

The momentum conservation principle DP describing static deflection of a transverse loaded beam, Figure 6.2, leads to the E-B and Timoshenko theories. Now assume the

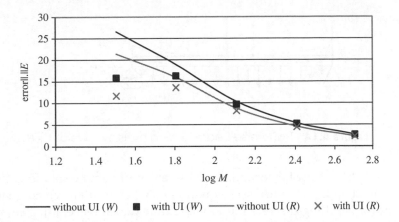

Figure 6.12 Timoshenko theory maximum displacement DOF error, with and without underintegration, $8 \le M \le 128$

beam of mass density ρ is deflected by a central point load P, Figure 6.14, which is suddenly removed. The beam will then initiate vibration in the transverse plane due to the load created by its own weight distribution $\rho g\, A$, where $A(x)$ is the cross-section area distribution.

The formulation starting point for the mechanical vibration GWS^h algorithm remains DP, now retaining the unsteady term

$$\text{DP: } \mathcal{L}(y) = \rho A(x)\frac{\partial^2 y}{\partial t^2} + \text{E}\frac{\partial^2}{\partial x^2}\left(I(x)\frac{\partial^2 y}{\partial x^2}\right) + p(x) + \text{P} = 0. \qquad (6.34)$$

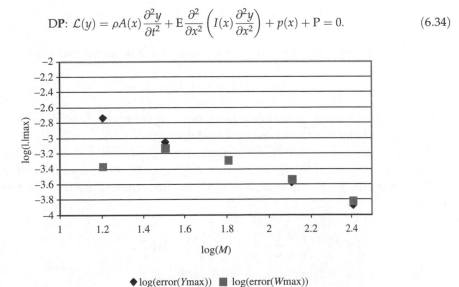

Figure 6.13 Normalized error in maximum displacement DOF for E-B and Timoshenko beam GWS^h solutions, $k = 1$ basis

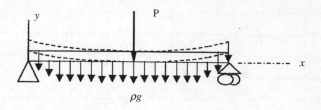

Figure 6.14 Transverse beam loaded by its own weight + P

Two GWS^h options exist for formulating discrete approximate solutions to equation (6.34). The direct approach addresses equation (6.34) as written which generates the requirement for handling discrete initial-value time-dependence, [3]. The more common practice alternative assumes beam motion is harmonic which leads to a *normal mode* analysis identifying the underlying spectra of natural frequencies and mode shapes fundamental to mechanical vibration.

Taking the latter tack, assume the transverse displacement dependence on x and t is separable, recall Chapter 3, as

$$y(x, t) \equiv Y(x)e^{i\omega t}. \tag{6.35}$$

Substitution of equation (6.35) into equation (6.34) produces

$$\text{DP: } \mathcal{L}(Y) = E\frac{\partial^2}{\partial x^2}\left(I(x)\frac{\partial^2 Y}{\partial x^2}\right) - \omega^2 \rho A(x)Y + P = 0, \tag{6.36}$$

where ω is the frequency of vibration for the equivalent distributed load

$$p(x) = -\rho\omega^2 A(x)Y(x). \tag{6.37}$$

The load distribution magnitude now depends on beam displacement and geometry, in sharp distinction to its being given data.

The process GWS^N ⇒ GWS^h for equation (6.36) proceeds as developed. Since the Hermite cubic basis $\{N_3^H(\zeta_\alpha)\}$ remains appropriate for the dominant biharmonic term in equation (6.36), its selection leads to the terminal matrix statement GWS^h ≡ $S_e\{WS\}_e = \{0\}$ with

$$\{WS\}_e = \int_{\Omega_e} \frac{d^2\{N_3^H\}}{dx^2} EI(x) \frac{d^2\{N_3^H\}^T}{dx^2} dx\{Q\}_e$$

$$- \int_{\Omega_e} \{N_3^H\}\omega^2 \rho A(x)\{N_3^H\}^T dx\{Q\}_e \tag{6.38}$$

$$V|_{\partial\Omega_e} + P\{\delta\} - \frac{d\{N_3^H\}}{dx} M|_{\partial\Omega_e} + N_{3'}\} \cdot$$

Interpolating $I(x)$ using $\{N_1\}$ and for element average area $A(x)$ the template is

$$\{WS\}_e = (E)(\ )\{IZZ\}(-3)[A3022LH]\{Q\}$$
$$-(\rho, \omega^2)(A)\{\ \}(1)[A200H]\{Q\}.$$
$$+P\{\delta\} + BCs$$
(6.39)

The matrix $[A3022LH]$ is identically $[STIFF]$, equation (6.10), which readily handles all useful beam moment of inertia distributions $I(x)$. The distributed load is the scalar $\rho\omega^2\bar{A}_e$ multiplying what the FE literature calls the (Hermite) basis *mass matrix* $[MASS]_e$. Specifically

$$\rho\omega^2\bar{A}_e[MASS]_e \equiv \int_{\Omega_e} \{N_3^H\}\rho A(x)\omega^2\{N_3^H\}^T \, dx$$

$$= \rho\omega^2\bar{A}_e \int_{\Omega_e} \{N_3^H\}\{N_3^H\}^T \, dx$$

$$= \frac{\rho\omega^2\bar{A}_e l_e}{420}
\begin{bmatrix}
156, & 22\ell_e, & 54, & -13\ell_e^2 \\
 & 4\ell_e^2, & 13\ell_e, & -3\ell_e^2 \\
 & & 156, & -22\ell_e \\
(\text{sym}) & & & 4\ell_e^2
\end{bmatrix}_e$$
(6.40)

$$= \rho\omega^2\bar{A}_e l_e[A200H]_e$$

An exercise suggests confirmation of $[A200H]_e$ in equation (6.40). Averaging the element area circumvents creation of a mixed basis *mass hypermatrix*. Another exercise suggests construction of $[A3000LH]_e$ as generated by assuming $A_e(x) = \{N_1\}^T\{A\}_e$.

The normal mode solution statement is equation (6.39) absent the last line, for point load P removed, and endpoint pin connections, hence slope, shear, and moment are not given BC data. Its assembly produces the *homogeneous* in $\{Q\}$ global matrix statement

$$([STIFF] - \omega^2[MASS])\{Q\} = \{0\}.$$
(6.41)

Recalling Chapter 3 and the basic Cramer's rule for solving a matrix statement, only two solution possibilities exist. If the *trivial* solution $\{Q\} = \{0\}$ is unacceptable the *sole* alternative is that the *determinant* (det) of the matrix $([STIFF] - \omega^2[MASS])$ must vanish. This occurrence identifies equation (6.41) as an *eigenvalue* statement, hence solving

$$det\left([MASS]^{-1}[STIFF] - \omega^2[I]\right) = \{0\}$$
(6.42)

where $[I]$ is the (diagonal) identity matrix, yields determination of the eigenvalues of $[MASS]^{-1}[STIFF]$. Taking the positive square root

$$\omega \Rightarrow \omega_i^h = \omega_1^h, \, \omega_2^h, \, \omega_3^h, \dots, \omega_n^h, \dots$$
(6.43)

are the GWS^h algorithm-generated beam vibration *natural frequencies*. Superscript h enforces that these are only *approximations* to the genuine (continuum) natural frequencies ω_i.

The number n of GWS^h solution eigenvalues that can be determined is bounded by the matrix order of $[MASS]^{-1}[STIFF]$. Their magnitudes range monotonically from smallest to largest unless there exists a repeated eigenvalue. An engineering normal-mode vibration analysis is usually interested in the few lowest natural frequencies. Hence, the eigenvalue portion of the GWS^h solution process generates the ordered set ω_i^h for $1 \leq i \leq n$ for n not too large. For each eigenvalue, the normal mode DOF array $\{Q\} \Rightarrow \{Q_i\} \equiv \{Q(\omega_i^h)\}$ is determined via the solution of

$$([STIFF] - \omega_i^2[MASS])_r\{Q_i\} = \{0\} \tag{6.44}$$

where subscript r denotes matrix order reduction to nontrivial rank.

Computer lab 6.3 details a regular mesh-refinement study generating GWS^h normal mode solutions for a spring loaded cantilever beam, Figure 6.15. The left-end BCs are zero displacement and zero slope while the unconstrained right end is subject to a spring restoring force (point load) with spring constant k. The *data* defining the problem include beam length L, Youngs modulus E, centroidal moment of inertia I, mass density ρ, cross-sectional area A, and k.

Hand illustration of the GWS^h algorithm formation process is quite informative. For a uniform $M = 2$ element discretization of L, the resulting Hermite cubic basis DOF for these two Ω_e are

$$\begin{aligned}
\{Q\}_{e=1} &= \{0, \quad 0, \quad Y2, \quad DY2\}^T \\
\{Q\}_{e=2} &= \{Y2, \quad DY2, \quad Y3, \quad DY3\}^T.
\end{aligned} \tag{6.45}$$

For these DOF assemble the two symmetric $[A200H]_e$ matrices (6.40).

$$[A200H]_e = S_e \left(\begin{bmatrix} 0 & 0 & 0 & 0 \\ & 0 & 0 & 0 \\ & & 156 & -22\ell_e \\ sym & & -22\ell_e & 4\ell_e^2 \end{bmatrix}, \begin{bmatrix} 156 & 22\ell_e & 54 & -13\ell_e \\ & 4\ell_e & 13\ell_e & -3\ell_e^2 \\ & & 156 & -22\ell_e \\ sym & & & 4\ell_e^2 \end{bmatrix} \right)$$

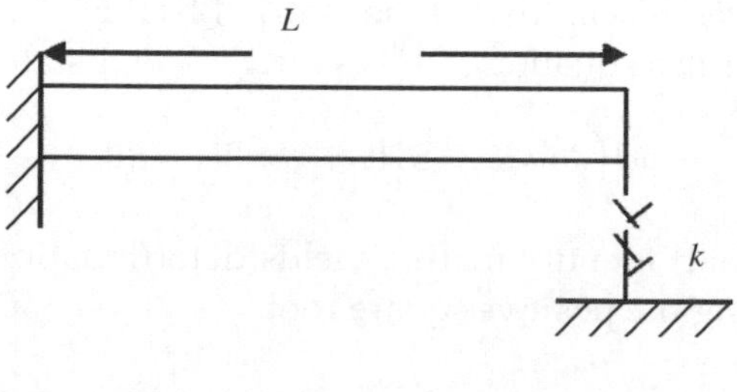

Figure 6.15 Spring restrained vibrating cantilever beam

then insert $l_e = L/2$ and the multiplier to generate global $\omega^2[\text{MASS}]$

$$\omega^2[\text{MASS}] = \frac{\rho\omega^2 AL}{840} \begin{bmatrix} 312 & 0 & 54 & -13(L/2) \\ & 8(L/2)^2 & 13(L/2) & -3(L/2)^2 \\ & & 156 & -22(L/2) \\ \text{sym} & & & 4(L/2)^2 \end{bmatrix} \begin{Bmatrix} Y2 \\ DY2 \\ Y3 \\ DY3 \end{Bmatrix}. \tag{6.46}$$

Do the identical operation on the two [STIFF] matrices, equation (6.10). Then inserting the point load spring restoring force $k\delta(Y3)$ produces

$$[\text{STIFF}] = \frac{EI}{(L/2)^3} \begin{bmatrix} 24 & 0 & -12 & 6L/2 \\ & 8(L/2)^2 & -6L/2 & 2(L/2)^2 \\ & & 12+k & -6L/2 \\ \text{sym} & & & 4(L/2)^2 \end{bmatrix}. \tag{6.47}$$

For a uniform beam and the following data: $L = 1$ m, $E = 2.1\text{E}09$ N/m^2, $I = 5\text{E-}05$ m^4, $\rho A = 200$ kg/m, $k = 2\text{E}06$ N/m, an analytical solution is published, [3]. This solution's four lowest natural frequencies are $\omega_1 = 829$, $\omega_2 = 5.05\text{E}3$, $\omega_3 = 1.41\text{E}4$, and $\omega_4 = 2.73\text{E}4$ rad/s. The $M = 2$ global matrix statement (6.41) is order four, and for the given data $\omega_1^h = 806$, $\omega_2^h = 5.09\text{E}3$, $\omega_3^h = 1.72\text{E}4$, $\omega_4^h = 5.00\text{E}4$ rad/s. The two lowest frequencies are "close" while the other two are truly inaccurate, not surprising realizing the coarseness of the $M = 2$ mesh.

Figures 6.16 and 6.17 graph the $M = 64$ GWS^h algorithm normal-mode solution *displacement* and *slope* DOF arrays, respectively, for the computer lab 6.3 solution first four natural frequencies. Note the eigenmode period decreases with increasing frequency. Figure 6.18 compares the normalized displacement solution magnitudes for the four frequencies; the slope solution relative magnitudes are similar in appearance.

As equation (6.36) is biharmonic, and neglecting the impact of the algebraic manipulations in equations (6.34)–(6.44), theory predicts asymptotic convergence in the energy norm is order four for the cubic Hermite basis, equation (6.14). However, natural frequency convergence is a more important design criteria and an appropriate norm is not obvious. For the first four natural frequencies, computer lab uniform mesh refinement *a posteriori* data clearly confirms convergence, Table 6.12. Of note, but not resolved, the lowest and highest frequencies do not converge to those generated by the analytical solution as do the middle two frequencies.

As smooth data drive this problem statement, Taylor series (TS) convergence estimation is appropriate. For the computer lab *a posteriori* data Figure 6.19 graphs TS-generated convergence for beam displacement at the spring constraint and natural frequency. The former is clearly (only!) quadratic for all $4 \le M \le 256$, in

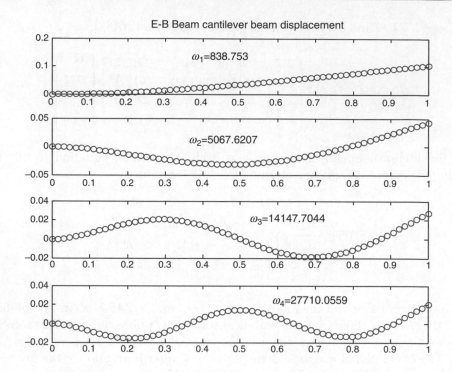

Figure 6.16 First four normal mode displacement DOF, $M = 64$

distinction to that predicted by equation (6.14). Convergence for the two lowest frequencies approximates quadratic while that for the next two is solidly quadratic to $M = 64$. Thereafter, the evidenced divergence is probably due to compute round off error. Factually, round-off pollutes the $M = 64$ lowest frequency, evidenced by the nonmonotonicity of these data in Table 6.12. This established quantitative convergence, TS-solid with caveats, indicates the cited analytical determinations for ω_1 and ω_4 are flawed.

Table 6.12 Natural frequency convergence (E04 rad/s)

M	ω_1^h	ω_2^h	ω_3^h	ω_4^h
4	0.0838220	0.5070094	1.4249880	2.8102408
8	0.0838704	0.5067519	1.4155006	2.7769875
16	0.0838752	0.5067581	1.4148080	2.7713884
32	0.0838754	0.5067615	1.4147716	2.7710265
64	0.0838753	0.5067621	1.4147705	2.7710054

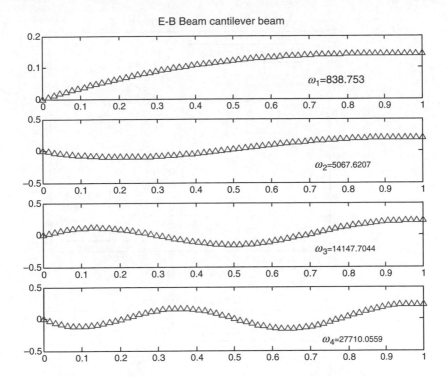

Figure 6.17 First four normal mode slope DOF, $M = 64$

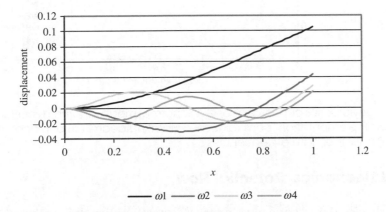

Figure 6.18 Normalized eigenmode displacement distributions, $M = 64$

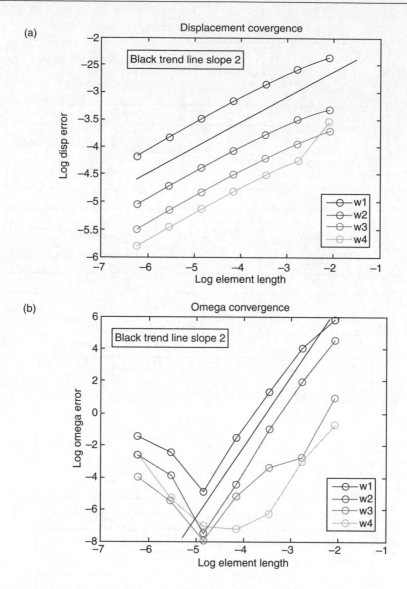

Figure 6.19 Taylor series convergence, GWS^h normal mode algorithm solution, cubic Hermite basis implementation, constrained E-B vibrating beam, (a) displacement at constraint, (b) first four natural frequencies

6.6 Fluid Mechanics, Potential Flow

In fluid mechanics, the aerodynamics *potential flow* assumption defines the velocity vector as $\mathbf{u} \equiv -\nabla\phi$, recall Chapter 2.4, which significantly simplifies the solution of the Navier–Stokes conservation principle PDE system. Potential flow simulation has found wide utility in aerodynamics performance prediction, as the resultant computed pressure distribution is imposed as *data* for laminar/turbulent boundary layer flow prediction.

The *venturi*, a duct of smoothly varying cross-sectional area $A(z)$, is a basic fluid flow rate measuring device. Always possessing an axis of symmetry, an $n = 1$ axisymmetric analysis is appropriate for the determination of the associated pressure distribution. Thereby, DM as expressed in equation (2.29) becomes altered to

$$\text{DM:}\; \mathcal{L}(\phi) = -\frac{d^2\phi}{dz^2} - \frac{d\phi}{dz}\frac{d(\log A)}{dz} = 0. \tag{6.48}$$

Multiplying equation (6.48) through by $A(z)$ clears the log yielding

$$\mathcal{L}(\phi) = -A(z)\frac{d^2\phi}{dz^2} - \frac{d\phi}{dz}\frac{dA}{dz} = 0 \tag{6.49}$$

as the ODE for the $\text{GWS}^N \Rightarrow \text{GWS}^h$ process. It is an exercise to verify that $\text{GWS}^h \equiv S_e$ $\{WS\}_e = \{0\}$ generates the three contribution template

$$\begin{aligned}
\{WS\}_e = \;&()(\,){AREA}(-1)[A3011d]\{Q\} \\
&+()(\,){AREA}(-1)[A3101d]\{Q\} \\
&-()(\,){AREA}(-1)[A3101d]\{Q\}
\end{aligned} \tag{6.50}$$

Thus the GWS^h integration-by-parts operation generates exact cancellation of the area derivative term in equation (6.49), and also identifies the Neumann BC as $\{N\}A(z)\frac{d\phi}{dz}|$ leading to algorithm template completion

$$\begin{aligned}
\{WS\}_e = \;&()(\,){AREA}(-1)[A3011d]\{Q\} \\
&+(UDOTN)(AREA)\{\,\}(0)[ONE]\{\,\}\,.
\end{aligned} \tag{6.51}$$

The velocity potential function is but a computational variable, with the solution of equation (6.51) required manipulated to generate the pressure distribution. Recalling classical fluid mechanics, the Bernoulli equation is a streamline integral of the momentum principle DP, equivalently DE

$$\text{DE:}\; p(x) = p_\infty - (1/2)\rho\mathbf{u}\cdot\mathbf{u}. \tag{6.52}$$

Hence, pressure can be postprocessed via a GWS^h written on

$$\mathcal{L}(p) = p - p_\infty + (1/2)\rho\nabla\phi\cdot\nabla\phi = 0 \tag{6.53}$$

where ρ is the (constant) fluid density. The GWS^h template is

$$\begin{aligned}
\{WS(P)\}_e = \;&()(\,){\,}(1)[A200d]\{P\} \\
&-()(\,){\,}(1)[A200d]\{P\,inf\} \\
&+(1/2, rho)(\,){Q}(-1)[A3101d]\{Q\}\,.
\end{aligned} \tag{6.54}$$

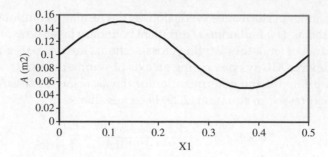

Figure 6.20 Venturi-type duct cross-section area distribution

One can also predict the distribution of velocity magnitude (speed) from the kinematics definition

$$\mathcal{L}(u) = u(z) + \nabla\phi \cdot \mathbf{k} = 0 \tag{6.55}$$

leading to the GWS^h template

$$\{WS(U)\}_e = (\ )(\ )\{\ \}(1)[A200d]\{U\}$$
$$+(\ )(\ )\{\ \}(0)[A201d]\{Q\} \tag{6.56}$$

Computer lab 6.4 defines a mesh-refinement study to compute the pressure distribution resulting from potential flow in a venturi device with cross-sectional area distribution $A(z) = A_o + B\ \sin(2\pi z/L)$, Figure 6.20. The onset speed is $U_\infty = 5\,\text{m/s}$ with fluid density $\rho = 480\,\text{kg/m}^3$ and geometric data $A_o = 0.1\,\text{m}^2$, $B = 0.05\,\text{m}^2$, and $L = 0.5\,\text{m}$.

The $k = 1$ basis GWS^h solutions for potential function, pressure and velocity distributions for uniform mesh refinements $8 \leq M \leq 64$ are overlaid in Figures 6.21–6.23.

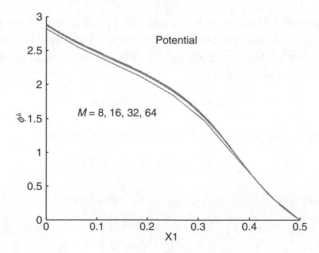

Figure 6.21 Venturi device potential distributions, $8 \leq M \leq 64$

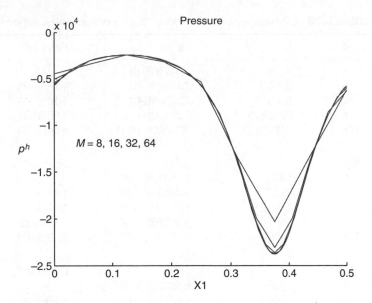

Figure 6.22 Venturi device pressure distributions, $8 \leq M \leq 64$

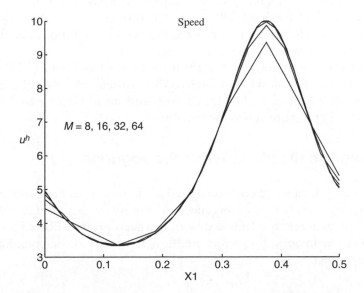

Figure 6.23 Venture device fluid speed distributions, $8 \leq M \leq 64$

Table 6.13 Convergence, GWS^{h}k = 1 basis venturi flow solutions

Mesh	M	l_e	$\|\phi^h\|_E$	Error$\|\phi^h\|_E$	Slope
H	8	0.125	7.04700E-01		
h/2	16	0.0625	7.17200E-01	4.166E-03	
h/4	32	0.03125	7.20500E-01	1.100E-03	1.921
h/8	64	0.015625	7.21400E-01	3.000E-04	1.874
h/16	128	0.0078125	7.21600E-01	6.666E-05	2.169
Pressure					
Mesh	M	l_e	$\|p^h\|_E$	Error$\|p^h\|_E$	Slope
H	8	0.125	2.58840E+00		
h/2	16	0.0625	3.09450E+00	1.687E-01	
h/4	32	0.03125	3.24360E+00	4.970E-02	1.763
h/8	64	0.015625	3.28260E+00	1.300E-02	1.934
h/16	128	0.0078125	3.29240E+00	3.266E-03	1.992
Speed					
Mesh	M	l_e	$\|u^h\|_E$	Error$\|u^h\|_E$	Slope
h	8	0.125	8.89330E+00		
h/2	16	0.0625	9.42320E+00	1.766E-01	
h/4	32	0.03125	9.57150E+00	4.943E-02	1.837
h/8	64	0.015625	9.60970E+00	1.273E-02	1.956
h/16	128	0.0078125	9.61930E+00	3.200E-03	1.992

Pressure, and to a lesser degree velocity, is quite sensitive to mesh density M in the deceleration section of the duct $0.3 < z < 0.4$. Interestingly, only *very subtle* slope changes with M occur in potential in this region, yet these solutions generate the large alterations recorded in velocity, hence pressure.

Quantitative error estimation is straight forward via the theory (6.20). The venturi data specification is very *smooth* and Table 6.13 confirms GWS^h solution adherence to theory for $2\gamma = 2k = 2$ for ϕ^h and u^h for $M \geq 32$, and for p^h for $M \geq 64$. The estimated error for the $M = 64$ pressure and velocity solutions is ~1%.

6.7 Electromagnetic Plane Wave Propagation

The conservation principles in electromagnetic field theory, upon constitutive closure, are called *Maxwell's equations*, a divergence-curl system of second-order PDEs, [4]. The dimensionally degenerate $n = 1$ ODE describing axial propagation of an electromagnetic plane wave in an optically perfect medium, for ϕ the electric potential (volts) and ω plane wave frequency, is

$$\mathcal{L}(\phi) = -\frac{d^2\phi}{dx^2} - \omega^2\phi = 0. \tag{6.57}$$

The typical BCs for equation (6.57) are $\phi = 0$ at $x = 0$ and $d\phi/dx = 0$ at $x = L$ for a wave tube of length L. For these homogeneous BCs, equation (6.57) defines an *eigenvalue* problem possessing solutions only for specific frequencies $\omega \Rightarrow \omega_i$. As fully detailed in Chapter 2, the analytical solution eigenvalues are $\omega_i \Rightarrow \omega_n = n\pi/L$ for all integers $n > 0$, and the corresponding eigenvectors (mode shapes) are $\phi_n = \sin(\omega_n x)$.

There exists no real need to define a GWS^h for equation (6.57) due to its linearity admitting the analytical solution. This could be altered by the wave tube medium being optically nonperfect or the geometry not rectangular cartesian. In either instance, equation (6.57) would inherit a coefficient k, analogous to diffusion in heat conduction, that would alter equation (6.57) to

$$\mathcal{L}(\phi) = -\frac{d}{dx}\left(k(x)\frac{d\phi}{dx}\right) - \omega^2\phi = 0. \tag{6.58}$$

The GWS^h for equation (6.58) is

$$\text{GWS}^h = ([\text{DIFF}(k)] - \omega^2[\text{MASS}])\{Q\} = \{0\} \tag{6.59}$$

hence, as with harmonic vibration, the GWS^h solution process starts by determining the first n eigenvalue approximations ω_i^h via

$$\det\left([\text{MASS}]^{-1}[\text{DIFF}(k)] - \omega^2[\text{I}]\right) = 0 \tag{6.60}$$

then solving for the companion n eigenmodes from the reduced-order GWS^h, equation (6.59), with template

$$\{WS\}_e = ()(){K}_e(-1)[A3011d]\{Q_i\}_e \\ -(\omega_i^h, \omega_i^h)(){ }(1)[A200d]\{Q_i\}_e \tag{6.61}$$

As specified, equations (6.60) and (6.61) may be implemented using any completeness degree $n = 1$ FE trial space basis $\{N_k(\zeta_\alpha)\}$.

6.8 Convection–Radiation Finned Cylinder Heat Transfer

Setting up *practical* simulation problem statements in the computational engineering sciences typically places significant demands on BC identification/application. An $n = 1$ statement supporting topical exposition is the finned cylinder heat exchanger, Figure 6.24.

The engineering design requirement might be to determine fin spacing and taper for minimum weight, with ease of manufacture, while meeting an overall heat-exchange quota criterion. The cylinder geometry is axisymmetric and conduction along the cylinder axis can be assumed of minor significance, hence ignorable.

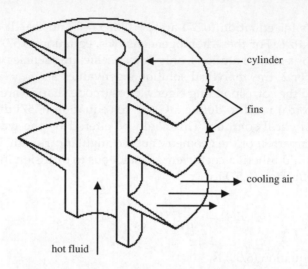

Figure 6.24 Finned cylinder heat exchanger

Hence, the prototype $n = 1$ model defines a geometry containing the *generic* fin connected to a section of cylinder of length equal to the fin spacing. Figure 6.25 graphs the $n = 1$ analysis geometry. The axisymmetric steady DE principle with convection BCs on the fins and cylinder wall is

$$\mathcal{L}(T) = -\frac{1}{r}\frac{d}{dr}\left[rk\frac{dT}{dr}\right] = 0, \text{ on } \Omega \subset \Re^1. \tag{6.62}$$

$$\ell(T) = k\frac{dT}{dn} + h(T - T_h) = 0, \text{ on } \partial\Omega_R \subset \Re^1. \tag{6.63}$$

In equation (6.63), $\partial\Omega_R$ denotes surfaces whereon *Robin* BCs are applied. Thereon, h is the thermal convection coefficient with T_h the exchange air temperature, with commensurate data for the fluid inside the cylinder heat exchange. There are no fixed temperatures and thickness variation due to fin taper is a design requirement. Finally, observe the Robin BCs are applied on surfaces of dimension identical to that of DE, equation (6.62).

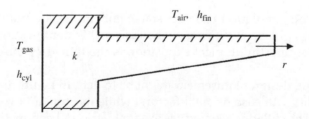

Figure 6.25 Geometry and BCs for generic fin-cylinder analysis

While equation (6.62) is an $n = 1$ ODE, the problem geometry suggests GWS^N should be cast on a multidimensional region to enable accurate BC definition/application. Hence, for all β

$$GWS^N = \int \Psi_\beta \mathcal{L}(T^N) r\,dr\,d\theta\,dz \equiv 0. \tag{6.64}$$

Inserting equation (6.62), integrating by parts, substituting equation (6.63) as appropriate, followed by $GWS^N \Rightarrow GWS^h = S_e\{WS\}_e = \{0\}$ and completing the suggested exercise will confirm that

$$
\begin{aligned}
\{WS\}_e &= \int_{\Omega_e} \frac{d\{N\}}{dr} k \frac{d\{N\}^T}{dr} r\,dr\,d\theta\,dz \{Q\}_e \\
&+ \int_{\partial\Omega_e \cap \partial\Omega_{Rfin}} \{N\}h\{N\}^T (\{Q\}_e - \{T_h\}_e) r\,dr\,d\theta\,dz . \\
&+ \int_{\partial\Omega_e \cap \partial\Omega_{Rcyl}} \{N\}h\{N\}^T (\{Q\}_e - \{T_h\}_e) r\,d\theta\,dz
\end{aligned}
\tag{6.65}
$$

The differential elements for the two heat-exchange surface integrals are distinct and the 2π from integration over $d\theta$ will cancel out.

While more detailed than the previous heat-transfer example, the template generated by integral completions in equation (6.65) remains of the form (5.13), that is,

$$\{WS\}_e = ([DIFF]_e + [BC]_e)\{Q\}_e - \{b(h, Tr)\}_e. \tag{6.66}$$

The suggested exercise will verify equation (6.66) takes the form

$$
\begin{aligned}
\{WS\}_e &= (COND)(THK)\{RAD\}(-1)[A3011d]\{Q\} \\
&+ (HC, RC)(THK)\{\ \}(\ )[ONE]\{Q\} \\
&- (HC, RC, THC)(THK)\{\ \}(\ )[ONE]\{\ \} \\
&+ (2HF)(\ )\{RAD\}(1)[A3000d]\{Q\} \\
&- (2HF)(\ )\{RAD\}(1)[A3000d]\{THF\}.
\end{aligned}
\tag{6.67}
$$

The notation in equation (6.67) is: cylinder convection and exchange temperature, $h \Rightarrow HC$ and $T_c \Rightarrow THC$, fin convection and exchange temperature, $h \Rightarrow HF$ and $T_c \Rightarrow TFC$, with the 2 signifying convection on both surfaces of the fin. THK denotes both the cylinder segment vertical span and the fin thickness (distribution), both handled as element average data. Finally, RC denotes the cylinder inner wall radius with $\{RAD\}$ the radial coordinate array of the discretization DOF.

For this proposed design optimization study, hand-assembly of $GWS^h = S_e\{WS\}_e = \{0\}$ for a coarse discretization M precisely illustrates how the finned cylinder heat exchanger *works*. Therefore, assume an $M = 2 + 5$, $k = 1$ basis FE discretization of the cylinder + fin, which generates an 8 DOF matrix statement. In equation (6.66) let d_{jk} and h_{jk} denote the matrix elements of $[\text{DIFF}]_e$ and $[\text{BC}]_e$. Then for superscripts signifying element index e, $1 \leq e \leq 7$, and subscripts matrix row–column indices, symbolic assembly produces

$$\text{GWS}^h = S_e\{\text{WS}\}_e = S_e([\text{DIFF}]_e + [\text{BC}]_e = \{b(h, T_c\}) \Rightarrow$$

$$
\begin{bmatrix}
0 & d_{11}^1 + h_{11}^1 & d_{12}^1 \\
d_{21}^1 & d_{22}^1 + d_{11}^2 & d_{12}^2 \\
d_{21}^2 & d_{22}^2 + d_{11}^3 + h_{11}^3 & d_{12}^3 + h_{12}^3 \\
d_{21}^3 + h_{21}^3 & d_{22}^3 + h_{22}^3 + d_{11}^4 + h_{11}^4 & d_{12}^4 + h_{12}^4 \\
d_{21}^4 + h_{21}^4 & d_{22}^4 + h_{22}^4 + d_{11}^5 + h_{11}^5 & d_{12}^5 + h_{12}^5 \\
\cdot & \cdot & \cdot \\
\cdot & \cdot & \cdot \\
d_{21}^7 + h_{21}^7 & d_{22}^7 + h_{22}^7 & 0
\end{bmatrix}
\begin{Bmatrix}
Q1 \\ Q2 \\ Q3 \\ Q4 \\ Q5 \\ Q6 \\ Q7 \\ Q8
\end{Bmatrix}
=
\begin{Bmatrix}
h_c T_1^1 \\
0 \\
h_f T_1^3 \\
h_f T_2^3 + h_f T_1^4 \\
h_f T_2^4 + h_f T_1^5 \\
\cdot \\
\cdot \\
h_f T_2^7
\end{Bmatrix}. \qquad (6.68)
$$

To enhance visualization, the global $[\text{DIFF}] + [\text{BC}]$ matrix in equation (6.68) is expressed in *shifted column format*, wherein the matrix principal diagonal terms are entered in the central vertical column. Hence, the matrix sub- and super-diagonals appear as vertical columns to the left and right of the central column. This format amounts to a shift of one column left for each succeeding row in a matrix.

The symbolic matrix statement (6.68) clearly illustrates how radial finned cylinder thermal diffusion d_{jk} is *weak form theory* enhanced by fin surface convection h_{jk} in the orthogonal z-direction! Similarly the hT_c term on the fin acts as a heat *sink*, an action mathematically identical with a (negative) distributed source s on Ω.

A more challenging GWS^h formulation, hence solution process, results upon interpretation of Figure 6.24 as the cylinder of a small air-cooled gasoline engine, for example, lawn tractor, motorcycle, and so on. In this case, the cylinder wall heat-transfer definition must include radiation as well as convection. The BC replacement for equation (6.63) is

$$\ell(T) = k\frac{dT}{dn} + h(T - T_h) + \varepsilon\sigma(T^4 - T_r^4) = 0, \text{ on } \partial\Omega_R \subset \Re^1 \qquad (6.69)$$

where ε is the product of surface emissivity with view factor and σ is the Stefan–Boltzmann constant.

The GWS^N integration-by-parts process leading to equation (6.65) is appropriately modified with net alteration of template (6.67) to

$$\{WS\}_e = (COND)(THK)\{RAD\}(-1)[A3011d]\{Q\}$$
$$+(HC, RC)(THK)\{\ \}(\ )[ONE]\{Q\}$$
$$-(HC, RC, THC)(THK)\{\ \}(\ )[ONE]\{\ \}$$
$$+(2HF)(\ )\{RAD\}(1)[A3000d]\{Q\} \qquad (6.70)$$
$$-(2HF)(\ )\{RAD\}(1)[A3000d]\{THF\}$$
$$+(EPS, SIG, RC)(THK)\{\ \}(\ )[ONE]\{Q^4\}$$
$$-(EPS, SIG, RC, TR^4)(THK)\{\ \}(\ )[ONE]\{\ \}.$$

The radiation BC addition has thus rendered $GWS^h = S_e\{WS\}_e = \{0\}$ very nonlinear, which generates the need for a matrix iterative solution procedure. The Newton algorithm remains the choice, as defined in equations (6.24) and (6.25), and the jacobian template is *analytically* generated from equation (6.70) as

$$[JAC]_e = (COND)(THK)\{RAD\}(-1)[A3011d][\ ]$$
$$+(HC, RC)(THK)\{\ \}(\ )[ONE][\ ]$$
$$+(2HF)(\ )\{RAD\}(1)[A3000d][\ ] \qquad (6.71)$$
$$+(4, EPS, SIG, RC)(THK)\{\ \}(\ )[ONE][Q^3].$$

Computer lab 6.5 implements equations (6.70) and (6.71) with appropriate small air-cooled engine data leading to a convergence study. The lab *a posteriori* data will verify the asymptotic error estimate (6.20) remains applicable. Drawing on the variational roots underlying the theory, the GWS^h DE principle energy norm definition for equation (6.62) with BCs (6.69) is

$$\|T\|_E \equiv \frac{1}{2}\int_\Omega \frac{dT}{dr}k\frac{dT}{dr}rdrd\theta dz + \frac{1}{2}\left(\int_{\partial\Omega_c} hT^2 rd\theta dz + \int_{\partial\Omega_f} hT^2 rdrd\theta dz\right). \qquad (6.72)$$

The discrete computed form remains as expressed in equation (5.65)

$$\|T^h\|_E = \frac{1}{2}\sum_{e=1}^{M}\{Q\}_e^T[DIFF + BC]_e\{Q\}_e. \qquad (6.73)$$

Suggested problem data are given in the computer lab directions. Figure 6.26 graphs the $GWS^h k=1$ basis fin-cylinder temperature distribution for uniform fin thickness and $M=8, 128$ uniform meshes. The temperature variation is linear in the cylinder and appears parabolic over the fin span. Even though the solutions appear visually insensitive to M, the *a posteriori* data clearly confirm quadratic asymptotic convergence in both energy and in Q_{max}, Tables 6.14 and 6.15.

Figure 6.27 graphs the $M=8, 128$ uniform mesh $k=1$ basis fin-cylinder distribution of temperature for the fin tapered to a point, as theoretically conjectured to promote optimal heat transfer, [5]. These solutions do not overlay, with the cylinder surface temperature slightly elevated and fin-tip temperature substantially lower than for

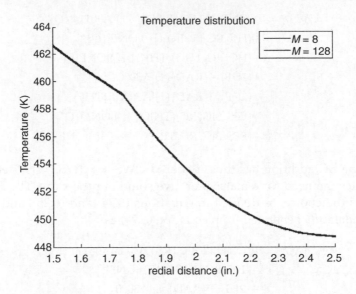

Figure 6.26 Temperature distribution, uniform thickness fin, $k = 1$

the uniform fin solution. The temperature distribution is piecewise continuous linear throughout with the change in slope remaining at the cylinder-fin interface.

This very modest alteration to fin geometry does not appear to constitute *nonsmooth* data, but indeed it reduces the $k = 1$ basis solution asymptotic convergence rate to

Table 6.14 Asymptotic convergence in energy

Mesh	M	l_e	$\|T\|_E$	$\|T\|_E$ error	Slope
h	8	0.125	24029.97307		
$h/2$	16	0.0625	24030.72805	0.25166	
$h/4$	32	0.03125	24030.91693	0.06296	1.9990
$h/8$	64	0.01563	24030.96416	0.01574	1.9997
$h/16$	128	0.00781	24030.97597	0.00394	1.9999

Table 6.15 Asymptotic convergence in Q_{max}

Mesh	M	l_e	Q_{max}	Q_{max} error	Slope
h	8	0.125	462.6049		
$h/2$	16	0.0625	462.6515	0.01552	
$h/4$	32	0.03125	462.6631	0.00388	1.9989
$h/8$	64	0.01563	462.6660	0.00097	1.9997
$h/16$	128	0.00781	462.6668	0.00024	1.9999

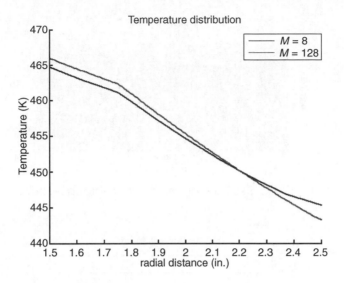

Figure 6.27 Temperature distribution, nonuniform thickness fin, $k = 1$

linear, Figure 6.28. Hence, the $r - 1$ term in equation (6.20) now dominates. If the linear theory is truly appropriate for this nonlinear problem, then this phenomenon must carry over to the $k = 2$ basis implementation, which it does, Figure 6.29. The $k = 2$ basis solution does exhibit improved relative accuracy, as experienced with the GWS^h beam formulations.

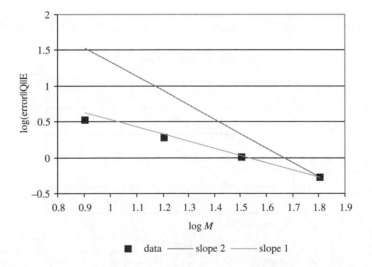

Figure 6.28 GWS^h $k = 1$ basis solution asymptotic convergence, nonuniform thickness fin tapered to a point

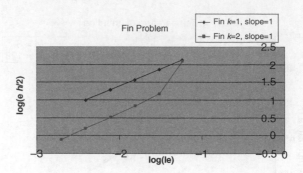

Figure 6.29 GWS^h k = 1,2 basis solution asymptotic convergence, nonuniform thickness fin tapered to a point

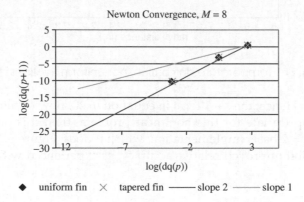

Figure 6.30 Newton algorithm iterative convergence rate, M=8

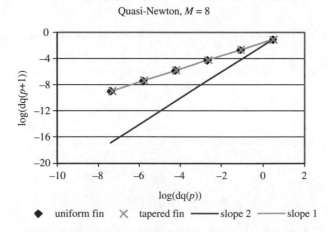

Figure 6.31 quasi-Newton algorithm iterative convergence rate, M=8

The Newton iterative convergence rate for the GWS^h algorithm, equations (6.70) and (6.71), is indeed quadratic for both uniform and tapered fin geometries for all M, see Figure 6.30 for $M = 8$. Eliminating the nonlinear radiation contribution in the jacobian, the last line in equation (6.71), generates a quasi-Newton algorithm and reduces the convergence rate to essentially linear, Figure 6.31. The number of matrix iterations required to achieve a given iterative convergence is about double that of the Newton algorithm.

6.9 Chapter Summary

The goal of this chapter is to expose key derivational issues for the weak form algorithm process GWS^N $\Rightarrow$ GWS^h $\Rightarrow$ S$_e${WS}$_e$ across a spectrum of subjects in the *computational engineering sciences*. Discipline statements become ODEs generated via simplifying assumptions on the fundamental conservation principles DM, DP, and DE, identified *well-posed* by pertinent boundary conditions (BCs) and constitutive closure.

Restriction to $n = 1$ cleanly supports theory generalization for *nonsmooth data*, which degrades *asymptotic convergence*, and algorithm nonlinearity. Via the *FEmPSE* toolbox and {WS}$_e$ templates, MATLAB®-enabled hands-on computer labs generate requisite *a posteriori* data for theory verifications. BC definitions are exposed without the encumbering coordinate transformations intrinsic to GWS^h formulation for $n > 1$ partial differential equations (PDEs), fully developed in subsequent chapters.

The structural mechanics discipline developments are via E-B and Timoshenko (T) beam theories for DP. The E-B classical form is a biharmonic ODE with 4th order derivative requiring two integration-by-parts for GWS^N. The cubic Hermite trial space basis was introduced as precisely appropriate for handling all admissible BCs, leading to theory generalization for differential order factor m

$$\|e^h\|_E \leq C l_e^{2\gamma} \|\text{data}\|_{L2}^2, \gamma = (k + 1 - m). \tag{6.14}$$

For *smooth data* the cubic Hermite basis implemented E-B beam theory performance is accurately predicted by equation (6.14). Recognizing the 4th order ODE is the combination of two 2nd order ODEs, which is also the standard form of the T beam theory, with its distinct state variable, led to GWS^h implementations using $k = 1,2$ Lagrange bases. Extension to mixed distributed and point loadings generated *a posteriori* data that initiated another theory generalization requirement for *data nonsmoothness* factor r

$$\|e^h\|_E \leq C l_e^{2\gamma} \|\text{data}\|_{L2}^2, \quad \gamma \equiv \min(k + 1 - m, r - m) \tag{6.20}$$

with term definitions:

k = FE trial space basis completeness degree
m = integer order (1,2) of the underlying variational principle
r = for specified *data*, solution differentiability order

Indeed, visually appearing smooth but in fact *nonsmooth data* definition led to degradation of asymptotic convergence to ∼ linear for any completeness degree *k* basis implementation for either beam theory, as predicted by equation (6.20). The T beam formulation also facilitated examination of the FE community "trick" of element matrix *under integration*. This operation was precisely identified as introducing an artificial *numerical diffusion* mechanism that improved prediction accuracy using coarse meshes.

The logical next development is *mechanical vibration* of a beam under the influence of its weight distribution. The starting point is dynamic DP leading to a separation of variables assumption of *harmonic motion*. The resultant ODE + BCs GWS^h algorithm is *homogeneous* generating a *normal mode* eigenvalue-eigenvector solution process. The net result is identification of the sequence of *natural frequencies* of vibration

$$\omega \Rightarrow \omega_i^h = \omega_1^h, \omega_2^h, \omega_3^h, \ldots, \omega_n^h, \ldots \tag{6.43}$$

with convergence under mesh-refinement a computer lab verification. A brief description of *electromagnetic* wave propagation in an optically perfect medium leads to the identical GWS^h solution process.

The fluid mechanics example results from the *potential flow* assumption, with DM generating an ODE with both 1st and 2nd order differential terms for a *venturi* device. The sought pressure distribution is post processed via a GWS^h algorithm written on the algebraic Bernoulli equation, an integral form of DE. The algorithm family is completed via a GWS^h written on the vector calculus definition

$$\mathcal{L}(u) = u(z) + \nabla\phi \cdot \mathbf{k} = 0. \tag{6.55}$$

Computer lab *a posteriori* data confirm asymptotic convergence is accurately predicted by equation (6.20) with completeness degree *k* dominance for all state variables in the problem statement.

The terminal topic is *convective-radiative* heat transfer in a finned cylinder heat exchanger. A multidimensional GWS^N enables precise definition of the range of BCs required for closure of an $n = 1$ simulation. Symbolic assembly on a coarse mesh of the resultant GWS^h algorithm generates physical insight into the manner of thermal conductivity enhancement by convection heat transfer in the orthogonal direction, equation (6.68).

Computer lab *a posteriori* data confirm pertinence of the theory (6.20) in predicting linear and *nonlinear* GWS^h algorithm performance, the latter benefiting from analytic formation of the Newton jacobian for matrix iterative solution. The tapered fin geometry constitutes *nonsmooth data* degrading asymptotic convergence to ∼ linear for any completeness degree *k* basis. This quantifies appropriateness of the *linear-theorized* asymptotic convergence (6.20) in predicting nonlinear algorithm performance. An added verification is iterative convergence degradation to linear upon removal of the nonlinear radiation term contribution in the jacobian.

Exercises

6.1 For the E-B beam theory biharmonic ODE (6.1), proceed through $\text{GWS}^N \Rightarrow \text{GWS}^h$ $\Rightarrow S_e\{\text{WS}\}_e$ hence verify the template (6.6).

6.2 Verify the matrix elements in the cubic Hermite trial space basis $\{N_3^H(\zeta_\alpha)\}$, equation (6.8).

6.3 For the E-B beam theory dual ODE resolution (6.15) and (6.16), verify the $\{\text{WS}\}_e$ template pair (6.18) and (6.19) including the BCs.

6.4 Verify the nonlinear E-B GWS^h Newton algorithm statement (6.24)–(6.26), also the matrix iterative template (6.27) and (6.28).

6.5 Proceed through the GWS^h process for the Timoshenko beam theory ODE pair (6.29) and (6.30) to confirm the template (6.33). Confirm the underintegrated $k = 1$ basis element matrix (6.32), constitutes a numerical diffusion mechanism.

6.6 For transverse beam harmonic vibration, verify the GWS^h algorithm template (6.39) for the separation of variables (6.35).

6.7 Using analytical integration, equation (5.28) confirm the $[A200H]_e$ matrix elements for equation (6.39) are defined by $[\text{MASS}]$, equation (6.40).

6.8 Define some of elements in the mixed basis matrix $[A3000LH]_e$ selecting $\{N_1(\zeta_\alpha)\}$ for $I(x)$ interpolation.

6.9 Verify the GWS^h template statement (6.51) including the Neumann BC (6.51) for quasi-one-dimensional potential flow.

6.10 Verify the GWS^h pressure and velocity postprocessing templates (6.54) and (6.56).

6.11 For convective heat transfer in a finned cylinder, proceed through equations (6.64) and (6.65) hence verify template (6.67).

6.12 For addition of radiation heat transfer, (6.69), verify the Newton template (6.70) and (6.71).

Computer Labs

The *FEmPSE* toolbox and all MATLAB® .m files required for conducting the computer labs are available for download at www.wiley.com/go/baker/finite.

6.1 Execute a convergence study, in displacement and energy, for the E-B beam problem using the cubic Hermite element formulation. Select a suitable distribution for I, and a practical (nonuniform) loading, hence confirm the mesh needed for 0.1% accuracy of max displacement.

6.2 Two forms for DP lead to $n = 1$ ODE pairs for prediction of the deflection of horizontal beams due to loading and various BCs. The MATLAB® .m file for the E-B beam theory GWS^h statement is available, and simple edits generate the companion Timoshenko beam theory template. Conduct mesh-refinement studies for both beam formulations using an M refinement factor of two. Hence determine the mesh M necessary for beam center deflection accuracy to be 0.1%. Confirm the theoretical asymptotic rate of convergence in energy.

6.3 Access the E-B beam normal mode GWS^h formulation MATLAB® .m file. Input BCs and other data specific for the cantilever beam example. Generate the first five natural frequencies and normal mode shapes for factor of two mesh refinements on $M = 4$ to 64. Estimate the uniform mesh M required for engineering accuracy of each natural frequency.

6.4 Aerodynamics weak interaction theory combines a far-field potential flow formulation of DM with laminar/turbulent boundary layer forms for DM and DP. Download the GWS^h algorithm .m file for predicting quasi-1D potential flow in a variable area duct. Execute the base case specification, then conduct a mesh refinement convergence study for ϕ, p, and u.

6.5 Small horsepower air-cooled gasoline engines achieve thermal balance via fins cast integral to the cylinder. The simplified DE principle reduces simulation to a 1-D axisymmetric ODE with a radiation BC at the cylinder wall and convection BCs on the wall and both fin surfaces. Access the MATLAB® .m file and enter BCs and other data pertinent for this analysis. Conduct a mesh-refinement study to determine the M necessary to generate the cylinder wall temperature to one-half degree accuracy.

References

[1] Clough, R.W. *et al.* (1960) Proceedings ASCE 2nd Conference on Electronic Computation.
[2] Oden, J.T. and Reddy, J.N. (1976) *An Introduction to the Mathematical Theory of Finite Elements*, Wiley-Interscience, New York, NY.
[3] Kelly, S.G. (2000) *Fundamentals of Mechanical Vibrations*, McGraw-Hill, New York, NY.
[4] Reitz, M. and Milford, J. (1962) *Electromagnetic Field Theory*, McGraw-Hill, New York, NY.
[5] Incropera, F.P. and Dewitt, D.P. (2004) *Fundamentals of Heat and Mass Transfer*, Wiley, New York, NY.

7

Steady Heat Transfer, $n > 1$: $n = 2, 3$ GWS^h for DE + BCs, FE bases, convergence, error mechanisms

7.1 Introduction

Essential theoretical and performance aspects of $1 \leq k \leq 3$ FE trial space basis discrete implementation of a Galerkin weak statement GWS^N are completed in Chapter 6 $n=1$ exposure. Extending to $n=2,3$ dimensional (n-D) DE + BCs statements build on established precepts with little added *theory* but with a substantial increase in the *details* of algorithm conversion to computable form.

The n-D steady heat transfer statement DE + BCs including convection, (*nonlinear*) radiation, and effflux remains the exposition vehicle

$$\text{DE}: \quad \mathcal{L}(T) = -\nabla \cdot (k\nabla T) - s = 0, \quad \text{on } \Omega \subset \mathfrak{R}^n \tag{7.1}$$

$$\text{BCs}: \quad \begin{aligned} \ell(T) &= k\nabla T \cdot \hat{\mathbf{n}} + h(T - T_c) + \mathbf{f} \cdot \hat{\mathbf{n}} \\ &\quad + f(vf, \varepsilon)\sigma(T^4 - T_r^4) = 0, \text{ on } \partial\Omega_R \subset \mathfrak{R}^{n-1}\cdot \end{aligned} \tag{7.2}$$

$$T(\mathbf{x}_b) = T_b(\mathbf{x}_b), \quad \text{on } \partial\Omega_D \subset \mathfrak{R}^{n-1}. \tag{7.3}$$

The DE *partial differential equation* (PDE) equation (7.1) holds *only* interior to the region Ω lying on n-D Euclidean space, symbolized as $\mathfrak{R}^n$. Robin efflux (7.2) and Dirichlet (7.3) BCs are enforced on the union of $(n-1)$ dimensional segments of the boundary

Finite Elements ⇔ Computational Engineering Sciences, First Edition. A. J. Baker.
© 2012 John Wiley & Sons, Ltd. Published 2012 by John Wiley & Sons, Ltd.

$\partial\Omega$ which *totally surrounds* Ω. In radiation heat transfer $f(vf,\varepsilon)$ is the to-be-determined function of *viewfactor* and surface *emissivity* with σ the Stefan–Boltzmann constant,

The weak form recipe for formulating and assessing performance of the FE basis discretely implemented GWS^h*approximate solution* for equations (7.1)–(7.3) remains:

$$\text{approximation}: \quad T(\mathbf{x}) \cong T^N(\mathbf{x}) \Rightarrow T^{\hat{h}}(\mathbf{x}) \equiv \cup_e T_e(\mathbf{x}) \tag{7.4}$$

$$\text{FE basis}: \quad T_e(\mathbf{x}) \equiv \{N_k(\mathbf{v}(\mathbf{x}))\}^T \{Q\}_e \tag{7.5}$$

$$\text{error extremization}: \quad \mathrm{GWS}^N \equiv \int_\Omega \Psi_\beta(\mathbf{x})\mathcal{L}(T^N)\mathrm{d}\tau = 0, \; \forall\beta \tag{7.6}$$

$$\Rightarrow \mathrm{GWS}^h = S_e\{WS\}_e \equiv \{0\}$$

$$\text{element contribution}: \quad \{WS\}_e = ([\mathrm{DIFF}]_e + [\mathrm{BC}]_e)\{Q\}_e - \{b(\mathrm{data})\}_e \tag{7.7}$$

$$\text{asymptotic convergence}: \quad \|e^h\|_E \leq Ch_e^{2\gamma}\left(\|\mathrm{data}\|^2_{\Omega,L2} + \|\mathrm{data}\|^2_{\partial\Omega,L2}\right), . \tag{7.8}$$

$$\gamma \equiv \min(k+1-m, \; r-m)$$

Viewing equations (7.4)–(7.8) the *new* implementation issue for all n is determination of FE trial space bases $\{N_k(\mathbf{v}(\mathbf{x}))\}$ with *degrees of freedom* (DOF), hence pertinent local coordinate systems $\mathbf{v}(\mathbf{x})$ spanning FE domain geometries, Figure 5.1. In equation (7.6)$\forall$ denotes *for all* and recall $m \equiv 1$ in equation (7.8) for the *elliptic* boundary value (EBV) PDE (7.1).

7.2 Multidimensional FE Bases and DOF

The DOF for multidimensional FE trial space bases are identified one-to-one with the geometric shapes for the domain Ω_e. The linear FE basis ($k=1$) is easiest to establish via geometric arguments. For the line/triangle/tetrahedron element family, Figure 5.1, the $k=1$ FE bases *define* the ζ_α*natural coordinate* (NC) system

$$\begin{aligned} n &= 1: \{N_1\}^T = \{\zeta_1, \zeta_2\} \\ n &= 2: \{N_1\}^T = \{\zeta_1, \zeta_2, \zeta_3\} \, . \\ n &= 3: \{N_1\}^T = \{\zeta_1, \zeta_2, \zeta_3, \zeta_4\} \end{aligned} \tag{7.9}$$

The NC system is *normalized* and *linearly dependent*, since $1 \leq \alpha \leq n+1$ by definition and

$$\sum_{\alpha=1}^{n+1} \zeta_\alpha = 1. \tag{7.10}$$

Each ζ_α is identically zero at every vertex node of Ω_e except *that* node with index α whereas ζ_α is unity.

Higher completeness degree FE bases $\{N_k(\zeta_\alpha)\}$ insert nonvertex DOF and involve logical extension on the ζ_α polynomials for the $n = 1$ basis, equation (7.10). Including equation (5.21) for completeness, the $k = 2$ Lagrange trial space basis $\{N_2(\zeta_\alpha)\}$ family for $1 \leq n \leq 3$ is

$$\{N_2(\zeta_\alpha)\} = \left\{ \begin{array}{c} \zeta_1(2\zeta_1 - 1) \\ 4\zeta_1\zeta_2 \\ \zeta_2(2\zeta_2 - 1) \end{array} \right\}_{n=1} , = \left\{ \begin{array}{c} \zeta_1(2\zeta_1 - 1) \\ \zeta_2(2\zeta_2 - 1) \\ \zeta_3(2\zeta_3 - 1) \\ 4\zeta_1\zeta_2 \\ 4\zeta_2\zeta_3 \\ 4\zeta_3\zeta_1 \end{array} \right\}_{n=2} , = \left\{ \begin{array}{c} \zeta_1(2\zeta_1 - 1) \\ \zeta_2(2\zeta_2 - 1) \\ \zeta_3(2\zeta_3 - 1) \\ \zeta_4(2\zeta_4 - 1) \\ 4\zeta_1\zeta_2 \\ 4\zeta_2\zeta_3 \\ 4\zeta_3\zeta_4 \\ 4\zeta_4\zeta_1 \end{array} \right\}_{n=3} . \tag{7.11}$$

Polynomial ordering in equation (7.11) for node numbering in Figure 7.1 is clearly visible. Nonvertex DOF are located at each line midpoint.

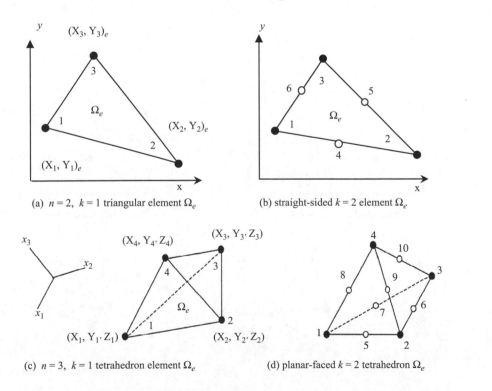

(a) $n = 2$, $k = 1$ triangular element Ω_e

(b) straight-sided $k = 2$ element Ω_e

(c) $n = 3$, $k = 1$ tetrahedron element Ω_e

(d) planar-faced $k = 2$ tetrahedron Ω_e

Figure 7.1 Natural coordinate $\{N_k(\zeta_\alpha)\}$ basis finite element domains, $n = 2, 3$

It's obvious that unions of triangles–tetrahedra readily admit arbitrary meshings. The quadrilateral–hexahedron element family, Figure 5.1, is similarly versatile when using boundary-fitted coordinate transformations [1]. The $n=2,3$, $k=1,2$ FE *tensor product* basis elements are graphed in Figure 7.2; the $k=1$ basis $\{N_1^+(\eta_k)\}$ defines the local cartesian coordinate system η_k spanning these element geometries.

$$\{N_1^+(\eta_k)\} = \frac{1}{4}\left\{\begin{array}{l}(1-\eta_1)(1-\eta_2)\\(1+\eta_1)(1-\eta_2)\\(1+\eta_1)(1+\eta_2)\\(1-\eta_1)(1+\eta_2)\end{array}\right\}_{n=2}, \quad = \frac{1}{8}\left\{\begin{array}{l}(1-\eta_1)(1-\eta_2)(1-\eta_3)\\(1+\eta_1)(1-\eta_2)(1-\eta_3)\\(1+\eta_1)(1+\eta_2)(1-\eta_3)\\(1-\eta_1)(1+\eta_2)(1-\eta_3)\\(1-\eta_1)(1-\eta_2)(1+\eta_3)\\(1+\eta_1)(1-\eta_2)(1+\eta_3)\\(1+\eta_1)(1+\eta_2)(1+\eta_3)\\(1-\eta_1)(1+\eta_2)(1+\eta_3)\end{array}\right\}_{n=3}. \tag{7.12}$$

The η_k coordinate system is *linearly independent* as $1 \leq k \leq n$, and is rectangular cartesian on the Ω_e geometry transformed to η-space.

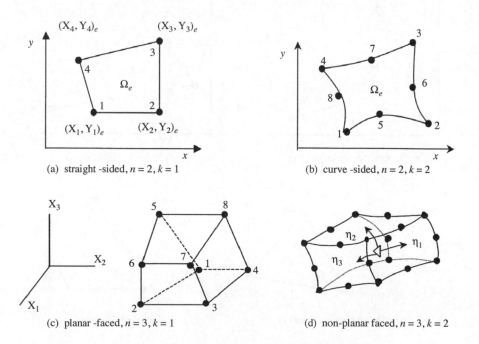

(a) straight-sided, $n = 2, k = 1$

(b) curve-sided, $n = 2, k = 2$

(c) planar-faced, $n = 3, k = 1$

(d) non-planar faced, $n = 3, k = 2$

Figure 7.2 Tensor product $\{N_k^+(\eta)\}$ basis finite element domains, $n=2,3$

The range of $k > 1$ bases $\{N_k^+(\eta_k)\}$ is widely reported in the FE structures literature, cf., [2]. For example, for DOF ordering defined in Figure 7.2b, the $n = 2 = k$ *serendipity* FE trial space basis is

$$\{N_2^+(\eta_k)\} = \frac{1}{4} \begin{Bmatrix} (1-\eta_1)(1-\eta_2)(-\eta_1-\eta_2-1) \\ (1+\eta_1)(1-\eta_2)(\eta_1-\eta_2-1) \\ (1+\eta_1)(1+\eta_2)(\eta_1+\eta_2-1) \\ (1-\eta_1)(1+\eta_2)(-\eta_1+\eta_2-1) \\ 2(1-\eta_1^2)(1-\eta_2) \\ 2(1+\eta_1)(1-\eta_2^2) \\ 2(1-\eta_1^2)(1+\eta_2) \\ 2(1-\eta_1)(1-\eta_2^2) \end{Bmatrix}. \tag{7.13}$$

The discrete implementation GWS^h for $DE + BCs$ statements in n-D again requires construction of the selected FE basis *matrix* library for diffusion, source, and BC terms in equations (7.1)–(7.3). While conceptually elementary, formation of the $[\text{DIFF}]_e$ matrix does involve considerable detail, as a transformation from global $x_i = (x,y,z)$ to the local (ζ_α or η_k) system is required to be implemented.

Via the chain rule, this calculus operation in index notation is

$$[\text{DIFF}]_e\{Q\}_e = \int_{\Omega_e} \nabla\{N_k\} \cdot k_e \nabla\{N_k\}^T d\tau\{Q\}_e$$

$$= \int_{\hat{\Omega}_e} \frac{\partial\{N_k\}}{\partial v_j} \left(\frac{\partial v_j}{\partial x_i}\right)_e k_e \frac{\partial\{N_k\}^T}{\partial v_k} \left(\frac{\partial v_k}{\partial x_i}\right)_e \det_e dv\{Q\}_e \tag{7.14}$$

where superscript *hat* signifies the *master* element $\hat{\Omega}_e$ of unit dimension in transform space spanned by the *intrinsic* coordinate system v_j, the placeholder for ζ_α or η_k. Further in equation (7.14), $\det_e$ is the determinant of the coordinate transformation jacobian $\partial v_j/\partial x_i$ with dv the n-D differential element for $\hat{\Omega}_e$.

7.3 Multidimensional FE Operations for $\{N_k(\zeta_\alpha)\}$

For $v_j \Rightarrow \zeta_\alpha$ in equation (7.14) the coordinate transformation $\zeta_\alpha = \zeta_\alpha(x_i)$ is linear and explicit. The jacobians $\partial v_j/\partial x_i$ and $\partial v_k/\partial x_j$ are thus known constant *data* on Ω_e, hence also is $\det_e$. Approximating the thermal conductivity k_e on Ω_e by the element average, equation (7.14) simplifies to

$$[\text{DIFF}]_e = \bar{k}_e \left(\frac{\partial \zeta_j}{\partial x_i}\right)_e \left(\frac{\partial \zeta_k}{\partial x_i}\right)_e \det_e \int_{\hat{\Omega}_e} \frac{\partial\{N_k\}}{\partial \zeta_j} \frac{\partial\{N_k\}^T}{\partial \zeta_k} d\zeta_1 d\zeta_2 d\zeta_3 d\zeta_4. \tag{7.15}$$

The element-dependent multipliers preceding the integral in equation (7.15) are *known data*. As developed shortly, the transformation jacobian contains algebraic differences of element vertex node coordinates $\{XI\}_e$, $1 \leq I \leq n$, divided by $\det_e$. Retaining the use of capital letters to denote discrete data label these elements of the jacobian matrix as

$$\left(\partial \zeta_j / \partial x_i \right)_e \equiv \text{ZETA}JI_e / \text{DET}_e. \tag{7.16}$$

A determinant cancellation thus occurs in equation (7.15), so for any basis degree k and $1 \leq I \leq n$ equation (7.15) is

$$[\text{DIFF}]_e \equiv \text{COND}_e \, \text{ZETA}JI_e \, \text{ZETA}KI_e \, \text{DET}_e^{-1} [\text{M2}JKd]. \tag{7.17}$$

Thereby, the NC $\{N_k(\zeta_\alpha)\}$ basis matrix library for the diffusion term in equation (7.1) is the $1 \leq (J, K) \leq n+1$ set of matrices $[\text{M2}JKd]$. Recall the convention $M \Rightarrow (A,B,C)$ for $n = (1,2,3)$.

What remains is the determination of the NC system $\zeta_\alpha(x_i)$, hence the coordinate transformation $\partial \zeta_j / \partial x_i$. The convenient starting point for all n is polynomials on $\mathbf{x}$ of completeness degree $k = 1$

$$\begin{aligned} n = 1: \, Te(x) &= a_1 + a_2 x \\ n = 2: \, Te(x,y) &= a_1 + a_2 x + a_3 y \\ n = 3: \, Te(x,y,z) &= a_1 + a_2 x + a_3 y + a_4 z \end{aligned} \tag{7.18}$$

Selecting $n = 2$ for exposition, evaluating that polynomial in equation (7.18) at the triangle vertex nodes, Figure 7.1, generates the algebraic system

$$\begin{aligned} Te(x = X_1, y = Y_1) &\equiv Q1 = a_1 + a_2 X_1 + a_3 Y_1 \\ Te(x = X_2, y = Y_2) &\equiv Q2 = a_1 + a_2 X_2 + a_3 Y_2 \\ Te(x = X_3, y = Y_3) &\equiv Q3 = a_1 + a_2 X_3 + a_3 Y_3 \end{aligned} \tag{7.19}$$

Solving this 3×3 matrix statement via Cramer's rule (an exercise) yields

$$\begin{aligned} a_1 &= [(X_2 Y_3 - X_3 Y_2)Q1 + (X_3 Y_1 - X_1 Y_3)Q2 + (X_1 Y_2 - X_2 Y_1)Q3]/2A_e \\ a_2 &= [(Y_2 - Y_3)Q1 + (Y_3 - Y_1)Q2 + (Y_1 - Y_2)Q3]/2A_e \\ a_3 &= [(X_3 - X_2)Q1 + (X_1 - X_3)Q2 + (X_2 - X_1)Q3]/2A_e \end{aligned} \tag{7.20}$$

$$\det[\cdot]_e = 2A_e = (X_1 Y_2 - X_2 Y_1) + (X_3 Y_1 - X_1 Y_3) + (X_2 Y_3 - X_3 Y_2). \tag{7.21}$$

The $\{N_1(\zeta_\alpha)\}$ trial space basis, hence coordinate transformation, is formed by extracting $\{Q\}_e$ in equation (7.20) as the common multiplier. The definition for $n = 2$ is

$$Te(x,y) \equiv \{N_1(\zeta_a)\}^T \{Q\}_e \Rightarrow \{\zeta_1, \zeta_2, \zeta_3\}\{Q\}_e \tag{7.22}$$

and completing the substitutions yields

$$\begin{aligned} \zeta_1 &= [\{X_2 Y_3 - X_3 Y_2\}_e + (Y_2 - Y_3)_e x + (X_3 - X_2)_e y]/2A_e \\ \zeta_2 &= [\{X_3 Y_1 - X_1 Y_3\}_e + (Y_3 - Y_1)_e x + (X_1 - X_3)_e y]/2A_e \\ \zeta_3 &= [\{X_1 Y_2 - X_2 Y_1\}_e + (Y_1 - Y_2)_e x + (X_2 - X_1)_e y]/2A_e \end{aligned} \tag{7.23}$$

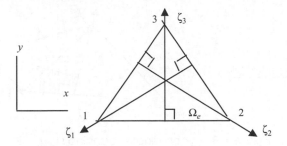

Figure 7.3 ζ_α system spanning Ω_e, $n=2$

Note that each ζ_α indeed is a linear polynomial in (x, y). The expression comparable to equation (7.23) for the $n=1$ basis is $\zeta_1 = (X_2 - x)/l_e$ and $\zeta_2 = (-X_1 + x)/l_e$, as results upon eliminating the $\bar{x}$ translation in Chapter 4.3. Figure 7.3 graphs the ζ_α system (7.23) spanning an $n=2$ Ω_e. Each ζ_α coordinate originates perpendicular to the pertinent triangle line segment becoming unity at "its" DOF. It is transparent the ζ_α system satisfies all $(0,1)$ requirements in equation (7.22) at each Ω_e vertex node.

The ζ_α system determination for $n=3$ requires solution of a 4×4 matrix statement, hence determinant expansions by cofactors. The trial space basis definition is

$$T_e(x,y,z) = \{N_1(\zeta_\alpha)\}^T\{Q\}_e = \{\zeta_1, \zeta_2, \zeta_3, \zeta_4\}\{Q\}_e \tag{7.24}$$

$$\text{with solution} \quad \{N_1(\zeta_\alpha)\} \equiv \begin{Bmatrix} \zeta_1 \\ \zeta_2 \\ \zeta_3 \\ \zeta_4 \end{Bmatrix} = \frac{1}{6V_e} \begin{Bmatrix} a_1 + b_1 + c_1 + d_1 \\ -(a_2 + b_2 + c_2 + d_2) \\ a_3 + b_3 + c_3 + d_3 \\ -(a_4 + b_4 + c_4 + d_4) \end{Bmatrix} \tag{7.25}$$

where V_e is the tetrahedron volume. The coefficient sets in equation (7.25) are the determinants

$$V_e \equiv \frac{\det}{6} \begin{bmatrix} 1 & X_1 & Y_1 & Z_1 \\ 1 & X_2 & Y_2 & Z_2 \\ 1 & X_3 & Y_3 & Z_3 \\ 1 & X_4 & Y_4 & Z_4 \end{bmatrix} \qquad b_i \equiv -\det \begin{bmatrix} 1 & Y_j & Z_j \\ 1 & Y_m & Z_m \\ 1 & Y_p & Z_p \end{bmatrix}_e$$

$$a_i \equiv \det \begin{bmatrix} X_j & Y_j & Z_j \\ X_m & Y_m & Z_m \\ X_p & Y_p & Z_p \end{bmatrix} \qquad c_i \equiv \det \begin{bmatrix} X_j & 1 & Z_j \\ X_m & 1 & Z_m \\ X_p & 1 & Z_p \end{bmatrix}_e \tag{7.26}$$

$$d_i \equiv -\det \begin{bmatrix} X_j & Y_j & 1 \\ X_m & Y_m & 1 \\ X_p & Y_p & 1 \end{bmatrix}_e$$

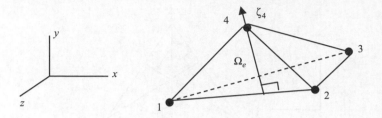

Figure 7.4 ζ_4 coordinate spanning Ω_e, $n=3$

wherein the indices (i, j, m, p) permute $(1, 2, 3, 4)$ over the vertex nodes of Ω_e, Figure 7.1. Figure 7.4 illustrates (only) the ζ_4 coordinate, readily verified to meet the $(0,1)$ requirement of equation (7.24) at all nodes of Ω_e, as thus do ζ_1, ζ_2, and ζ_3.

7.4 The NC$k=1,2$ Basis FE Matrix Library

The $n=1,2,3$ NC $k=1,2$ FE trial space bases $\{N_k(\zeta)\}$ are identified. Their identity enables construction of GWS^h for all n via the developed formalism

$$\text{GWS}^h = S_e\{\text{WS}\}_e = \{0\}, \{\text{WS}\}_e = ([\text{DIFF}]_e + [\text{BC}]_e)\{Q\}_e - \{b\}_e. \qquad (7.27)$$

The lead term in $\mathcal{L}(q)$, equation (7.1), again generates $[\text{DIFF}]_e$ and its construction involves the greatest expenditure of effort. For the $k=1$ trial space basis, the gradient operators can be extracted from the integral since $\{N_1(\zeta_\alpha\}$ is linear in **x**. $[\text{DIFF}]_e$ thus becomes

$$[\text{DIFF}]_e = \int_{\Omega_e} \nabla\{N_1\} \cdot k_e \nabla\{N_1\}^T d\tau = \nabla\{N_1\} \cdot \nabla\{N_1\}^T \int_{\Omega_e} k_e d\tau \qquad (7.28)$$

with the remaining integral (as usual) producing $\bar{k}_e$ times l_e, A_e, or V_e.
The chain rule leads to the accounting for ζ_α dependence as

$$\begin{aligned}
\nabla\{N_1\} &\equiv \hat{\mathbf{i}}\frac{\partial\{N_1\}}{\partial x} + \hat{\mathbf{j}}\frac{\partial\{N_1\}}{\partial y} + \hat{\mathbf{k}}\frac{\partial\{N_1\}}{\partial z} \\
&= \hat{\mathbf{i}}\frac{\partial\{N_1\}}{\partial \zeta_\alpha}\frac{\partial \zeta_\alpha}{\partial x} + \hat{\mathbf{j}}\frac{\partial\{N_1\}}{\partial \zeta_\alpha}\frac{\partial \zeta_\alpha}{\partial y} + \hat{\mathbf{k}}\frac{\partial\{N_1\}}{\partial \zeta_\alpha}\frac{\partial \zeta_\alpha}{\partial z}, \quad 1 \leq \alpha \leq n+1
\end{aligned} \qquad (7.29)$$

Via equation (7.29) the gradient dot product in equation (7.28) expanded for $1 \leq I \leq n$ and $1 \leq \alpha, \beta \leq n+1$ yields

$$\nabla\{N_1\} \cdot \nabla\{N_1\}^T = \left(\frac{\partial\{N_1\}}{\partial \zeta_\alpha}\frac{\partial\{N_1\}^T}{\partial \zeta_\beta}\right)\left(\frac{\partial \zeta_\alpha}{\partial x_i}\frac{\partial \zeta_\beta}{\partial x_i}\right)_e \qquad (7.30)$$

and the spatial derivatives for $1 \leq \alpha \leq 4$ generate the matrices

$$\frac{\partial \{N_1\}}{\partial \zeta_\alpha} = \{1,0,0,0,\}^T, \{0,1,0,0,\}^T, \{0,0,1,0,\}^T, \{0,0,0,0,1\}^T. \tag{7.31}$$

Each matrix product in equation (7.30) is thus a pointer placing the corresponding index scalar metric data product into the correct row–column location in $[\text{DIFF}]_e$. Therefore, for $k = 1$ GWS^h implementations on triangles and tetrahedra, the matrices $[M2JKL]$ defined in equation (7.16) for $[\text{DIFF}]_e$ are the set of "Kronecker delta" matrices

$$[M2JKL] \Rightarrow [M\delta_{jk}] \equiv \begin{bmatrix} 1 & \text{at matrix element location } j,k \\ 0 & \text{everywhere else} \end{bmatrix}. \tag{7.32}$$

With equation (7.32), the completion step for equation (7.28) is the determination of the $\zeta_\alpha = \zeta_\alpha(\mathbf{x})$ coordinate transformation metric data array $\partial \zeta_\alpha / \partial x_i$, an order $(n+1) \times n$ matrix of constants solely dependent on the global nodal coordinates of Ω_e. The transformations are analytical, via equations (7.22), (7.25), and the $n = 2$ suggested exercise produces

$$\left[\frac{\partial \zeta_\alpha}{\partial x_i}\right]_e = \frac{1}{2A_e} \begin{bmatrix} Y_2 - Y_3, & X_3 - X_2 \\ Y_3 - Y_1, & X_1 - X_3 \\ Y_1 - Y_2, & X_2 - X_1 \end{bmatrix}_e. \tag{7.33}$$

The $n = 3$ transformation matrix is more detailed, as it involves determinantal expansions, equations (7.25) and (7.26). The essential result is

$$\left[\frac{\partial \zeta_\alpha}{\partial x_i}\right]_e = \frac{1}{6V_e}[f(\Delta X \Delta Y, \quad \Delta X \Delta Z, \quad \Delta Y \Delta Z]_e. \tag{7.34}$$

Metric data products in equation (7.30) require summation on $1 \leq i \leq n$

$$\left(\frac{\partial \zeta_\alpha}{\partial x_i}\frac{\partial \zeta_\beta}{\partial x_i}\right) = \left(\frac{\partial \zeta_\alpha}{\partial x_1}\frac{\partial \zeta_\beta}{\partial x_1} + \frac{\partial \zeta_\alpha}{\partial x_2}\frac{\partial \zeta_\beta}{\partial x_2} + \frac{\partial \zeta_\alpha}{\partial x_3}\frac{\partial \zeta_\beta}{\partial x_3}\right)_e. \tag{7.35}$$

For $n = 2$, introducing the definition $\zeta_{\alpha i} \equiv 2A_e \partial \zeta_\alpha / \partial x_i$ to signify *only* the $[\Delta X, \Delta Y]_e$ matrix in equation (7.33), and noting $[M\delta_{jk}] \Rightarrow [M\delta_{\alpha\beta}]$, the suggested exercise will verify

$$[M\delta_{\alpha\beta}]\left(\frac{\partial \zeta_\alpha}{\partial x_i}\frac{\partial \zeta_\beta}{\partial x_i}\right)_e = \frac{1}{4A_e^2}\begin{bmatrix} \zeta_{11}^2 + \zeta_{12}^2, & \zeta_{11}\zeta_{21} + \zeta_{12}\zeta_{22}, & \zeta_{11}\zeta_{31} + \zeta_{12}\zeta_{32} \\ \zeta_{21}\zeta_{11} + \zeta_{22}\zeta_{12}, & \zeta_{21}^2 + \zeta_{22}^2, & \zeta_{21}\zeta_{31} + \zeta_{22}\zeta_{32} \\ \zeta_{31}\zeta_{11} + \zeta_{32}\zeta_{12}, & \zeta_{31}\zeta_{21} + \zeta_{32}\zeta_{22}, & \zeta_{31}^2 + \zeta_{32}^2 \end{bmatrix}_e. \tag{7.36}$$

After cancellation of one $A_e = \det_e/2$ by the integral in equation (7.28), the terminal $k = 1$ basis, $n = 2$ [DIFF]$_e$ element matrix is

$$[\text{DIFF}]_e = \frac{\bar{k}_e}{2\,\det_e} \begin{bmatrix} \zeta_{11}^2 + \zeta_{12}^2, & \zeta_{11}\zeta_{21} + \zeta_{12}\zeta_{22}, & \zeta_{11}\zeta_{31} + \zeta_{12}\zeta_{32}, \\ \zeta_{21}\zeta_{11} + \zeta_{22}\zeta_{12}, & \zeta_{21}^2 + \zeta_{22}^2, & \zeta_{21}\zeta_{31} + \zeta_{22}\zeta_{32}, \\ \zeta_{31}\zeta_{11} + \zeta_{32}\zeta_{12}, & \zeta_{31}\zeta_{21} + \zeta_{32}\zeta_{22}, & \zeta_{31}^2 + \zeta_{32}^2, \end{bmatrix}_e . \tag{7.37}$$

For any $k > 1$ basis triangle/tetrahedron implementation, the [M2JKd] is another full set of element independent matrices with entries the ratio of integers. For $k = 2$ and after a rather lengthy calculus process the symmetric matrices forming [B2JKQ] are

$$[B211Q] = \frac{1}{3} \begin{bmatrix} 3 & 0 & 0 & 0 & 0 & 0 \\ & 0 & 0 & 0 & 0 & 0 \\ & & 0 & 0 & 0 & 0 \\ & & & 8 & 0 & 4 \\ & & & & 0 & 0 \\ & & & & & 8 \end{bmatrix}, \qquad [B222Q] = \frac{1}{3} \begin{bmatrix} 0 & 0 & 0 & 0 & 0 & 0 \\ & 3 & 0 & 0 & 0 & 0 \\ & & 0 & 0 & 0 & 0 \\ & & & 8 & 4 & 0 \\ & & & & 8 & 0 \\ & & & & & 0 \end{bmatrix}$$

$$[B233Q] = \frac{1}{3} \begin{bmatrix} 0 & 0 & 0 & 0 & 0 & 0 \\ & 0 & 0 & 0 & 0 & 0 \\ & & 3 & 0 & 0 & 0 \\ & & & 0 & 0 & 0 \\ & & & & 8 & 4 \\ & & & & & 8 \end{bmatrix}$$

$$[B212Q] + [B221Q] = \frac{1}{3} \begin{bmatrix} 0 & -1 & 0 & 4 & 0 & 0 \\ & 0 & 0 & 4 & 0 & 0 \\ & & 0 & 0 & 0 & 0 \\ & & & 8 & 4 & 4 \\ & & & & 0 & 8 \\ & & & & & 0 \end{bmatrix} \tag{7.38}$$

$$[B213Q] + [B231Q] = \frac{1}{3} \begin{bmatrix} 0 & 0 & -1 & 0 & 0 & 4 \\ & 0 & 0 & 0 & 0 & 0 \\ & & 0 & 0 & 0 & 4 \\ & & & 0 & 8 & 4 \\ & & & & 0 & 4 \\ & & & & & 8 \end{bmatrix}$$

$$[B223Q] + [B232Q] = \frac{1}{3} \begin{bmatrix} 0 & 0 & 0 & 0 & 0 & 0 \\ & 0 & -1 & 0 & 4 & 0 \\ & & 0 & 0 & 4 & 0 \\ & & & 0 & 4 & 8 \\ & & & & 8 & 4 \\ & & & & & 0 \end{bmatrix}$$

Table 7.1 Natural coordinate $k = 1$ FE basis matrix library, $1 \leq n \leq 3$

$n = 1$:

$$[DIFF]_e = \frac{\bar{k}_e}{l_e} \begin{bmatrix} 1 & -1 \\ -1 & 1 \end{bmatrix}, \quad \{b\}_e = \frac{l_e}{6} \begin{bmatrix} 2 & 1 \\ 1 & 2 \end{bmatrix} \{SRC\}_e, \quad [BC]_e = h \begin{bmatrix} \delta_{e1} \\ & \delta_{eM} \end{bmatrix}$$

$n = 2$:

$$[DIFF]_e = \frac{\bar{k}_e}{4 A_e} \begin{bmatrix} \zeta_{11}^2 + \zeta_{12}^2, & \zeta_{11}\zeta_{21} + \zeta_{12}\zeta_{22}, & \zeta_{11}\zeta_{31} + \zeta_{12}\zeta_{32}, \\ \zeta_{21}\zeta_{11} + \zeta_{22}\zeta_{12}, & \zeta_{21}^2 + \zeta_{22}^2, & \zeta_{21}\zeta_{31} + \zeta_{22}\zeta_{32}, \\ \zeta_{31}\zeta_{11} + \zeta_{32}\zeta_{12}, & \zeta_{31}\zeta_{21} + \zeta_{32}\zeta_{22}, & \zeta_{31}^2 + \zeta_{32}^2, \end{bmatrix}_e$$

$$\{b(s)\}_e = \frac{A_e}{12} \begin{bmatrix} 2 & 1 & 1 \\ 1 & 2 & 1 \\ 1 & 1 & 2 \end{bmatrix} \{SRC\}_e, \quad [BC]_e = \frac{\bar{h}_e l_e}{6} \begin{bmatrix} 2 & 1 \\ 1 & 2 \end{bmatrix}$$

$n = 3$:

$$[DIFF]_e = \frac{\bar{k}_e}{36 V_e} \begin{bmatrix} \zeta_{11}^2 + \zeta_{12}^2 + \zeta_{13}^2, & \zeta_{11}\zeta_{21} + \zeta_{12}\zeta_{22} + \zeta_{13}^2, & \zeta_{23}^2, \ldots \\ & \cdot \quad \cdot \quad \cdot & \zeta_{21}^2 + \zeta_{22}^2 + \zeta_{23}^2, \ldots \text{etc.} \end{bmatrix}_e$$

$$\{b(s)\}_e = \frac{V_e}{20} \begin{bmatrix} 2 & 1 & 1 & 1 \\ 1 & 2 & 1 & 1 \\ 1 & 1 & 2 & 1 \\ 1 & 1 & 1 & 2 \end{bmatrix} \{SRC\}_e, \quad [BC]_e = \frac{\bar{h}_e A_e}{12} \begin{bmatrix} 2 & 1 & 1 \\ 1 & 2 & 1 \\ 1 & 1 & 2 \end{bmatrix}$$

In completing the NC basis FE matrix library, the source term in equation (7.1), along with Robin BCs (7.2) simply require integrals of products of the NCs. For all n, these integrals are analytically evaluable via

$$\int_{\Omega_e} \zeta_1^p \zeta_2^q \zeta_3^r \zeta_4^s = \det_e \frac{p! \, q! \, r! \, s!}{(n + p + q + r + s)!}. \tag{7.39}$$

Table 7.1 summarizes the $DE + BCs$ principle GWS^h algorithm linear NC $\{N_1(\zeta_\alpha)\}$ basis element matrix library for $1 \leq n \leq 3$.

7.5 NC Basis {WS}$_e$ Template, Accuracy, Convergence

For the process $GWS^N \Rightarrow GWS^h \Rightarrow S_e\{WS\}_e$ the algorithm object keyed $\{WS\}_e$ *template* definition remains as in equation (6.5)

$$\{WS\}_e = (const) \begin{pmatrix} elem \\ const \end{pmatrix}_e \begin{Bmatrix} distr \\ data \end{Bmatrix}_e^T (\text{metric data; det}) [\text{Matrix}] \begin{Bmatrix} Q \text{ or} \\ data \end{Bmatrix}_e. \tag{7.40}$$

Eliminating the $\bar{k}_e$ simplification and interpolating all *data*, the GWS^h algorithm template for (7.1)–(7.3) for any n and any basis degree k is

$$
\begin{aligned}
\{WS\}_e &= ([DIFF]_e + [BC]_e)\{Q\}_e - \{b(data(s,h,T_c,f_n))\}_e \\
&= (\ )(\ )\{COND\}(ZETAJI, ZETAKI; -1)[M30JKd]\{Q\} \\
&\quad -(\ )(\ )\{\ \}(;1)[M200d]\{SRC\} \\
&\quad +(\ )(\ )\{H\}(;1)[N3000d](\{Q\} - \{TC\}) \\
&\quad +(SIG)(\ )\{f(vf,\varepsilon\}(;1)[N3000d]\left(\{Q\}^4 - \{TR\}^4\right) \\
&\quad -(\ )(\ )\{\ \}(;1)[N200d]\{\mathbf{F} \bullet \mathbf{n}\}
\end{aligned}
\tag{7.41}
$$

The linear algebra jacobian template is

$$
\begin{aligned}
[JAC]_e &= (\ )(\ )\{COND\}(ZETAJI, ZETAKI; -1)[M30JKd][\ ] \\
&\quad +(\ )(\ )\{H\}(;1)[N3000d][\ ] \\
&\quad +(4, SIG)(\ )\{f(vf,\varepsilon\}(;1)[N3000d]diag[Q^3]
\end{aligned}
\tag{7.42}
$$

The summation indices range $1 \leq I \leq n$, $1 \leq J$, $K \leq n+1$. Element matrix prefix M is replaced with N for the BC matrices which are $n-1$ dimensional. All *data* are assumed interpolated in terms of nodal DOF. The radiation implementation specifics must account for view factor; this derivation is deferred pending development of the tensor product basis GWS^h implementation.

The MATLAB® environment *FEmPSE* toolbox enables $k=1$ basis implementations on $n=2$. Computer lab 7.1 seeks solution of equation (7.1) with BC equation (7.3) for a solution domain containing a high-conductivity heat sink embedded in a low-conductivity dielectric subject to external heating (think of a PC card). The dielectric/conductor domains share DOF, hence $\Omega = \Omega_c \cup \Omega_d$ with data definitions $k=1.0, s=0$ in Ω_c while $k=0.1, s=200$ in Ω_d.

MATLAB® functionality includes generation of Delauney triangulations on $n=2$ domains Ω. Figure 7.5 left graphs the result with mesh nonuniformity pertinent to (anticipated) temperature gradient distribution. Figure 7.5 right graphs the GWS^h temperature distribution perspective for one of the computer lab 7.1 data specifications.

The mathematically pertinent DE + BCs simulation supporting accuracy/ convergence assessment is conduction in a cylindrical pipe subject to internal flow convective heat transfer with Dirichlet outer-wall BCs. The problem geometry and given data are symmetric hence any pie-shaped domain definition is appropriate. For a quarter circle, Figure 7.6, BCs must be applied on the exposed x and y equal constant boundary segments. Problem symmetry dictates *adiabatic* is required; it is *most* convenient that via GWS^h this form of equation (7.2) is free!

This statement possesses an analytical solution, logarithmic in radius r which, via k $(x) \Rightarrow k \times r$, is functionally identical to the $n=1$ accuracy/convergence computer lab 5.1. Hence, M-dependent GWS^h solution surface temperatures are known for comparison. For the $M=32$ discretization the $k=1$ basis solution is axisymmetric but

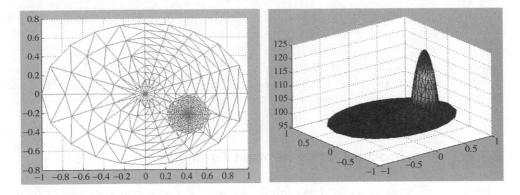

Figure 7.5 Conductor-dielectric triangulation and GWS^h solution

underpredicts (!?) the inner surface temperature of the $n = 1$, $M = 4$ (identical radial resolution) solution, Table 5.2, by several degrees. Why?

Computer lab 7.2 azimuthal mesh-refinement study will verify the additional error mechanism at play is inaccurate computation of the flux BC term $\|\text{data}\|^2_{\partial\Omega,L2}$ in the asymptotic error estimate equation (7.8). Specifically, since the cylinder curved boundary segments are $k = 1$ basis approximated by the chord, clearly illustrated in Figure 7.6, the surface area over which the equation (7.2) convective heating BC acts is underestimated.

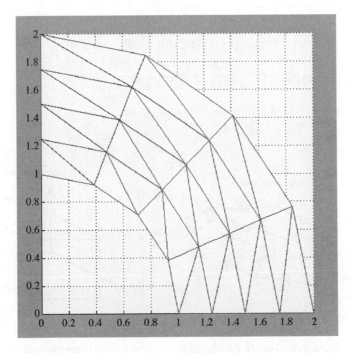

Figure 7.6 Quarter cylinder domain Ω with triangulation

This error mechanism is mathematically similar to that exposed in computer lab 5.3 wherein convergence of $k=1$ basis solutions did not adhere to theory for coarse meshes M. That error source was underprediction of the domain data norm $\|data\|^2_{\Omega,L2}$. In either instance as noted, data-interpolation error goes to zero faster than solution approximation error, as will be verified in completing this lab. As the problem data are smooth, convergence rate is dominated by the basis factor k rather than the $r-1$ term, equation (7.8).

7.6 The Tensor Product Basis Element Family

Moving to the FE tensor product basis family, Figure 7.2, the calculus operations forming $[DIFF]_e$ become more detailed for quadrilateral and hexahedron shaped elements that depart significantly from parallelograms. This results from the local–global coordinate transformation $\eta_j(x_i)$ not being explicitly available. (Note: this issue is also true for NC basis triangles/tetrahedra possessing curved boundaries.)

The resolution is to *interpolate* the x_i coordinate system spanning Ω_e via the basis $\{N_k^+(\boldsymbol{\eta})\}$ and the element geometric node global coordinate data $\{XI\}_e$, $1 \leq I \leq n$. The result is transformation of an arbitrary curvilinear Ω_e in physical (x_i) space to a unit cartesian parallelepiped in transform (η_j) space, Figure 7.7. This interpolation handles elements possessing straight and/or curved boundaries, $n=2,3$ sides/surfaces, independent of basis completeness degree k.

The coordinate interpolation statement for all n is

$$(x_i)_e = \{N_k^+(\eta)\}^T \{XI\}_e, \tag{7.43}$$

which produces the explicit forward coordinate transformation

$$\left(\frac{\partial x_i}{\partial \eta_j}\right)_e = \frac{\partial}{\partial \eta_j} \{N_k^+(\eta)\}^T \{XI\}_e. \tag{7.44}$$

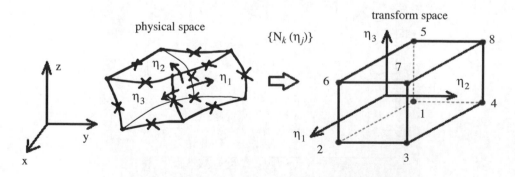

Figure 7.7 Tensor product basis coordinate transformation

Substituting $v_j \Rightarrow \eta_j$ in equation (7.14) and introducing the summation index m (to avoid confusion with conductivity k) yields

$$
\begin{aligned}
[\text{DIFF}]_e &= \int_{\Omega_e} \nabla\{N_k^+\} \cdot k_e \{N_k^+\}^T d\tau \\
&= \int_{\Omega_e} \frac{\partial \{N_k^+\}}{\partial \eta_j} k_e \frac{\partial \{N_k^+\}^T}{\partial \eta_m} \left(\frac{\partial \eta_j}{\partial x_i} \frac{\partial \eta_m}{\partial x_i} \right)_e d\tau
\end{aligned}
\tag{7.45}
$$

For any completeness degree $k \geq 1$ the spatial derivative of $\{N_k^+\}$ in equation (7.45) retains functions of η_j. For example, for $n = 2$ via equation (7.12)

$$
\frac{\partial \{N_1^+\}}{\partial \eta_j} = \frac{1}{4} \frac{\partial}{\partial \eta_j} \left\{ \begin{array}{l} (1-\eta_1)(1-\eta_2) \\ (1+\eta_1)(1-\eta_2) \\ (1+\eta_1)(1+\eta_2) \\ (1-\eta_1)(1+\eta_2) \end{array} \right\} = \frac{1}{4} \left\{ \begin{array}{l} -(1-\eta_2) \\ (1-\eta_2) \\ (1+\eta_2) \\ -(1+\eta_2) \end{array} \right\}_{j=1} , \frac{1}{4} \left\{ \begin{array}{l} -(1-\eta_1) \\ -(1+\eta_1) \\ (1+\eta_1) \\ (1-\eta_1) \end{array} \right\}_{j=2} .
\tag{7.46}
$$

Clearly matrix products of equation (7.46) in $[\text{DIFF}]_e$ cannot be extracted from the integrand, as allowed for the ζ_α element family.

The terminal complication for $[\text{DIFF}]_e$ integral evaluations is absence of the explicit form of the transformation jacobian matrix with elements $(\partial \eta_j / \partial x_i)_e$. Hence it must be generated from the forward transformation (7.43) via a matrix inversion. Again using $n = 2$ for illustration

$$
\left[\frac{\partial \eta_j}{\partial x_i} \right]_e = \left[\frac{\partial x_i}{\partial \eta_j} \right]_e^{-1} = \begin{bmatrix} \dfrac{\partial x_1}{\partial \eta_1} & \dfrac{\partial x_1}{\partial \eta_2} \\ \dfrac{\partial x_2}{\partial \eta_1} & \dfrac{\partial x_2}{\partial \eta_2} \end{bmatrix}_e^{-1} = \frac{1}{\det[J]_e} \begin{bmatrix} \dfrac{\partial x_2}{\partial \eta_2} & -\dfrac{\partial x_1}{\partial \eta_2} \\ -\dfrac{\partial x_2}{\partial \eta_1} & \dfrac{\partial x_1}{\partial \eta_1} \end{bmatrix}_e
\tag{7.47}
$$

for $[J]_e$ the jacobian matrix of the forward transformation (7.44). The $n = 3$ construction is a direct but more detailed as the inverse contains matrix cofactors with multiple embedded determinants.

Continuing the $k = 1$, $n = 2$ exposition, the elements of the inverse transformation matrix in equation (7.47) using equation (7.44) are

$$
\left(\frac{\partial x_1}{\partial \eta_1} \right)_e = \frac{\partial}{\partial \eta_1} \{N_1^+(\boldsymbol{\eta})\}^T \{XI\}_e
$$

$$
= \frac{1}{4} \left[(1-\eta_2)(X2 - X1)_e + (1+\eta_2)(X3 - X4)_e \right] \equiv eta_{22}
$$

$$
\left(\frac{\partial x_1}{\partial \eta_2} \right)_e = \frac{1}{4} \left[(1-\eta_1)(X4 - X1)_e + (1+\eta_1)(X3 - X2)_e \right] \equiv -eta_{12}
\tag{7.48}
$$

$$
\left(\frac{\partial x_2}{\partial \eta_1} \right)_e = \frac{1}{4} \left[(1-\eta_2)(Y2 - Y1)_e + (1+\eta_2)(Y3 - Y4)_e \right] \equiv -eta_{21}
$$

$$
\left(\frac{\partial x_2}{\partial \eta_2} \right)_e = \frac{1}{4} \left[(1-\eta_1)(Y4 - Y1)_e + (1+\eta_1)(Y3 - Y2)_e \right] \equiv eta_{11}
$$

which introduces the *notation* eta$_{ji}$ for later use. Therefore, even for the elementary $k=1$ TP basis every matrix element in the coordinate transformation is η_j-dependent, hence also is det$[J]_e$ in the denominator of equation (7.47). Then admitting conductivity k_e may be variable data, *no* terms defining the [DIFF]$_e$ integrand (7.45) may *a priori* be extracted to simplify the integration process.

7.7 Gauss Numerical Quadrature, $k=1$ TP Basis Library

Precise resolution of the [DIFF]$_e$ formation issue exists, but becomes computationally intense for large TP basis meshings M. As all integrand data are functions of η_j after transformation, the domain of the integral (7.45) is really the *rectangular cartesian master* element $\hat{\Omega}_e$, Figure 7.7 right. The requisite differential element transformation is

$$d\tau = \det[J]_e d\eta_1 d\eta_2 d\eta_3 \equiv \det[J]_e d\boldsymbol{\eta}, \qquad (7.49)$$

hence one det$[J]_e$ from the transformation product $\left(\frac{\partial \eta_j}{\partial x_i} \frac{\partial \eta_m}{\partial x_i}\right)_e$ in equation (7.45), see equation (7.47), is canceled. Label each cofactor matrix element in the inverse as (eta$_{ji}$)$_e$, as illustrated for $n=2$ in equation (7.48), by extracting the remaining det$[J]$. Then denoting the column–row matrix product elements of [DIFF]$_e$ as (diff$_{\alpha\beta}$)$_e$, $1 \leq (\alpha,\beta) \leq f(k, n)$ for $1 \leq i \leq n$, the generic integral required evaluated in equation (7.45) is

$$\left(\text{diff}_{\alpha\beta}\right)_e \equiv \iiint_{\hat{\Omega}_e} k_e \frac{\partial N_\alpha}{\partial \eta_j} \frac{\partial N_\beta}{\partial \eta_m} \left(\text{eta}_{ji}\text{eta}_{mi}\right)_e \det{}_e^{-1} d\boldsymbol{\eta}. \qquad (7.50)$$

The essence of equation (7.50) is

$$\left(\text{diff}_{\alpha\beta}\right)_e \equiv \int_{\hat{\Omega}} \left(\text{integrand}_{j,m,i}^{\alpha,\beta}(\boldsymbol{\eta})\right) d\boldsymbol{\eta}, \quad 1 \leq j,m,i \leq n \qquad (7.51)$$

and to predictable accuracy equation (7.51) can be evaluated via Gauss numerical quadrature as the dense summation operation

$$\left(\text{diff}_{\alpha\beta}\right)_e \cong \sum_{i=1}^{n}\sum_{j=1}^{n}\sum_{m=1}^{n}\sum_{p=1}^{P}\sum_{q=1}^{Q}\sum_{r=1}^{R} H_p H_q H_r \left(\text{integrand}_{j,m,i}^{\alpha,\beta}(\boldsymbol{\eta} \Rightarrow \boldsymbol{\eta}_{j,m}^{p,q,r})\right). \qquad (7.52)$$

The theory for *Gauss quadrature* predicts a Pth order quadrature rule (P,Q,R for $n=3$) will exactly integrate a polynomial of degree $2P-1$ spanning the *cartesian* element $\hat{\Omega}_e$, [2]. Hence, the summation limits (P,Q,R) are chosen to meet accuracy requirements dependent on the highest degree polynomial in equation (7.50). This theory is not exact

Table 7.2 Gauss quadrature rule data, $n=2$

P, Q	Coordinates η_1^q	Coordinates η_2^q	Weights H_p	Weights H_q
1	0.0	0.0	1.0	1.0
2	$\pm 1/\sqrt{3}$	$\pm 1/\sqrt{3}$	1.0	1.0
3	0.0	0.0	8/9	8/9
	$\pm\sqrt{0.6}$	$\pm\sqrt{0.6}$	5/9	5/9

for equation (7.50) however, as the integrand is a *rational* polynominal because of the embedded $\det_e^{-1}$.

Table 7.2 lists the $n=2$ rule *Gauss point* coordinates with P,Q coefficients and weights H. Figure 7.8 presents geometric coordinates and the $n=3$ data specifications are the obvious extension.

Completing the Gauss quadrature loop for matrix $[DIFF]_e$ for basis degree k augments the summation operation (7.52) to

$$[DIFF]_e \cong \sum_{\alpha,\beta}^{k,n} \sum_i^n \sum_j^n \sum_m^n \sum_p^P \sum_q^Q \sum_r^R H_p H_q H_r \left(integrand_{j,m,i}^{\alpha,\beta} (\boldsymbol{\eta}_{j,m}^{p,q,r}) \right) \qquad (7.53)$$

which confirms that code implementation of Gauss quadrature generates a deep nesting of very short FOR loops, a *compute-intense* operation.

Certainly seeking a simplification that at least admits extracting the transformation data $(eta_{ji})_e$, hence also $\det_e$, from the integrand in equation (7.50) is appropriate. An accurate approximation does exist for the $k=1$ TP basis *only* (!) implementation of $[DIFF]_e$. For the TP domain Ω_e a *parallelepiped* and for $n=2,3$, Figure 7.9, the coordinate transformation metric data are constants on $\hat{\Omega}_e$.

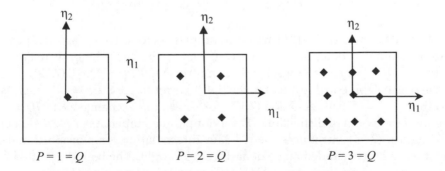

Figure 7.8 Gauss quadrature coordinates, $n=2$

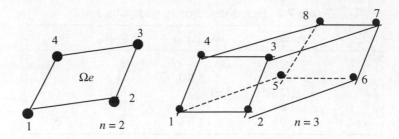

Figure 7.9 Bilinear TP basis parallelepiped Ω_e, $n = 2, 3$

Continuing the $n = 2$ exposition, for this restriction terms in (7.48) simplify to, for example,

$$\left(\frac{\partial x_2}{\partial \eta_2}\right)_e = \frac{1}{4}(Y4 - Y1 + Y3 - Y2)_e = \Delta Y_e/2 \equiv \text{eta}_{11}$$

$$\left(\frac{\partial x_1}{\partial \eta_2}\right)_e = \frac{1}{4}(X4 - X1 + X3 - X2)_e = \Delta X_e/2 \equiv -\text{eta}_{12} \tag{7.54}$$

The same result occurs for $n = 3$ via direct geometric extension. Further, since the metric data are now constants on $\hat{\Omega}_e$ so is det $[J]_e$.

For the TP basis parallelepiped element restriction and assuming average conductivity $\bar{k}_e$, the *exact* restatement of equation (7.45) is

$$[\text{DIFF}]_e = \det{}_e^{-1}(\text{eta}_{ji}\text{eta}_{ki})_e\bar{k}_e \int_{\Omega^e} \frac{\partial\{N_1^+\}}{\partial \eta_j}\frac{\partial\{N_1^+\}^T}{\partial \eta_k}\,d\boldsymbol{\eta}. \tag{7.55}$$

All element-dependent *data* are now extracted from the integrand, which admits writing equation (7.55) for TP parallelepiped Ω_e in the form of equation (7.17)

$$[\text{DIFF}]_e \equiv \text{COND}_e\ \text{ETA } JI_e\ \text{ETA } KI_e\ \text{DET}_e^{-1}[\text{M } 2JKL]. \tag{7.56}$$

The elements of matrix [M2*JKL*] are ratios of integers for all n. All are independent of Ω_e with their number determined by $1 \leq (I, J, K) \leq n$. This is *the* TP basis major distinction with the NC $k = 1$ basis Kronecker delta matrix [M2*JKL*] $\Rightarrow$ [Mδ_{jk}], (7.32).

For the parallelepiped Ω_e restriction, the [DIFF]$_e$*template* for the $k = 1$ TP basis is identical to the NC $k = 1$ basis, equation (7.17), for $d \Rightarrow L$, the exchange ZETA*JI*$_e$ $\Leftrightarrow$ ETA*JI*$_e$, and noting j index range distinctions. For a nonparallelepiped *rhomboidal* element, the approximation (7.55) introduces metric data interpolation *error* proportional to the cosine of extremum included angle departure from right. The boundary fitted coordinate transformations engendering TP basis geometric versatility, [1], rarely generate elements sufficiently distorted to invalidate (7.56).

To symbolize the Gauss quadrature *option* for arbitrary k TP basis GWS^h formulations $[\text{M2}JKd] \Rightarrow [\text{M2}JKE]$ denotes generation of the *Element-dependent* matrix library resulting from quadrature. It must be emphasized that GWS^h implementation using FE TP bases $\{N_k^+(\eta)\}$, $k > 1$, or $k = 1$ basis meshes containing substantially nonparallelepiped elements Ω_e, *always* requires Gauss quadrature to accurately generate element matrices [M . . .] containing spatial derivatives!

Formation of $[\text{DIFF}]_e$ for TP basis GWS^h implementations is complete for all k and n. The $k \geq 1$ TP basis element-*ind*ependent matrix library is easily completed for source and BC terms in equations (7.1)–(7.2) for any n. Table 7.3 is the $k = 1$ basis summary comparable to Table 7.1 for the $k = 1$ NC basis.

Comparing the $k = 1$ NC/TP basis data, Tables 7.1 and 7.3 confirm the significant regularity pervading element matrix constructions. Now including BCs and recalling

Table 7.3 Parallelepiped tensor product $k = 1$ basis FE matrix library

$n = 1$:

$$[\text{DIFF}]_e = \frac{\bar{k}_e}{2\det_e}\begin{bmatrix} 1 & -1 \\ -1 & 1 \end{bmatrix}, \ \{b\}_e = \frac{\det_e}{3}\begin{bmatrix} 2 & 1 \\ 1 & 2 \end{bmatrix}\{\text{SRC}\}_e, \ [\text{BC}]_e = h_e\begin{bmatrix} \delta_{e1} & \\ & \delta_{eM} \end{bmatrix}$$

$n = 2$:

$$[\text{DIFF}]_e \equiv \text{COND}_e \ \text{ETA} \ JI_e \ \text{ETA} \ KI_e \ \text{DET}_e^{-1} \ [\text{B } 2JKL]$$
$$= \text{COND}_e \ (\text{ETA } 11_e^2 + \text{ETA } 12_e^2)\text{DET}_e^{-1}[\text{B } 211L] + \dots$$
$$+ \text{COND}_e \ (\text{ETA } 21_e^2 + \text{ETA } 22_e^2)\text{DET}_e^{-1}[\text{B } 222L]$$

$$[\text{B } 211L] = \frac{1}{6}\begin{bmatrix} 2 & -2 & -1 & 1 \\ -2 & 2 & 1 & -1 \\ -1 & 1 & 2 & -2 \\ 1 & -1 & -2 & 1 \end{bmatrix}, \dots, [\text{B } 222L] = \frac{1}{6}\begin{bmatrix} 2 & 1 & -1 & -2 \\ 1 & 2 & -2 & -1 \\ -1 & -2 & 2 & 1 \\ -2 & -1 & 1 & 2 \end{bmatrix}$$

$$\{b\}_e = \frac{\det_e}{9}\begin{bmatrix} 4 & 2 & 1 & 2 \\ 2 & 4 & 2 & 1 \\ 1 & 2 & 4 & 2 \\ 2 & 1 & 2 & 4 \end{bmatrix}\{\text{SRC}\}_e, \quad [\text{BC}]_e = \frac{\bar{h}_e\det_e}{3}\begin{bmatrix} 2 & 1 \\ 1 & 2 \end{bmatrix}$$

$n = 3$:

$$[\text{DIFF}]_e \equiv \text{COND}_e \ \text{ETA} \ JI_e \ \text{ETA} \ KI_e \ \text{DET}_e^{-1} \ [\text{C } 2JKL]$$
$$= \text{COND}_e(\text{ETA } 11_e^2 + \text{ETA } 12_e^2 + \text{ETA } 13_e^2)\text{DET}_e^{-1}[\text{C } 211L] + \dots$$
$$+ \text{COND}_e(\text{ETA } 31_e^2 + \text{ETA } 32_e^2 + \text{ETA } 33_e^2)\text{DET}_e^{-1}[\text{C } 233L]$$

$$\{b_e\} = \frac{\det_e}{27}\begin{bmatrix} 8 & 4 & 2 & 4 & 4 & 2 & 1 & 2 \\ & 8 & 4 & 2 & 4 & 4 & 2 & 1 \\ & & 8 & 4 & 2 & 4 & 4 & 2 \\ & & & & & & & \cdot \\ (\text{sym}) & & & & & & & 8 \end{bmatrix}\{\text{SRC}\}_e, \quad [\text{BC}]_e = \bar{h}_e\frac{\det_e}{9}\begin{bmatrix} 4 & 2 & 1 & 2 \\ 2 & 4 & 2 & 1 \\ 1 & 2 & 4 & 2 \\ 2 & 1 & 2 & 4 \end{bmatrix}$$

definition (7.40), the GWS^h TP basis template for all k is the appearance alteration to the NC template (7.41)

$$
\begin{aligned}
\{WS\}_e &= \left([DIFF]_e + [BC]_e\right)\{Q\}_e - \{b(data(s, h, T_c, T_r, f_n))\}_e \\
&= (\ )(\ )\{COND\}(ETAJI, ETAKI; -1)[M30JKE]\{Q\} \\
&\quad -(\ )(\ )\{\ \}(; 1)[M200d]\{SRC\} \\
&\quad +(\ )(\ )\{H\}(; 1)[N3000d](\{Q\} - \{TC\}) \\
&\quad +(SIG)(\ )\{f(vf, \varepsilon\}(; 1)[N200d]\left(\{Q\}^4 - \{TR\}^4\right) \\
&\quad -(\ )(\ )\{\ \}(; 1)[N200d]\{\mathbf{F} \bullet \mathbf{n}\}
\end{aligned}
\tag{7.57}
$$

Restricting $\{WS\}_e$ (7.57) to parallelogram $k=1$ TP basis amounts to moving $\{COND\}$ to the second bracket, replacing $[M30JKE]$ with $[M2JKL]$, and the exchange $d \Rightarrow L$.

Simulations can certainly use a mixture of NC/TP elements within a single problem analysis. Select $k > 1$ basis matrices have been identified; texts focused in FE structural mechanics applications present very complete details [2]. Commercial codes admit mixed-k basis meshes using interpolation theory to handle the generated *hanging* DOF. These code-implementation topics lie beyond the scope of this text.

7.8 Convection–Radiation BC GWS^h Implementation

Genuine DE statements rarely have Dirichlet BCs as given data, as this occurs *only* when the exchange medium undergoes phase change, for example, boiling/freezing liquids. That the GWS^h algorithm precisely embeds all pertinent Robin gradient BC statements as *natural* warrants exposition.

For the fourth line in template (7.57), convective heat transfer is categorized into three modes, *natural, mixed, and forced*. Encompassing convection heat transfer coefficient correlations are available in the form $h = h(k, Re, Gr, Pr, x)$ for definitions [3]:

$k \Rightarrow$ heat exchange device thermal conductivity
$Re \Rightarrow$ exchange fluid flow Reynolds number, $Re = \rho UL/\mu$
$Gr \Rightarrow$ operating mode Grashoff number, $Gr = \rho^2 \beta g \Delta T\, x^3/\mu^2$
$Pr \Rightarrow$ exchange fluid Prandtl number, $Pr = \mu c_p/k$
$x \Rightarrow$ distance from onset flow impingement on device

The heat exchange medium non-D groups involve *flow* reference velocity U and length scale L with exchange *fluid* properties density ρ, dynamic viscosity μ, compressibility β, specific heat c_p, and thermal conductivity k. The temperature difference between fluid and heat exchange device is ΔT with g the magnitude of gravity.

The convection coefficient correlations for the three convective heat transfer modes are:

$$\begin{aligned}
&forced: \ h = C_1 k Pr^{1/3} Re^{1/2}/x, \ C_1 = 0.664 \\
&natural: \ h = C_2 k (GrPr)^n, \ C_2 = 0.540 \\
&\qquad\qquad Gr = \rho\beta\Delta T x^3/\mu^2 \\
&\qquad\qquad n = (0.25, 0.3) \text{ for } \Delta T < 0, \ \Delta T > 0 \\
&mixed: \ \text{smooth transition between the two correlations}
\end{aligned} \tag{7.58}$$

The existence of natural/mixed/forced convection heat transfer mode is determined by Gr/Re^2 being of O ($\gg 1$, ~ 1, $\ll 1$). The convection coefficient data generated via equation (7.58) enter the $\{WS\}_e$ template as nodal DOF $\{H\}_e$, third bracket in the fourth line in equation (7.57).

Specification of radiation heat transfer is much more detailed with the GWS^h algorithm yielding a precise implementation statement. Radiation environments typically involve surface facets that are a *gray body*, defined as possessing emissivity ε less than unity. Radiation involves the sum total of all surface facet pairs with geometrical orientation requiring determination of the line of sight.

Figure 7.10 illustrates the geometric essential. Lambert's cosine law identifies the radiation received at differential area dA_i emitted from differential area dA_k in terms

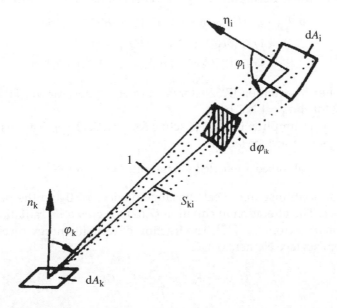

Figure 7.10 Surface facets in radiation energy exchange

of their respective normal unit vectors and separation distance r leading to the definition of *viewfactor*

$$F_{k\to i} \equiv \frac{1}{A_k} \int_{A_k} \int_{A_i} \frac{\cos\phi_k \cos\phi_i}{\pi r_{k\to i}^2} \, dA_k dA_i. \tag{7.59}$$

For a linear/bilinear NC/TP basis discrete implementation surface facets are planar, hence the *kernel* in equation (7.59) is symmetric

$$K_{k\to i} = \frac{\cos\phi_k \cos\phi_i}{\pi r_{ki}^2} = K_{i\to k} \equiv K_{ik}. \tag{7.60}$$

The fact that kernel (7.60) is element distributed data is traditionally ignored in view factor evaluation codes, [4], engendering the simplification $F_{k\to i} = A_i K_{i-k}$ and $F_{i\to k} = A_k K_{ik}$. Thus the view factors are equal if and only if (*iff*) the facet areas are equal.

For the $k=1$ basis restriction, viewfactors enter the GWS^h algorithm as *element data* in the template second bracket. Thus the exchange surface integral for BC (7.2) is

$$\int_{\partial\Omega_R \cap \partial\Omega} \Psi_\beta f(vf, \varepsilon) \sigma \left[(T^N)^4 - T_{rad}^4 \right] d\sigma$$
$$\Rightarrow \sigma\varepsilon \int_{\partial\Omega_e \cap \partial\Omega_R} \{N\}\{N\}^T \left(F_{i\to k}\{Q^4\}_{e=i} - F_{k\to i}\{Q^4\}_{e=k} \right) d\sigma \tag{7.61}$$

The element subscripts in equation (7.61) enforce that the integrals are on *distinct* domains Ω_e. For a perfect emitter, called a *black body*, $\varepsilon = 1$ and the Stefan–Boltzmann constant is $\sigma = 5.67 \times 10^{-8}$ [W/m^{3}K]. For the kernel K_{ik} a constant and via the text relationship following equation (7.60), the GWS^h BC statement for black-body radiation exchange between the generic pair of $k=1$ TP basis element $n=2$ surfaces is

$$\sigma\varepsilon \int_{\partial\Omega_e \cap \partial\Omega_{rad}} \{N\} K_{ki} \{N\}^T \left(A_k\{Q^4\}_{e=i} - A_i\{Q^4\}_{e=k} \right) d\sigma$$
$$= 4\sigma K_{ki} (\det_k \det_i) [B200L] (\{Q^4\}_i - \{Q^4\}_k) \tag{7.62}$$
$$= 4\sigma [B200L] (\det_i F_{i\to k}\{Q^4\}_i - \det_k F_{k\to i}\{Q^4\}_k)$$

The 4 multiplier in equation (7.62) stems from $A_e = 4\det_e$ for the TP basis; the constant becomes 2 for the $k=1$ NC basis.

The black-body assumption is too restrictive for practical applications. The energetic balance on the generic surface facet is

$$\text{absorbed} + \text{reflected} + \text{transmitted} = \varepsilon + \rho + \delta = 1 \tag{7.63}$$

For the typical nontransparent heat exchange device, $\delta = 0$ and the portion reflected becomes $\rho = 1 - \varepsilon$. The alteration to equation (7.62) for *gray body* radiation is expressed in terms of Gebhart factors $G_{k\to i}$ [5]. The fraction of radiant energy received at surface element dA_i from surface element dA_k is

$$G_{k\to i} \equiv \varepsilon_i F_{k\to i} + \sum_{j=1}^{n} \rho_j F_{k\to j} G_{j\to i} \tag{7.64}$$

which reverts to black-body radiation in the absence of reflection.

For n interacting surfaces, equation (7.64) generates the order n-square full matrix statement in Gebhart factor coupling

$$
\begin{bmatrix}
(F_{11}\rho_1 - 1), & \rho_2 F_{12}, & \cdots, & \rho_n F_{1n} \\
\rho_1 F_{21}, & (F_{22}\rho_2 - 1), & \cdots, & \rho_n F_{2n} \\
& & \cdot & \\
\rho_1 F_{n1}, & & \cdots, & , & (F_{nn}\rho_n - 1)
\end{bmatrix}
\begin{Bmatrix}
G_{1i} \\
G_{2i} \\
\cdot \\
G_{ni}
\end{Bmatrix}
= \varepsilon_i
\begin{Bmatrix}
-F_{1\to i} \\
-F_{2\to i} \\
\cdot \\
-F_{n\to i}
\end{Bmatrix}.
\tag{7.65}
$$

All self viewfactors F_{kk} are zero hence the matrix diagonal is -1. There exist as many equations (7.65) as radiation interaction surface facets. Note that only a single LU decomposition is required to solve n right-hand sides.

For this gray body formulation, equation (7.62) for the $k=1$ TP basis implementation becomes

$$
\int_{\partial\Omega_{rad}\cap\partial\Omega} \Psi_{\beta} f(vf, \varepsilon) \sigma \left[(T^N)^4 - T_{rad}^4 \right] d\sigma
$$

$$
\Rightarrow \sigma \int_{\partial\Omega_e\cap\partial\Omega_{rad}} \{N\}\{N\}^T \left(\varepsilon_i \{Q^4\}_{e=i} - \sum_{k=1}^{n} \varepsilon_k G_{k\to i} \{Q^4\}_{e=k} \right) d\sigma.
\tag{7.66}
$$

$$
= 4\sigma [B200L] \left(\varepsilon_i \det_i \{Q^4\}_i - \sum_{k=1}^{n} \varepsilon_k \det_k G_{k\to i} \{Q^4\}_k \right)
$$

The resultant *very nonlinear* contribution to the GWS^h algorithm jacobian (7.21) is

$$
[JAC(\{Q\})]_e \equiv \frac{\partial}{\partial\{Q\}_e} \left[4\sigma[B200L] \left(\varepsilon_i \det_i \{Q^4\}_i - \sum_{k=1}^{n} \varepsilon_k \det_k G_{k\to i} \{Q^4\}_k \right) \right].
\tag{7.67}
$$

$$
= 16\sigma [B200L] \left(\varepsilon_i \det_i \mathrm{diag}[Q^3]_i - \sum_{k=1}^{n} \varepsilon_k \delta\tau_k G_{k\to i} \mathrm{diag}[Q^3]_k \right)
$$

For the n gray-body GWS^h radiation formulation (7.66), the replacement for the radiation BC line in template (7.57) is

$$
\{WS\}_e = \ldots + (SIG)(\,)\{f(vf, \varepsilon)\}(; 1)[N200d]\left(\{Q\}^4 - \{TR\}^4 \right)
$$

$$
\Rightarrow (4SIG)(EPS(i))\{\ \}(0; 1)[B200L]\{Q^4\}_{e=i}
\tag{7.68}
$$

$$
- \sum_{k=1}^{n} [(4SIG)(EPS(k), G(k \to i))\{\ \}(0; 1)[B200L]\{Q^4\}_{e=k}]
$$

The suggested exercise will verify the companion contribution to the Newton jacobian template is

$$[JAC]_e \Rightarrow (16SIG)(EPS(i))\{\ \}(0;1)[B200L]\text{diag}[Q^3]_{e=i}$$

$$- \sum_{k=1}^{n}[(16SIG)(EPS(k),G(k \to i))\{\ \}(0;1)[B200L]\text{diag}[Q^3]_{e=k}]. \tag{7.69}$$

The GWS^h formulation (7.66)–(7.69) is valid for black-body radiation by replacing distinct emissivity with $\varepsilon = 1$ and $G_{k \to i}$ with $F_{k \to i}$. The $\Sigma_k [\cdots]$ operation in equation (7.66) is dependent on the number n of exchange surfaces, hence equation (7.68) can be an *exceptionally long* element loop. Thereby the radiation contribution to $[JAC]_e$, equation (7.69), *greatly expands* the bandwidth of assembled global $[JAC]$. These issues can be moderated by zeroing out Gebhart factors $G_{k \to i}$, equivalently viewfactors $F_{k \to i}$, with magnitude less than a specified threshold.

7.9 Linear Basis GWS^h Template Unification

Unification of the NC/TP $k = 1$ basis GWS^h algorithm $[DIFF]_e$ templates completes compute organization for arbitrary n DE + BCs, equations (7.1)–(7.3). The theory organization (7.6) and (7.7) template is

$$\{WS\}_e = (\text{const})\begin{pmatrix} \text{elem} \\ \text{const} \end{pmatrix}_e \begin{Bmatrix} \text{distr} \\ \text{data} \end{Bmatrix}_e^T (\text{metric data}; \det)[\text{Matrix}]\begin{Bmatrix} Q \text{ or} \\ \text{data} \end{Bmatrix}_e. \tag{7.40}$$

Alter the metric data prefix to "ET" for NC and parallelogram TP basis coordinate transformations (7.41) and (7.56). The resulting $[DIFF]_e$ expression for average conductivity is

$$[DIFF]_e = (\ )(COND)\{\ \}(\text{ET } JI, \text{ET } KI; -1)[M2JKL]\{Q\}, \tag{7.70}$$

where $M \Rightarrow (A, B, C)$ for $n = 1, 2, 3$ and $1 \le (I, J, K) \le n, n+1$.

The last data organization issue is implementing the *very* short n-dependent repeated index loops in equation (7.70). An efficiency results by transforming the matrix $ETJI_e$ to a single-subscript array. For NC/TP basis coordinate systems for $n = 2, 3$ the suggested organizations are

$$\text{ET } JI_e \Rightarrow \begin{bmatrix} 1 & 2 \\ 3 & 4 \\ 5 & 6 \end{bmatrix}, \quad \begin{bmatrix} 1 & 2 \\ 3 & 4 \end{bmatrix}_{n=2} \quad \text{or} \quad \begin{bmatrix} 1 & 2 & 3 \\ 4 & 5 & 6 \\ 7 & 8 & 9 \\ 10 & 11 & 12 \end{bmatrix}, \quad \begin{bmatrix} 1 & 2 & 3 \\ 4 & 5 & 6 \\ 7 & 8 & 9 \end{bmatrix}_{n=3}. \tag{7.71}$$

Letting index association denote multiplication, for example, $12 \equiv 1 \times 2$, the compute sequences for the $k = 1$ NC and TP $[DIFF]_e$ contribution to $\{WS\}_e$ for $n = 2$ are

$$
\begin{aligned}
[DIFF]_e = \;&(\ )(COND)\{\ \}(11 + 22; -1)[B211L]\{Q\} \\
+ &(\ )(COND)\{\ \}(33 + 44; -1)[B222L]\{Q\} \\
+ &(\ )(COND)\{\ \}(55 + 66; -1)[B233L]\{Q\} \\
+ &(\ )(COND)\{\ \}(13 + 24; -1)[B212L]\{Q\} \\
+ &(\ )(COND)\{\ \}(24 + 13; -1)[B221L]\{Q\}. \\
+ &(\ )(COND)\{\ \}(15 + 26; -1)[B213L]\{Q\} \\
+ &(\ )(COND)\{\ \}(26 + 15; -1)[B231L]\{Q\} \\
+ &(\ )(COND)\{\ \}(35 + 46; -1)[B223L]\{Q\} \\
+ &(\ )(COND)\{\ \}(46 + 35; -1)[B232L]\{Q\}
\end{aligned}
\tag{7.72}
$$

$$
\begin{aligned}
[DIFF]_e = \;&(\ )(COND)\{\ \}(11 + 22; -1)[B211L]\{Q\} \\
+ &(\ )(COND)\{\ \}(33 + 44; -1)[B222L]\{Q\} \\
+ &(\ )(COND)\{\ \}(13 + 24; -1)[B212L]\{Q\} \\
+ &(\ )(COND)\{\ \}(24 + 13; -1)[B221L]\{Q\}
\end{aligned}
\tag{7.73}
$$

The comparative $k = 1$ $n = 3$ TP template compute sequence is

$$
\begin{aligned}
[DIFF]_e = \;&(\ )(COND)\{\ \}(11 + 22 + 33; -1)[C211L]\{Q\} \\
+ &(\ )(COND)\{\ \}(44 + 55 + 66; -1)[C222L]\{Q\} \\
+ &(\ )(COND)\{\ \}(77 + 88 + 99; -1)[C233L]\{Q\} \\
+ &(\ )(COND)\{\ \}(14 + 25 + 36; -1)[C221L]\{Q\} \\
+ &(\ )(COND)\{\ \}(47 + 58 + 69; -1)[C223L]\{Q\}. \\
+ &(\ )(COND)\{\ \}(14 + 25 + 36; -1)[C212L]\{Q\} \\
+ &(\ )(COND)\{\ \}(17 + 28 + 39; -1)[C213L]\{Q\} \\
+ &(\ )(COND)\{\ \}(17 + 28 + 39; -1)[C231L]\{Q\} \\
+ &(\ )(COND)\{\ \}(47 + 58 + 69; -1)[C232L]\{Q\}
\end{aligned}
\tag{7.74}
$$

7.10 Accuracy, Convergence Revisited

Computer lab 7.3 further examines the impact of $\|data\|^2_{\partial\Omega, L2}$ on $\gamma = \min[k, r - 1]$ in the $m = 1$ asymptotic error estimate (7.8). The benchmark problem is steady conduction in an L-shaped region heated by a source. For the $n = 2, 3$ geometries, Figure 7.11, the non-convex corner boundary singularity will dominate convergence rate hence accuracy.

Figure 7.12 graphs the illustrative NC $k = 1$ basis solution in perspective. The computer lab 7.3 regular mesh-refinement study generates *a posteriori* data confirming that both TP and NC $k = 1$ basis GWS^h solutions converge at a *suboptimal* rate approaching 5/3. As the $DE + BCs$ solution domain geometric closure also constitutes *data*, non-convexity engenders $r - 1$ dominance over k in equation (7.8). The suggested follow-up

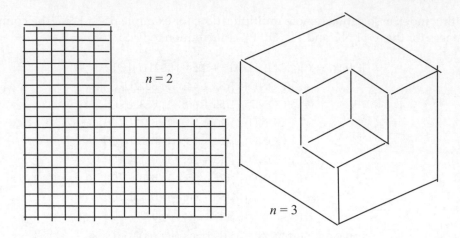

Figure 7.11 L-shaped regions, $n = 2, 3$

to computer lab 7.3 removes the nonconvex $\partial\Omega$ vertex: the *a posteriori* data confirms convergence returns to $2k = 2$ dominance.

Bottom line, *nonconvex* domain closures $\partial\Omega$ along with *nonsmooth* data on Ω and/or $\partial\Omega$ will adversely affect solution convergence to engineering accuracy. Further, in such *practical* situations, replacement of the $k = 1$ basis GWS^h implementation with a $k > 1$ basis is unlikely to improve algorithm convergence rate, but for any M, hence additional DOF, will likely generate improved accuracy.

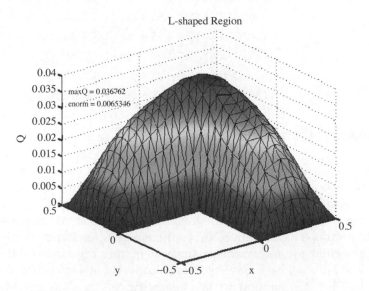

Figure 7.12 L-shaped region GWS^h solution T^h, $n = 2$

7.11 Chapter Summary

Exposition of the FE discrete implementation GWS^h of GWS^N for n-D DE + BCs, hence by extension a wide range of steady n-D, $m = 1$ EBV PDE + BCs statements in the *engineering sciences*, is complete. For the expository development statement

$$\text{DE}: \quad \mathcal{L}(T) = -\nabla \cdot (k\nabla T) - s = 0, \quad \text{on } \Omega \subset \mathfrak{R}^n \tag{7.1}$$

$$\text{BCs}: \quad \begin{aligned} \ell(T) &= k\nabla T \cdot \hat{\mathbf{n}} + h(T - T_r) + f \cdot \hat{\mathbf{n}} \\ &+ f(vf, \varepsilon)\sigma(T^4 - T_r^4) = 0, \text{ on } \partial\Omega_R \subset \mathfrak{R}^{n-1} \end{aligned} \tag{7.2}$$

$$T(\mathbf{x}_b) = T_b(\mathbf{x}_b), \quad \text{on } \partial\Omega_D \subset \mathfrak{R}^{n-1} \tag{7.3}$$

the FE trial space basis implementation GWS^h for all n and completeness degree k can be summarized as

$$\text{approximation}: \quad T(\mathbf{x}) \cong T^N(\mathbf{x}) \Rightarrow T^h(\mathbf{x}) \equiv \cup_e T_e(\mathbf{x}) \tag{7.4}$$

$$\text{FE basis}: \quad T_e(\mathbf{x}) \equiv \{N_k(v(\mathbf{x}))\}^T \{Q\}_e \tag{7.5}$$

$$\text{error extremization}: \quad \text{GWS}^N \equiv \int_\Omega \Psi_\beta(\mathbf{x})\mathcal{L}(T^N)d\tau = 0, \forall\beta$$
$$\Rightarrow \text{GWS}^h = S_e\{\text{WS}\}_e \equiv \{0\} \tag{7.6}$$

$$\text{element contribution}: \quad \{\text{WS}\}_e = ([\text{DIFF}]_e + [\text{BC}]_e)\{Q\}_e - \{b(\text{data})\}_e. \tag{7.7}$$

Solutions generated by the GWS$^N \Rightarrow$ GWS$^h \Rightarrow \{\text{WS}\}_e$ process are theoretically guaranteed to be *optimal* in their basis completeness degree-equivalent peer group. The rate at which GWS^h solutions converge to engineering accuracy is predictable via

$$\|e^h\|_E \leq Ch_e^{2\gamma}\left(\|\text{data}\|_{\Omega,L2}^2 + \|\text{data}\|_{\partial\Omega,L2}^2\right), \gamma \equiv \min(k+1-m, r-m). \tag{7.8}$$

A quite complete range of $k \geq 1$ trial space basis finite element geometries, spanned by both natural and tensor product local coordinate systems, is developed. The terminal computable form is

$$\text{GWS}^h = S_e\{\text{WS}\}_e = \{0\}, \{\text{WS}\}_e = ([\text{DIFF}]_e + [\text{BC}]_e)\{Q\}_e - \{b\}_e. \tag{7.27}$$

The *template* facilitates conversion of every nuance of equation (7.27) into executable syntax via the object-oriented data organization

$$\{\text{WS}\}_e = (\text{const})\begin{pmatrix}\text{elem} \\ \text{const}\end{pmatrix}_e \begin{Bmatrix}\text{distr} \\ \text{data}\end{Bmatrix}_e^T (\text{metric data; det})[\text{Matrix}]\begin{Bmatrix}Q \text{ or} \\ \text{data}\end{Bmatrix}_e. \tag{7.40}$$

The template for the n-D DE + BCs statement, for NC/TP $k = 1$ basis implementation, is the combination of equations (7.57), (7.66)–(7.70)

$$
\begin{aligned}
\{WS\}_e &= \left([DIFF]_e + [BC]_e \right) \{Q\}_e - \left\{ b\left(data(s, h, T_c, T_r, f_n) \right) \right\}_e \\
&= (\ )(COND)\{\ \}(ET\,JI, ET\,KI; -1)[M\,2JKL]\{Q\} \\
&\quad -(\ )(\ )\{\ \}(; 1)[M\,200L]\{SRC\} \\
&\quad +(\ )(\ )\{H\}(; 1)[N\,3000L](\{Q\} - \{TC\}) \\
&\quad -(\ )(\ )\{\ \}(; 1)[N\,200d]\{\mathbf{F} \bullet \mathbf{n}\} \\
&\quad +(4\,SIG)(EPS(i))\{\ \}(0; 1)[N\,200L]\{Q^4\}_{e=i} \\
&\quad -\sum_{k=1}^{n} \left((4\,SIG)(EPS(k), G(k \to i))\{\ \}(0; 1)[N\,200L]\{Q^4\}_{e=k} \right)
\end{aligned}
\qquad (7.75)
$$

$$
\begin{aligned}
[JAC]_e &= (\ )(COND)\{\ \}(ET\,JI, ET\,KI; -1)[M\,2JKL] \\
&\quad +(\ )(\ )\{H\}(; 1)[N\,3000L] \\
&\quad +(16\,SIG)(EPS(i))\{\ \}(0; 1)[N\,200L]diag[Q^3]_{e=i} \\
&\quad -\sum_{k=1}^{n} \left((16\,SIG)(EPS(k), G(k \to i))\{\ \}(0; 1)[N\,200L]diag[Q^3]_{e=k} \right)
\end{aligned}
$$

Exercises

Note: Symbolic softwares, for example, Maple™, Mathematica®, will assist completion of select exercises.

7.1 For a triangle verify the solution for (a_1, a_2, a_3), equation (7.20), hence equation (7.21).

7.2 For tetrahedra, highlight the a_i–d_i, solution process (7.26), then verify the $n = 3$ FE linear basis $\{N_1(\zeta_\alpha)\}$ satisfies the (0,1) Kronecker delta requirement at all geometric nodes of Ω_e.

7.3 Verify entries in the natural coordinate transformation matrix $[\partial \zeta_\alpha / \partial x_j]$, equation (7.33).

7.4 Verify correctness of the NC basis elements in the $n = 2$ diffusion term matrix $[DIFF]_e$, equation (7.37).

7.5 Verify the distributed source NC basis element matrices $\{b(s)\}_e$ for $n = 2, 3$, Table 7.1.

7.6 For quadrilateral and hexahedron elements, verify the TP basis $\{N_1^+\}$, equation (7.12), satisfies the (0,1) Kronecker delta requirement at all geometric nodes of Ω_e.

7.7 Verify that several members of TP serendipity basis $\{N_2^+\}$, equation (7.13), satisfy the nodal (0,1) conditions required of an FE basis.

7.8 Fill in the elements of the TP $n = 2$ coordinate transformation $(\partial n_j / \partial x_i)$, equation (7.47), using equation (7.48).

7.9 Generate the matrix elements of the $n = 3$ TP basis coordinate transformation replacement for equation (7.47).

7.10 Confirm the distributed source TP basis element matrices $\{b(s)\}_e$ for $n = 2, 3$, Table 7.3.

7.11 Verify the template (7.69) for the gray body radiation term contribution to the Newton jacobian (7.67).

Computer Labs

The *FEmPSE* toolbox and all MATLAB® .m files required for conducting the computer labs are available for download at www.wiley.com/go/baker/finite.

7.1 This computer lab defines an $n = 2$ heat sink embedded in a low thermal conductivity dielectric. The .m file instructs MATLAB® to generate an NC basis solution-adapted mesh via Delauney triangulation of a macro element definition. Review the template script then generate GWS^h solutions for a range of distinct thermal conductivities for various Dirichlet BCs. Visualize the resultant data in perspective.

7.2 Theory *validation* is for heat conduction through a pipe flowing a hot fluid. The analytical solution is radially logarithmic, hence all approximate solutions will have quantifiable error. Compare the MATLAB® template script to the algorithm statement. Execute a GWS^h NC $k = 1$ basis convergence study for *regular* azimuthal, then radial discretization refinement for given Robin and Dirichlet BCs. Graph the energy norm data, hence confirm the theory prediction accuracy.

7.3 The L-shaped region with uniform source $n = 2$ *benchmark* problem confirms solution accuracy dominated by the boundary geometry nonconvexity. Verify the MATLAB® .m file template for the $k = 1$ TP basis GWS^h algorithm. Execute a regular mesh-refinement study to verify suboptimal convergence rate in the max and energy norms. Then alter the .m file to fill in the omitted region and repeat the experiment. Then edit the .m file to specify the NC $k = 1$ basis algorithm (take guidance from lab 7.1) and repeat both experiments.

7.4 The DM PDE for an unconfined aquifer is equation (7.1) with thermal conductivity k replaced by the dependent variable, recall equation (2.32). Using as guidance the nonlinear radiation formulation (7.75), derive and state the associated nonlinear GWS^h algorithm. Then convert the lab 7.3 template .m file into the iterative Newton algorithm (recall computer lab 6.5 for guidance). Conduct a regular mesh-refinement study, hence assess iterative and asymptotic convergence rates compared to theory.

References

[1] Thompson, J.F., Thames, F., and Mastin, W. (1977) TOMCAT — A code for numerical generation of boundary-fitted curvilinear coordinate systems on fields containing any number of arbitrary two-dimensional bodies. *J. Comput. Phys.*, **24**, 274–302.

[2] Zienkiewicz, O.C. and Taylor, R.L. (1989) *The Finite Element Method*, McGraw-Hill, New York, NY.

[3] Incropera, F.P. and Dewitt, D.P. (2004) *Fundamentals of Heat and Mass Transfer*, Wiley, New York, NY.

[4] Walton, G.N. (1992) Calculation of obstructed viewfactors by adaptive integration, Technical Report NISTIR-6925, National Institute of Standards and Technology.

[5] Miller, F.P., Vandome, A.F., and Brewster, J.M. (2010) *Gebhart Factors*, VDM Publishing House Ltd., England, ISBN: 6134096326.

8

Finite Differences of Opinion: FE GWS^h connections to FD, FV methods

8.1 The FD–FE Correlation

Finite difference (FD) discrete methodology preceded both the *integral* finite volume (FV) and weak form finite element (FE) implementations for PDE transformation to algebraic (computable) form. Hence the question, "Is there a circumstance wherein the FE discrete implementation GWS^h embodies an FD construction?" The answer is yes, specifically interior node second-order accurate FD stencils for linear spatial derivatives on uniform cartesian meshes in n-dimensions are *exactly* recovered by linear NC basis $\{N_1(\zeta_\alpha)\}$ GWS^h implementations.

The verification is straightforward. Consider the linear elliptic boundary value (EBV) PDE written on $q(\mathbf{x})$

$$\mathcal{L}(q) = -\nabla^2 q - s = 0, \quad on \ \Omega \subset \Re^n. \tag{8.1}$$

Recalling Chapter 3 developments, equations (3.24)–(3.29), a Taylor series written symmetrically at the *generic* node with n-D coordinate $X_{j,k,i}$ produces the $n = 1, 2, 3$ sequence of *second-order* accurate *stencils*

$$-\frac{d^2 q}{dx^2}\bigg|_{X_j} \Rightarrow -\frac{Q_{j-1} - 2Q_j + Q_{j+1}}{\Delta x^2} + O(\Delta x^2). \tag{8.2}$$

$$-\nabla^2 q\big|_{n=2} \Rightarrow -\frac{1}{\Delta x^2}\left[Q_{j-1,k} \quad \begin{matrix} Q_{j,k+1} \\ -4Q_{j,k} \quad Q_{j+1,k} \\ Q_{j,k-1} \end{matrix}\right] + O(\Delta x^2). \tag{8.3}$$

$$-\nabla^2 q\big|_{n=3} \Rightarrow -\frac{1}{\Delta x^2}\left[-6Q_{j,k,i} + Q_{j\pm1,k\pm1,i\pm1}\right] + O(\Delta x^2). \tag{8.4}$$

FD *convention* evaluates $s(x)$ in equation (8.1) at the stencil node, hence $s(x_j) \Rightarrow SRC_j$, $SRC_{j,k}$, $SRC_{j,k,i}$. The uniform mesh restriction requires $\Delta x \equiv \Delta y \equiv \Delta z$ with Δx used as the mesh measure placeholder.

The $GWS^N \Rightarrow GWS^h \Rightarrow \{WS\}_e$ process to this point in the text terminates with the template

$$\{WS\}_e \equiv ([DIFF]_e + [BC]_e)\{Q\}_e - \{b\}_e \tag{8.5}$$

as the computable form for assembly of *all* FE domain contributions to the global matrix statement. While not commonly appreciated, assembly over all FE domains Ω_e sharing the generic mesh node with n-D coordinate $X_{j,k,i}$ directly forms an FE stencil.

The $k=1$ basis process is patently transparent on $n=1$ as only two FE domains share node X_j, Figure 8.1. The $\{WS\}_e$ stencil at X_j results by adding the $[DIFF]_e$ matrix lower row on Ω_e to that for the Ω_{e+1} matrix upper row. Indexing the DOF $\{Q\}$ on j and for $l_e \equiv l_{e+1} \equiv h$

$$\begin{aligned} S_j([DIFF]_e\{Q\}_e) &= S_j\left(\frac{1}{l_e}\begin{bmatrix} 1 & -1 \\ -1 & 1 \end{bmatrix}\begin{Bmatrix} Q_{j-1} \\ Q_j \end{Bmatrix}, \frac{1}{l_{e+1}}\begin{bmatrix} 1 & -1 \\ -1 & 1 \end{bmatrix}\begin{Bmatrix} Q_j \\ Q_{j+1} \end{Bmatrix}\right) \\ &= \frac{1}{h}\left(-Q_{j-1} + 2Q_j - Q_{j+1}\right) \end{aligned} \tag{8.6}$$

and indeed $S_j([DIFF]_e\{Q\}_e)$ exactly reproduces the $n=1$ FD stencil for $-d^2q/dx^2$, equation (8.2), to within the sign and a factor of $1/\Delta x$, due to integration by parts, then an integration. Specifically, note that equation (8.6) is identically the middle row–column matrix product in the expository assembled global GWS^h statement (4.27).

FD methodology requires separate explicit handling of a Robin BC for equation (8.1) which is not central to the discussion. The arbitrary FD decision $s(X_j) \rightarrow SRC_j$ is replaced via $\{WS\}_e$ with interpolation stencil

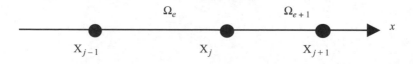

Figure 8.1 $(DIFF)_e\ k=1$ basis assembly at X_j, $n=1$

$$S_j(\{b\}_e) = S_j\left(\frac{l_e}{6}\begin{bmatrix}2 & 1\\ 1 & 2\end{bmatrix}\begin{Bmatrix}\mathrm{SRC}_{j-1}\\ \mathrm{SRC}_j\end{Bmatrix}, \frac{l_{e+1}}{6}\begin{bmatrix}2 & 1\\ 1 & 2\end{bmatrix}\begin{Bmatrix}\mathrm{SRC}_j\\ \mathrm{SRC}_{j+1}\end{Bmatrix}\right)$$

$$= \frac{l_e}{6}\left(\mathrm{SRC}_{j-1} + 4\mathrm{SRC}_j + \mathrm{SRC}_{j+1}\right) \qquad (8.7)$$

readily proven a higher (order of) accuracy interpolation.

Tedious algebraic operations are required to construct the $k=1$ NC basis $\{WS\}_e$ stencil for the $n=2,3$ linear laplacian. The suggested exercise asks to prove the $n=2$ $[DIFF]_e$ stencil for *arbitrary* orientation of the NC basis triangle diagonals, see Figure 8.2, is

$$S_{j,k}([DIFF]_e\{Q\}_e) \Rightarrow \begin{bmatrix} & -Q_{j,k+1} & \\ -Q_{j-1,k} & 4Q_{j,k} & -Q_{j+1,k} \\ & -Q_{j,k-1} & \end{bmatrix}. \qquad (8.8)$$

Similarly, for absolutely *arbitrary* orientation of the tetrahedron element facets, Figure 8.2, the $n=3$ $\{WS\}_e$ stencil for $[DIFF]_e$ is

$$S_{j,k,i}([DIFF]_e\{Q\}_e) \Rightarrow h\left[6Q_{j,k,i} - Q_{j\pm1,k\pm1,i\pm1}\right]. \qquad (8.9)$$

The $\{WS\}_e$ $n=2,3$ stencils for $\{b\}_e$ assembly at $X_{j,k,i}$ are direct extensions on the higher order $n=1$ stencil (8.7).

Clearly, GWS^h $\{N_1(\zeta_\alpha)\}$ basis implementations exactly reproduce uniform cartesian mesh second-order accurate FD stencils for all n, to within the integration influence on mesh measure h exponent and the term sign. And this results for absolutely arbitrary *orientation* of diagonals (surface facets) of $\partial\Omega_e$. Finally, the *arbitrary* FD choice $s(x) \Rightarrow$ SRC$_{j,k}$, SRC$_{j,k,i}$ is replaced in GWS^h by a higher order accurate interpolation of distributed (nonconstant) data.

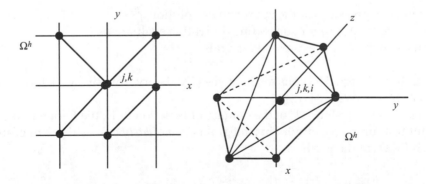

Figure 8.2   Generic uniform cartesian Ω^h for GWS^h nodal assembly

In summary:

- assembly of the $k=1$ NC basis discrete implementation GWS^h identically reproduces second-order accurate FD stencils for the linear laplacian at the generic interior node on a uniform *cartesian* mesh for all n
- the $k=1$ NC basis GWS^h node assembly for domain *data* generates a higher order accurate interpolation
- for all n, generic node assembly of the GWS^h could be *cumbersomely* expressed as a stencil
- maintaining FD stencil order-of-accuracy is challenging for *any* noncartesian meshing $\Omega^h \subset \Re^n$.
- in distinction, for arbitrary data and *regular* mesh refinement, for all $k \geq 1$ bases the GWS^h energy norm error estimate *guarantees* convergence of $O(2k, 2(r-1))$ for equation (8.1) for *arbitrary* Ω^h and Robin BCs!

The conclusion from this brief exposé is that no need persists for the FD approximation process for PDE systems in the engineering sciences, of which equation (8.1) is representative. Further, since implementing Robin BCs and noncartesian meshes places significant additional burdens on the FD process, this legacy methodology is *truly obsolete* in the present context.

8.2 The FV–FE Correlation

Difficulty in maintaining FD stencil order-of-accuracy for $\mathcal{L}(q)$, equation (8.1), led the CFD community (in particular) to develop FV *integral* discrete methodology. The variety of FV constructions can be interpreted as a non-Galerkin direct WS^N ⇒ WS^h ⇒ stencil formulation implemented in the absence of

- a functional expression for the approximate solution $q^h(\mathbf{x})$
- a basis function $\{N_k(\mathbf{v})\}$ for supporting derivative evaluations
- calculus to evaluate the expressed integrals

Each of these design decisions distinguishes FV theory from the weak form formality WS^N ⇒ GWS^N ⇒ GWS^h ⇒ {WS}$_e$.

For an arbitrary meshing Ω^h of the domain of equation (8.1), the FV integral employs a constant test function (unity). This integral is *immediately* cast as the sum of integrals over each *FV* Ω_v of the mesh as

$$\text{FV}^N \equiv \int_\Omega \mathcal{L}(q^N) d\tau \Rightarrow \sum_{\Omega^h = \cup \Omega_v} \left(\int_{\Omega_v} (-\nabla \cdot \nabla q^N - s) d\tau \right) \equiv 0. \tag{8.10}$$

Comparing equation (8.10) to GWS^N clearly confirms the FV^N distinction for test function. The subsequent FV^N ⇒ FV^h step immediately implements the divergence theorem generating *closed* surface integrals surrounding each FV in $\Omega^h = \cup\Omega_v$ as

$$\mathrm{FV}^h \equiv \sum_{\Omega^h} \left(-\int_{\partial\Omega_v} \nabla q^h \bullet \hat{\mathbf{n}}\,d\sigma - \int_{\Omega_v} s\,d\tau \right) \equiv 0. \tag{8.11}$$

Note the FV theory departure point (8.10) implies the integral of $\mathcal{L}(q^N)$ over Ω vanishes identically. This contradicts the fundamental WS^N precept $q(\mathbf{x}) = q^N(\mathbf{x}) + e^N(\mathbf{x})$! In truth, without immediately defining the discretization Ω^h, FV$^N \equiv 0$ is computationally intractable. Coupling definition of Ω^h with the divergence theorem generating FV^h circumvents this FV^N theoretical anomaly, which of course never entered into the original derivation of an FV algorithm.

The source term remains an integral on Ω_v centrally evaluated as in FD methodology. The *key* consequence of equation (8.11) is the requirement for a *staggered mesh* nodalization for evaluating the integrals about $\partial\Omega_v$ and on Ω_v in Ω^h, Figure 8.3.

As always $n = 1$ admits the transparent illustration for equation (8.11) formation. Absence of a functional form for q^h requires use of FD procedures for surface integral efflux evaluation. Figure 8.4 graphs the generic FV Ω_v centered at the node X_j and the two bounding surfaces $\partial\Omega_{vL}$ and $\partial\Omega_{vR}$, where L/R denotes "left/right."

The FV^h closed surface integral is FD approximated as the sum of left and right-end fluxes with sign determined by the dot product with outwards pointing $\hat{\mathbf{n}}$. Hence, to lowest-order accuracy

$$-\int_{\partial\Omega_v} \nabla q^h \cdot \hat{\mathbf{n}}\,d\sigma \approx [-(Q_{j+1} - Q_j)/\Delta x_R + (Q_j - Q_{j-1})]/\Delta x_L + O(\Delta x^2) \tag{8.12}$$

where Δx_L and Δx_R denote the appropriate geometric node spacing.

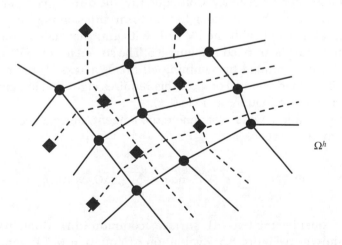

Figure 8.3 FV domain discretization $\Omega^h = \cup\Omega_v$ illustrating staggered nodalization on Ω_v

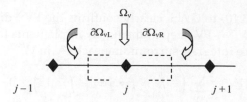

Figure 8.4 FV^h illustration for $n=1$

As with FD methods, the source term integral is FV^h-defined the average, hence $\int_{\Omega_v} s\,dx \Rightarrow \Delta x\,SRC_j$. Summing Ω_v and $\partial\Omega_v$ contributions and assuming $\Delta x_R = \Delta x = \Delta x_L$ the $n=1$ FV^h theory stencil

$$\begin{aligned}
FV_j^h &= -\sum_j \int_{\partial\Omega_{vj}} \nabla q^h \cdot \hat{n}\,d\sigma - \int_{\Omega_{vj}} s\,d\sigma \\
&= \frac{-1}{\Delta x}\left(Q_{j-1} - 2Q_j + Q_{j+1}\right) - \Delta x\,SRC_j
\end{aligned} \tag{8.13}$$

exactly reproduces the second-order FD stencil, (8.2), multiplied by Δx. As an improvement over FD, an exercise suggests deriving the nonuniform mesh alternative to equation (8.13). Then performing this operation on equation (8.6), the FV^h and $k=1$ basis GWS^h stencils are identical! Hence, for $n=1$ the FV^h algorithm (8.11) retains nonuniform mesh second-order accuracy for smooth data as predicted by the GWS^h asymptotic error estimate.

For $n>1$ the FV^h theory for equation (8.1) encounters head-on the coordinate transformation issue dominating n-D GWS^h integrals. Calculus is not available to support this requirement, and the FV^h formulation process instead employs algebraic FD methods and *insight* for order consistency. Consequently, the directly comparative $n=2$ FV^h stencil can be derived *only* by restriction to a uniform cartesian meshing.

The Ω_v index associated with the FV cell, with unit normals $\hat{n}$ and node lexicographic (j,k) numbering is graphed in Figure 8.5. The stencil for the FV^h surface integral in equation (8.11) is generated by summing all fluxes across Ω_v boundary segments that connect to the generic node. For $n=2$ this requires eight line integrals for the four $\partial\Omega_v$ surface segment terminations at node (j,k).

In FD operator notation, where s/n denote the tangent/normal coordinates, the surface integral formation essence is

$$FV_{j,k}^h = -\sum_{j,k}\int_{\partial\Omega_v} \nabla q^h \cdot \hat{n}\,d\sigma \Rightarrow \sum_{\alpha=1}^{8}\left(\pm\Delta Q_\alpha \Delta s/\Delta n\right)_{j,k}. \tag{8.14}$$

For exposition, consider the two $\partial\Omega_v$ surfaces common to Ω_v domains 1 and 4 with unit normals shown in Figure 8.5. Evaluation of equation (8.14) generates the two contributions

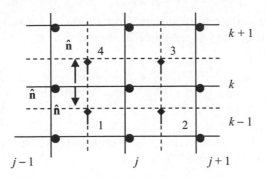

Figure 8.5 Cartesian staggered mesh for FV^h stencil at node (j, k)

$$-\int_{\partial\Omega_{1-4}} (\nabla q^h \cdot \hat{n}) d\sigma \cong -\frac{\Delta x}{2\Delta y}\left[Q_{j,k} + Q_{j-1,k} - (Q_{j,k-1} + Q_{j-1,k-1})\right]$$
$$-\int_{\partial\Omega_{4-1}} (\nabla q^h \cdot \hat{n}) d\sigma \cong -\frac{\Delta x}{2\Delta y}\left[Q_{j-1,k+1} + Q_{j,k+1} - (Q_{j-1,k} + Q_{j,k})\right]$$

Using these as guides, for a uniform mesh $\Delta x = \Delta y$ and summing the eight terms defined by equation (8.14), the suggested exercise will confirm the resultant $n = 2$ laplacian FV^h stencil is

$$\sum_{j,k}\left(-\int_{\partial\Omega_v} \nabla q^h \cdot \hat{n} d\sigma\right) \Rightarrow \frac{1}{2}\begin{pmatrix} -2 & 0 & -2 \\ 0 & 8 & 0 \\ -2 & 0 & -2 \end{pmatrix}_{j,k} = \begin{pmatrix} -Q_{j-1,k+1} & 0 & -Q_{j+1,k+1} \\ 0 & 4Q_{j,k} & 0 \\ -Q_{j-1,k-1} & 0 & -Q_{j+1,k-1} \end{pmatrix}. \quad (8.15)$$

Aside from FV^h integral (8.11) generating a Δx^2 multiplier for $n = 2$, equation (8.15) is distinct from the FD laplacian stencil (8.3) in involving the out-rigger $(j \pm 1, k \pm 1)$ corner nodal DOF rather than the (j,k) row/column DOF.

GWS^h correspondence with the FV^h formulation is via the tensor product (TP) $k = 1$ $\{N_1^+(\eta)\}$ basis implementation. Assembly at node (j,k) of the $n = 2$ [DIFF]$_e$ matrix, [B2JKL], Table 7.3, on the identical uniform cartesian mesh (an exercise) produces

$$S_{j,k}([DIFF]_e\{Q\}_e) \Rightarrow \frac{1}{6}\begin{pmatrix} -2 & -2 & -2 \\ -2 & 16 & -2 \\ -2 & -2 & -2 \end{pmatrix} Q_{j,k} \quad (8.16)$$

which exhibits complete nearest-neighbor coupling in distinction to equations (8.15) and (8.3). Interestingly, since the FD and FV^h stencils are both second-order accurate, so is their linear combination, hence the alternative stencil for the laplacian at node (j,k) that exhibits the coupling in equation (8.16) is

$$\frac{1}{2}(FD + FV^h)_{j,k} = \frac{1}{4}\begin{pmatrix} -2 & -2 & -2 \\ -2 & 16 & -2 \\ -2 & -2 & -2 \end{pmatrix} Q_{j,k}. \quad (8.17)$$

The FV^h algorithm source term in equation (8.11), a constant on each Ω_v, when summed at (j,k) produces the wider (than FD) stencil

$$\sum_{j,k}\int_{\Omega_n} s\, d\tau \Rightarrow \frac{\Delta x \Delta y}{16}\begin{pmatrix}1 & 2 & 1\\2 & 4 & 2\\1 & 2 & 1\end{pmatrix}\text{SRC}_{j,k} \tag{8.18}$$

as the data are now in fact interpolated. The comparative GWS^h stencil, the (j,k) assembly of the $n=2\,\{b\}_e$ matrix in Table 7.3, is

$$S_{j,k}(\{b(s)\}_e) \Rightarrow \frac{\Delta x \Delta y}{36}\begin{pmatrix}1 & 4 & 1\\4 & 16 & 4\\1 & 4 & 1\end{pmatrix}\text{SRC}_{j,k}. \tag{8.19}$$

This exhibits the same DOF spread as equation (8.18) but is weighted more heavily at the assembly node (j,k).

8.3 Chapter Summary

Legacy FD and FV^h discrete constructions for the EBV PDE (8.1), typically written as *stencils*, predate the GWS^h matrix-generating construction. The mission of this chapter is establishing mathematical correlations with the goal to distinguish functional differences and similarities. Coincidentally, it provides the opportunity to formalize GWS^h algorithm stencil construction, of significant theoretical utility in the following chapters. Because integral vector calculus is not involved in stencil formation, the generated comparisons are restricted to uniform rectangular cartesian meshes.

Somewhat surprising perhaps is that linear NC basis $\{N_1(\zeta_\alpha)\}$ GWS^h implementation on triangles/tetrahedra precisely reproduces FD stencils for the laplacian for $n=2,3$. GWS^h interpolation of source data is higher order than the traditional FD assumption of nodal. Since triangle/tetrahedron FE meshings can be arbitrarily nonregular, no real need remains for developing FD constructions. Of key pertinence, the theoretical asymptotic error estimate precisely predicts NC basis GWS^h algorithm performance for all k and n and for arbitrary data on any discretization.

The FD successor FV method is interpretable as a non-Galerkin weak formulation. Due to the absence of a trial space and vector integral calculus, FV^h implementation uses FD methods for flux approximation. In distinction to FD, the FV^h algorithm is directly applicable to use with arbitrary quadrilateral/hexahedron meshings.

The FV^h comparable GWS^h algorithm is discrete implementation using the TP $k=1$ $\{N_1^+(\eta)\}$ basis. A linear combination of the FD and FV^h laplacian stencils emulates that produced by this GWS^h algorithm. FV^h interpolation of source data is of DOF span comparable to that produced by GWS^h.

Since FE trial space basis implementation of GWS^h is so richly founded in theory and calculus, one must seriously question the desirability of continued use of *legacy* FD and FV^h methodologies in the computational engineering sciences.

Exercises

8.1 Verify that nodal assembly of the $\{N_1(\zeta_\alpha)\}$ basis GWS^h algorithm for the $n = 2, 3$ laplacian produces equations (8.8) and (8.9) for more than one cartesian triangle/tetrahedron diagonal/facet arrangement.

8.2 Make nonuniform the $n = 1$ mesh in Figure 8.4, then confirm that the FV^h and $k = 1$ basis GWS^h laplacian stencils are identical.

8.3 Verify the accuracy of the uniform rectangular cartesian mesh FV^h stencil for the laplacian, equation (8.15).

8.4 Repeat exercise 8.3 for the GWS^h laplacian stencil (8.16).

8.5 Confirm the correctness of the FV^h and $k = 1$ basis GWS^h stencils for data interpolation, equations (8.18) and (8.19).

9

Convection–Diffusion, $n=1$: unsteady energy transport, accuracy/convergence, dispersion error, numerical diffusion

9.1 Introduction

The weak form procedure for formulating and assessing performance of the n-dimensional GWS^h FE trial space basis implemented algorithm for the steady $DE + BCs$ conservation principle PDE system is complete. The algorithm strategy essence is:

$$\textit{approximation}: \quad T(\mathbf{x}) \cong T^N(\mathbf{x}) \Rightarrow T^{\hat{h}}(\mathbf{x}) \equiv \cup_e T_e(\mathbf{x}) \tag{9.1}$$

$$\textit{FE trial space basis}: \quad T_e(\mathbf{x}) \equiv \{N_k(\upsilon(\mathbf{x}))\}^T \{Q\}_e \tag{9.2}$$

$$\textit{error extremization}: \mathrm{GWS}^N \equiv \int_{\Omega} \Psi_\beta(\mathbf{x}) \mathcal{L}(T^N) \mathrm{d}\tau = 0, \forall \beta$$
$$\Rightarrow \mathrm{GWS}^h = S_e \{WS\}_e \equiv \{0\} \tag{9.3}$$

$$\textit{element contribution}: \quad \{WS\}_e = ([\mathrm{DIFF}]_e + [\mathrm{BC}]_e)\{Q\}_e - \{\mathrm{b(data)}\}_e \tag{9.4}$$

$$\textit{asymptotic convergence}: \quad \|e^h\|_E \leq Ch_e^{2\gamma} \left(\|\mathrm{data}\|_{\Omega,L2}^2 + \|\mathrm{data}\|_{\partial\Omega,L2}^2 \right),$$
$$\gamma \equiv \min(k+1-m, r-m) \tag{9.5}$$

Finite Elements ⇔ Computational Engineering Sciences, First Edition. A. J. Baker.
© 2012 John Wiley & Sons, Ltd. Published 2012 by John Wiley & Sons, Ltd.

This chapter extends the content of Chapter 7 to unsteadiness and the first-order spatial derivative associated with fluid convection. For clarity in development and assessment of expanded theoretical characterizations for newly identified error mechanisms, attention is restricted to the $n = 1$ DE + BCs + IC (*initial condition*) PDE system characterizing unsteady transport of heat.

$$\mathcal{L}(T) = \frac{\partial}{\partial t}(\rho c_p T) + \frac{\partial}{\partial x}\left(\rho c_p u T - k \frac{\partial T}{\partial x}\right) - s = 0, \text{ on } \Omega \times t \subset \Re^1 \times R^+ \tag{9.6}$$

$$\ell(T) = k dT/dn + h(T - T_c) + \mathbf{f} \cdot \hat{\mathbf{n}} = 0, \text{ on } \partial\Omega_R \times t \subset \Re^0 \times R^+ \tag{9.7}$$

$$T(x_b) = T_b(x_b), \quad \text{on } \partial\Omega_D \times t \subset \Re^0 \times R^+ \tag{9.8}$$

$$T(x, t_0) = T_0(x), \quad \text{on } \Omega \cup \partial\Omega \times t_0. \tag{9.9}$$

In equation (9.6), ρ is the density and c_p is specific heat at constant pressure, both assumed constant, with $u(x,t)$ the imposed fluid velocity magnitude. The remaining variables are familiar, the BCs may be functions of time, equation (9.9) states the IC and R^+ is the positive real number axis.

To identify fundamental scales, nondimensionalize (non-D) equations (9.6) and (9.7) by a reference length (L), speed (U), and timescale (L/U). Defining the non-D *potential temperature*

$$\Theta = (T - T_{\min})/(T_{\max} - T_{\min}) \tag{9.10}$$

the resultant non-D form for equations (9.6) and (9.7) is

$$\mathcal{L}(\Theta) = \frac{\partial\Theta}{\partial t} + u\frac{\partial\Theta}{\partial x} - \frac{1}{\text{Pe}}\frac{\partial}{\partial x}\left(k_\Theta \frac{\partial\Theta}{\partial x}\right) - s_\Theta = 0 \tag{9.11}$$

$$\ell(\Theta) = k_\Theta \frac{d\Theta}{dn} + \text{Nu}(\Theta - \Theta_c) + \mathbf{f}_\Theta \cdot \hat{\mathbf{n}} = 0 \tag{9.12}$$

defining non-D thermal conductivity $k_\Theta \equiv k(T)/k_{ref}$ to admit temperature dependence.

The non-D velocity $u \equiv u(x,t)/U$ is assumed variable and the subscript Θ on the *data* denotes the non-D forms. The key characterizing non-D group is the Peclet number, $\text{Pe} \equiv \rho c_p UL/k_{ref} = \text{RePr}$ where Pr and Re are the fluid (flow) Prandtl and Reynolds numbers respectively. The BC non-D group is Nusselt number, $\text{Nu} = hL/k_{ref}$.

9.2 The Galerkin Weak Statement

The key weak form feature in equation (9.11) is the time derivative that requires the appropriate decision in forming $\text{GWS}^N \Rightarrow \text{GWS}^h$. The addition of the first spatial derivative is not of major formulation consequence. The parent GWS^N statement for equations (9.11) and (9.12) is

$$\mathrm{GWS}^N \equiv \int_\Omega \Psi_\beta \mathcal{L}(\Theta^N)\,dx = 0, \forall \beta$$

$$= \int_\Omega \Psi_\beta \left(\frac{\partial \Theta^N}{\partial t} + u \frac{\partial \Theta^N}{\partial x} - s_\Theta \right) dx + \frac{1}{Pe} \int_\Omega \frac{d\Psi_\beta}{dx} k_\Theta \frac{\partial \Theta^N}{\partial x}\,dx$$

$$+ \frac{Nu}{Pe}\Psi_\beta(\Theta^N - \Theta_c)\Big|_{\partial\Omega_R} + \frac{1}{Pe}f_\Theta \cdot \hat{n}\Big|_{\partial\Omega_R} = 0 \qquad (9.13)$$

performing as usual integration by parts on the diffusion term.

Hence, $\mathrm{GWS}^N \Rightarrow \mathrm{GWS}^h$ requires a decision on where to place functional support for time dependence since

$$\Theta(x, t) \approx \Theta^N(x, t) \equiv \Theta^h(x, t) = \cup_e \Theta_e(x, t). \qquad (9.14)$$

Augmenting the developed FE trial space bases for time generates a totally new library of *space–time* finite elements. The alternative defines the approximation degrees-of-freedom (DOF) as time-dependent

$$\Theta_e(x, t) \equiv \{N_k(\zeta_\alpha)\}^T \{Q(t)\}_e. \qquad (9.15)$$

This decision facilitates direct GWS^h formulation/compute practice and further preserves trial space basis development in its entirety.

Taking the time partial derivative of equation (9.15) generates the approximation DOF *ordinary* derivative matrix

$$\frac{\partial \Theta_e}{\partial t} = \frac{\partial}{\partial t}\left(\{N_k(\zeta)\}^T \{Q(t)\}_e \right) = \{N_k\}^T \frac{d\{Q\}_e}{dt}. \qquad (9.16)$$

Combining equations (9.13)–(9.16) generates and summarizes, in symbolic notation, the process $\mathrm{GWS}^N \Rightarrow \mathrm{GWS}^h$

$$\mathrm{GWS}^h = S_e \left([\mathrm{MASS}]_e \frac{d\{Q\}_e}{dt} + ([\mathrm{VEL}]_e + [\mathrm{DIFF}]_e + [\mathrm{BC}]_e)\{Q\}_e - \{b\}_e \right)$$

$$\equiv [\mathrm{MASS}]\frac{d\{Q\}}{dt} + \{\mathrm{RES}\} = \{0\}. \qquad (9.17)$$

In equation (9.17) $[\mathrm{DIFF}]_e$ and $[\mathrm{BC}]_e$ are familiar matrix names. $[\mathrm{MASS}]_e$ and $[\mathrm{VEL}]_e$ are similar placeholders for the element matrices resulting from the time and velocity convection terms in equation (9.13). The second line in equation (9.17) expresses the *assembled* global matrix statement GWS^h.

9.3 GWS^h Completion for Time Dependence

For the elliptic boundary value (EBV) PDE systems considered to now, the GWS^h has directly produced the algebraic equation system for computing. For the *initial value-elliptic boundary value* (I-EBV) DE + BCs + IC system, GWS^h has instead produced a *matrix ordinary differential equation* (ODE) system, fully coupled in the time derivative by the global [MASS] matrix.

Mathematical issue resolution starts with symbolically solving equation (9.17) for the global time derivative DOF matrix

$$\frac{d\{Q\}}{dt} \equiv \{Q\}' = -[MASS]^{-1}\{RES\} \tag{9.18}$$

which also introduces *superscript prime* notation. The utility of the time-derivative is to enable evaluation of a Taylor series (TS) expansion in *time*. Let subscript n denote time level, hence $t_n = t_0 + n\Delta t$ and think back about Chapter 3.4 developments on TS order of accuracy. For $0 \leq \theta \leq 1.0$, a variable truncation error order TS describing time-evolution of the DOF matrix $\{Q(t)\}$ in equation (9.18) is

$$\{Q(t_{n+1})\} \equiv \{Q\}_{n+1} = \{Q\}_n + \Delta t \left(\theta\{Q\}'_{n+1} + (1-\theta)\{Q\}'_n\right) + O(\Delta t^2, \Delta t^3)\big|_\theta \tag{9.19}$$

The IC (9.9) provides the *data* $T(x,t_0) \Rightarrow \Theta(x,t_0) \Rightarrow \{Q(t_0)\}$ to initiate equation (9.19), and Figure 9.1 illustrates the role of θ in time-derivative location for this single-stage *Euler* ODE algorithm family.

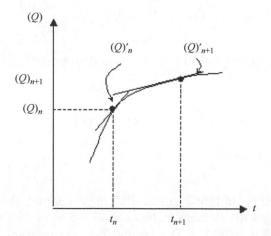

Figure 9.1 Variable implicit Euler ODE algorithm

Substituting GWS^h (9.18) into equation (9.19) produces

$$\{Q\}_{n+1} = \{Q\}_n - \Delta t[\text{MASS}]^{-1}(\theta\{\text{RES}\}_{n+1} + (1-\theta)\{\text{RES}\}_n) + \text{TE} \qquad (9.20)$$

where TE denotes truncation error. Noting $\{\text{RES}\}_{n+1}$ involves $\{Q\}_{n+1}$, the unknown *implicit* parts of equation (9.20) are now coalesced and the matrix inverse in equation (9.18) removed by through multiplication with [MASS] in equation (9.20). Then defining $\{\Delta Q\} \equiv \{Q\}_{n+1} - \{Q\}_n$ as the change in $\{Q(t)\}$ over Δt the terminal algebraic equation GWS$^N \Rightarrow$ GWS$^h + \theta$TS is

$$([\text{MASS}] + \theta\Delta t([\text{VEL}] + [\text{DIFF}] + [\text{BC}])_{n+1})\{\Delta Q\} = -\Delta t\{\text{RES}\}_n$$
$$\{\text{RES}\}_n = ([\text{VEL}] + [\text{DIFF}] + [\text{BC}])_n\{Q\}_n - \{b(s_\Theta, \Theta_c, \mathbf{f}_\Theta \cdot \hat{\mathbf{n}}\}_{n+\theta} \qquad (9.21)$$

In equation (9.21) θ remains a decision and time-dependent $\{b(data)\}$ is evaluated at $t_n + \theta\Delta t$. The selected θ-implicit Euler ODE algorithm is but one option for algebraic completion of the GWS^h. Other potential candidates include Runge-Kutta, leap-frog, predictor-corrector, and so on, and the choice is *yours*! Finally $\theta = {}^1/_2$ defines the *trapezoidal* or mid-point rule that possesses the optimal $O(\Delta t^3)$ TE.

9.4 GWS$^h + \theta$TS Algorithm Templates

One need only define the *new* element matrices in GWS$^h + \theta$TS for the $n=1$ unsteady DE + BCs + IC statement to start computing. For equations (9.11) and (9.12) defining a linear problem (9.21) *is* the Newton statement

$$[\text{JAC}]\{\Delta Q\} = -\Delta t\{\text{RES}\}_n. \qquad (9.22)$$

The matrices defined in equation (9.22) are the assembly of element matrix statements over the $n=1$ spatial discretization Ω^h

$$[\text{JAC}]_e = l_e[\text{A200}d] + \theta\Delta t\left(u_e[\text{A201}d] + (l_e\text{Pe})^{-1}[\text{A211}d] + (\text{Nu/Pe})\{\delta\}\right)$$
$$\{\text{RES}\}_e = \left(u_e[\text{A201}d] + (l_e\text{Pe})^{-1}[\text{A211}d] + (\text{Nu/Pe})\{\delta\}\right)\{Q\}_e - \{b\}_e \qquad (9.23)$$
$$\{b\}_e = l_e[\text{A200}d]\{\text{SRC}\}_e + (-\{\text{FN}\}_e + (\text{Nu/Pe})\{\text{TC}\}_e)\{\delta\}$$

No new $n=1$ element matrices are defined in equation (9.23) except for $\{\delta\}$, shorthand for the Kronecker delta matrix defined in equation (4.22). However, if the applied velocity is spatially or temporally variable, that is, $u_e = u_e(x,t)$, the altered flow convection term in $\{\text{RES}\}_e$ and $[\text{JAC}]_e$ is

$$u_e[\text{A201}d] \Rightarrow \{\text{U}(t)\}_e^T[\text{A3001}d]. \qquad (9.24)$$

Noting that $GWS^h + \theta TS \Rightarrow S_e\{WS\}_e$ and recalling

$$\{WS\}_e = (\text{const})\begin{pmatrix} \text{elem} \\ \text{const} \end{pmatrix}_e \left\{ \begin{matrix} \text{distr} \\ \text{data} \end{matrix} \right\}_e^T (\text{metric data; det})[\text{Matrix}]\left\{ \begin{matrix} Q \text{ or} \\ \text{data} \end{matrix} \right\}_e \qquad (7.40)$$

for data names $PEI = (Pe)^{-1}$, $NU = Nu$, $DT = \Delta t$, $TDT = \theta \Delta t$ the *template* for algorithm statement (9.23) is

$$\begin{aligned}
\Delta t\{RES\}_e = \ & (DT)(\ )\{U\}(;0)[A3001d]\{Q\} \\
& +(PEI, DT)(\ )\{\ \}(;-1)[A211d]\{Q\} \\
& -(DT)(\ )\{\ \}(;1)[A200d]\{SRC\} \\
& +(PEI, DT, FN)(\ )\{\ \}(;0)[ONE]\{\} \\
& +(PEI, NU, DT)(\ )\{\ \}(;0)[ONE]\{Q\} \\
& -(PEI, NU, DT, TC)(\ )\{\ \}(;0)[ONE]\{\ \} \qquad (9.25)
\end{aligned}$$

$$\begin{aligned}
[JAC]_e = \ & (\ )(\ )\{\ \}(;1)[A200d][\ ] \\
& +(TDT)(\ )\{U\}(;0)[A3001d][\ ] \\
& +(PEI, TDT)(\ )\{\ \}(;-1)[A211d][\ ] \\
& +(PEI, NU, TDT)(\ )\{\ \}(;0)[ONE][\ ]
\end{aligned}$$

For the case where equation (9.11) describes the *steady* $n = 1$ DE problem statement, the GWS^h algorithm template replacing equation (9.25) is

$$\begin{aligned}
\{RES\}_e = \ & (\ )(\ )\{\ \}(;1)[A200d]\{SRC\} \\
& +(PEI, FN)(\ )\{\ \}(;0)[ONE]\{\ \} \\
& +(PEI, NU, DT)(\ )\{\ \}(;0)[ONE]\{Q\} \\
& -(PEI, NU, TC)(\ )\{\ \}(;0)[ONE]\{\ \}
\end{aligned}$$
$$\qquad (9.26)$$

$$\begin{aligned}
[JAC]_e = \ & (\ )(\ )\{U\}(;0)[A3001d][\ ] \\
& +(PEI)(\ )\{\ \}(;-1)[A211d][\ ] \\
& +(PEI, NU)(\ )\{\ \}(;0)[ONE][\ ]
\end{aligned}$$

Anticipating nonlinearity, for example, temperature-dependent thermal conductivity, the Newton algorithm replacement for equation (9.22) is

$$[JAC]^p\{\delta Q\}^{p+1} = -\{FQ\}_{n+1}^p$$
$$\{FQ\}_{n+1}^p = [MASS](\{Q\}_{n+1}^p - \{Q\}_n) + \Delta t(\theta\{RES\}_{n+1}^p + (1-\theta)\{RES\}_n)$$
$$[JAC]^p = \partial\{FQ\}/\partial\{Q\} = [MASS] + \Delta t\theta\, \partial\{RES\}_{n+1}^p/\partial\{Q\} \qquad (9.27)$$

$$\{Q\}_{n+1}^{p+1} = \{Q\}_{n+1}^p + \{\delta Q\}^{p+1} = \{Q\}_n + \sum_{\alpha=0}^{p}\{\delta Q\}^{\alpha+1}$$

for p the iteration index, an integer > 0. For notation $QP = Q^p_{n+1}$, $QN = Q_n$, $TDT = \theta\Delta t$, and $TDT1 \equiv (1-\theta)\Delta t$, with $\{RES\}_e$ terms now doubled due to θ-dependence, the iterative $GWS^h + \theta TS$ algorithm template is

$$
\begin{aligned}
\{FQ\}_e = \ &(\,)(\,)\{\ \}(;1)[A200d]\{QP\} - (\,)(\,)\{\ \}(;1)[A200d]\{QN\}\\
&+(TDT)(\,)\{U\}(;0)[A3001d]\{QP\}\\
&+(TDT1)(\,)\{U\}(;0)[A3001d]\{QN\}\\
&+(PEI,TDT)(\,)\{\ \}(;-1)[A211d]\{QP\}\\
&+(PEI,TDT1)(\,)\{\ \}(;-1)[A211d]\{QN\}\\
&-(TDT)(\,)\{\ \}(;1)[A200d]\{SRC\}\\
&-(TDT1)(\,)\{\ \}(;1)[A200d]\{SRCN\}\\
&+(NU,PEI,TDT)(\,)\{\ \}(;0)[ONE]\{QP\}\\
&+(NU,PEI,TDT1)(\,)\{\ \}(;0)[ONE]\{QN\}\\
&-(NU,PEI,TC,DT)(\,)\{\ \}(;0)[ONE]\{\ \}\\
&-(NU,PEI,TC,TDT1)(\,)\{\ \}(;0)[ONE]\{\ \}\\
&-(PEI,FN,DT)(\,)\{\ \}(;0)[ONE]\{\ \}
\end{aligned}
\tag{9.28}
$$

$$
\begin{aligned}
[JAC]_e \quad &(\,)(\,)\{\ \}(;1)[A200d][\ ]\\
&+(TDT)(\,)\{U\}(;0)[A3001d][\ ]\\
&+(PEI,TDT)(\,)\{\ \}(;1)[A211d][\ ]\\
&+(NU,PEI,TDT)(\,)\{\ \}(;0)[ONE][\ ]
\end{aligned}
$$

9.5 $GWS^h + \theta TS$ Algorithm Asymptotic Error Estimates

The $GWS^h + \theta TS$ algorithm for $n=1$ unsteady/steady, linear, and nonlinear convective $DE + BCs + IC$ statement (9.6)–(9.12) is complete. Decisions remaining are θ and theoretical prediction of performance for the range of $\{N_k(\zeta_\alpha)\}$ $n=1$ FE trial space basis implementations.

For the unsteady statement and finite Peclet number $Pe^{-1} > 0$, and realizing $m=1$, the generalization of the theoretical *asymptotic* error estimate (9.5) is

$$
\|e^h(n\Delta t)\|_E \leq Cl_e^{2\gamma}\left(\|data\|^2_{\Omega,L2} + \|data\|^2_{\partial\Omega,L2}\right) + C_t\Delta t^{f(\theta)}\|T_0\|_E,
\tag{9.29}
$$

$$
\gamma = \min(k, r-1), f(\theta) = (2,3)
$$

where $\|T_0\|_E$ is the energy norm of projection of the IC, equation (9.9), onto the DOF of Ω^h. Hence, asymptotic convergence under mesh refinement remains dependent on basis completeness degree k for smooth data. The impact of truncation error for the selected ODE algorithm time-step size Δt, as a function of θ, is also clearly identified, with C_t another constant and 3 associated *only* with $\theta = 0.5$.

For most practical fluid convection situations Pe is large, up to the $O(10^5)$, hence Pe^{-1} is typically but modestly larger than zero. The convergence benefit of using a higher

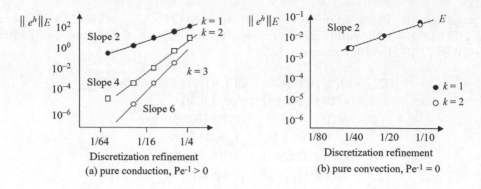

Figure 9.2 GWS^h + θTS algorithm solution convergence as a function of FE basis completeness degree k, (a) Pe^{-1} > 0, (b) Pe^{-1} = 0

degree $k > 1$ FE basis is lost in the limit Pe$^{-1} \Rightarrow 0$, as the significant mesh-dependent term in the alternative to equation (9.29) is independent of basis degree k, [1]. In this case, the replacement asymptotic error estimate is

$$\left\| e^h(n\Delta t) \right\|_E \leq Cl_e^2 \int_{t_0}^{n\Delta t} \|T(x,\tau)\|_{H^{k+1}}^2 d\tau + C_t \Delta t^{f(\theta)} \|T_0\|_E, \tag{9.30}$$

where $\int_{t_0}^{n\Delta t} \|T(x,\tau)\|_{H^{k+1}}^2 d\tau$ is the H^{k+1} norm squared of evolution of the *exact solution* over time span $t_0 \leq \tau \leq n\Delta t$. For the H^{k+1} norm to exist requires the exact solution $T(x,\tau)$ be sufficiently *smooth* such that its $k+1^{st}$ spatial derivative is square-integrable over time evolution to $n\Delta t$.

The precision with which asymptotic error estimates (9.29) and (9.30) characterize GWS^h + θTS algorithm performance is summarized in Figure 9.2. The left graph presents measured convergence under mesh refinement for GWS^h + θTS implemented using $1 \leq k \leq 3$ Lagrange trial space bases, equations (5.8, 5.21, 5.22), for an unsteady *pure conduction* statement. Thereby, Pe^{-1} corresponds to thermal diffusivity κ which of course is finite.

The right graph presents convergence for the GWS^h + θTS algorithm implemented using $1 \leq k \leq 2$ Lagrange bases for an unsteady *pure convection* statement, hence Pe^{-1} = 0. Obviously, no benefit accrues to use of a $k > 1$ basis. The Figure 9.2 data are in excellent agreement with theory and all solutions were generated time accurate using $\theta = 0.5$.

9.6 Performance Verification Test Cases

The classic $n = 1$ steady heat conduction verification is the *Peclet problem*, solutions of which amply illustrate convection term discretization error dominance as the diffusion term multiplier Pe^{-1} becomes small. The subject ODE is the source-free, constant speed

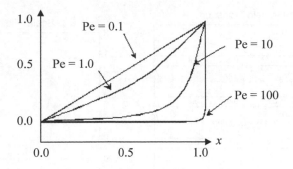

Figure 9.3 Peclet problem analytical solutions as $f(\text{Pe})$

and conductivity stationary form of equation (9.11)

$$\mathcal{L}(\Theta) = \frac{d\Theta}{dx} - \frac{1}{\text{Pe}}\frac{d^2\Theta}{dx^2} = 0, \quad \text{on } \Omega \subset \Re^1 \in (0,1). \tag{9.31}$$

For the Dirichlet BCs $\Theta(0) = 0$ and $\Theta(1) = 1$ the analytical solution for equation (9.31) is

$$\Theta(x) = \frac{1 - \exp(x\text{Pe})}{1 - \exp(\text{Pe})} \tag{9.32}$$

which exhibits a progressively pronounced wall-layer character as Pe becomes large, Figure 9.3.

Computer lab 9.1 suggests regular and solution-adapted mesh-refinement studies of the Peclet problem for $10 \leq \text{Pe} \leq 1000$. The inaugural $\text{Pe} = 100$ uniform mesh refinement *a posteriori* data for $10 \leq M \leq 80$ confirms inadequate mesh resolution introduces a node-to-node $(2\Delta x)$ *oscillatory error mode* into the $k=1$ basis GWS^h solution, Figure 9.4a. Lab continuation confirms identical $k=2$ basis results on inadequate M. Factually, the DOF in a uniform discretization Ω^h must exceed $\sim \text{Pe}/2$ to achieve a *monotone*, hence accurate solution.

Since in reality $\text{Pe} \leq O(10^5)$ is typical, uniform meshings are unrealistic and one must move to *solution-adapted* regular nonuniform meshing. A *regular* distribution of mesh *non-uniformity* is generated by a geometric factor P_r for progressive element measure alteration in the form $\ell_{e+1} \equiv P_r \ell_e$. For the lab continuation Figure 9.4b–d graph $k=1$ basis GWS^h DOF distributions generated on select P_r $M=20$ solution-adapted nonuniform meshes for $\text{Pe} = 100, 1000$.

The companion classic unsteady $n=1$ verification statement is pure convection of an energetic packet in complete absence of diffusion, that is, $\text{Pe}^{-1} = 0$. The pertinent differential equation (PDE) is the source-free, constant speed zero conductivity unsteady form of equation (9.11)

$$\mathcal{L}(\Theta) = \frac{\partial\Theta}{\partial t} + u\frac{\partial\Theta}{\partial x} = 0, \quad \text{on } \Omega \times t \subset \Re^1 \times R^+ \tag{9.33}$$

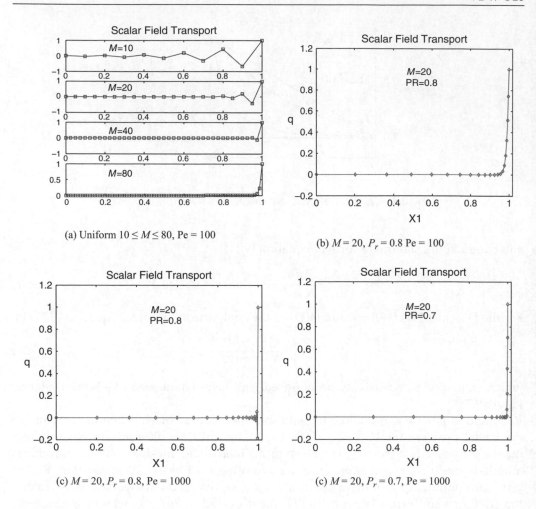

(a) Uniform $10 \leq M \leq 80$, Pe = 100

(b) $M = 20$, $P_r = 0.8$ Pe = 100

(c) $M = 20$, $P_r = 0.8$, Pe = 1000

(c) $M = 20$, $P_r = 0.7$, Pe = 1000

Figure 9.4 Peclet problem $k = 1$ GWS^h DOF distributionsfor select meshes M: (a) uniform $10 \leq M \leq 80$, Pe = 100; (b) $M = 20$, $P_r = 0.8$, Pe = 100; (c) $M = 20$, $P_r = 0.8$, Pe = 1000; (d) $M = 20$, $P_r = 0.7$, Pe = 1000

For the IC (9.9) and BC $\Theta(x = 0) = 0$ the analytical (characteristic) solution for equation (9.33) is

$$\Theta(x, t) = \Theta_0 \exp i(x - ut) \tag{9.34}$$

for i the imaginary unit.

The solution (9.34) guarantees *exact preservation* of the IC as the thermal packet is convected parallel to the x axis by u, Figure 9.5. Solution quality generated by any spatially discrete algorithm is strongly dependent on the integration time step size Δt, literature characterized by *Courant number*, the non-D time step $C \equiv u\Delta t / \Delta x$, explicitly dependent on mesh measure $\Delta x = h$.

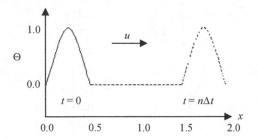

Figure 9.5 Exact solution, pure convective transport of energetic packet

The classic *smooth* IC is a gaussian or sine wave. For the gaussian, Figure 9.6 summarizes the *a posteriori* data generated in computer lab 9.2 for various M, C, and θ. In each graph, the IC is plotted over the first $L/4$ DOF of a uniform meshing of $0 \le x \le L$ for $M = 20$, 40, and 60. The number of time steps $n\Delta t$ to exactly propagate the IC over three wavelengths for each specified Courant number C must be computed. Each graph presents the solution at times corresponding to 1, 2, and 3 IC wavelength displacements.

For $C = 0.5 = \theta$ the coarsest $M = 20\,k = 1$ basis $GWS^h + \theta TS$ solution suffers significant peak distortion and profile corruption due to the $2\Delta x$ dispersion error wake trailing the solution, Figure 9.6a. As DOF interpolation of the IC is refined, hence increasing the mesh spanning L, solution quality improves, Figure 9.6b and c. Increasing C aggravates the dispersion error mechanism, Figure 9.6d and e.

Selecting any $0.5 < \theta \le 1.0$ introduces a *numerical diffusion* mechanism. The result is artificial diffusion of the solution peak coupled with dissipation of the dispersion error, Figure 9.6f. The choice $\theta = 0$ alters the truncation error to $O(\Delta t^2)$ and any $0 < \theta < 0.5$ defines an unstable time integration algorithm.

9.7 Dispersive Error Characterization

These $n = 1$ validation *a posteriori* data clearly illustrate that altering $DE + BCs$ with the fluid convection first derivative significantly impacts $GWS^h + \theta TS$ solution fidelity. In fact the convection term dominates for practical Pe and *all* discrete approximate solution methods inherit the *dispersive error* mode (*modulo* $2\Delta x$) that can *totally compromise* solution acceptability!

The legacy discrete algorithm community response to dispersion error is to add a strictly artificial (numerical) diffusion mechanism to the algorithm. The key issue thereby becomes, "Is the simulation process running on genuine (physics-based) diffusive closure model or is it dominated, hence polluted, by strictly *numerical diffusion*?" It is thus imperative to upgrade the theoretical basis for characterizing dispersion error cause and effect mechanisms.

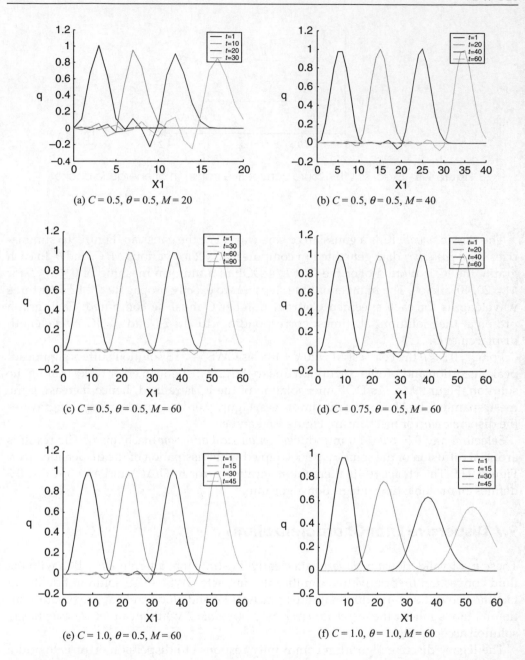

Figure 9.6 GWS^h + θTS $k = 1$ basis solutions for convected energetic packet: (a) $C = 0.5$, $\theta = 0.5$, $M = 20$; (b) $C = 0.5$, $\theta = 0.5$, $M = 40$; (c) $C = 0.5$, $\theta = 0.5$, $M = 60$; (d) $C = 0.75$, $\theta = 0.5$, $M = 60$; (e) $C = 1.0$, $\theta = 0.5$, $M = 60$; (f) $C = 1.0$, $\theta = 1.0$, $M = 60$

The precisely appropriate analytical framework is Fourier *modal analysis*. Writing equation (9.33) on the generic variable q

$$\mathcal{L}(q) = \frac{\partial q}{\partial t} + u \frac{\partial q}{\partial x} = 0 \tag{9.35}$$

the Fourier mode resolution of its characteristic solution is, [2]

$$q_\kappa(x, t) = e^{i\kappa(x-ut)} \tag{9.36}$$

for $\kappa \equiv 2\pi/\lambda$ the *wave number* of the Fourier mode of wavelength λ.

Recall from Figure 1.4 that *no* spatially discrete approximate solution method on Ω^h can resolve $\lambda = 2\Delta x$ wavelength solution content, as the DOF are identically zero. It is precisely this liability that initializes the dispersive error mode. Therefore, error quantification seeks to resolve how well short wavelength $\lambda = n\Delta x$, $2 \leq n \leq 2+\varepsilon$, Fourier content is propagated.

The analysis assumption is that any spatially semi-discrete WS^h solution to equation (9.35) possesses a Fourier mode resolution functionally identical to equation (9.36). Replacing x with its discrete equivalent

$$q_\kappa^h(x \Rightarrow j\Delta x, t) = e^{i\kappa(j\Delta x - \tilde{U}t)} \tag{9.37}$$

identifies the *mode velocity* $\tilde{U}(\kappa, \Delta x) \equiv \tilde{U}_{\text{real}} + i\tilde{U}_{\text{imag}}$. This *complex function* will likely differ from the input (real only) *data u*, equation (9.36), due to approximations underlying WS^h.

Fourier mode representation following time integration is also required. Any fully discrete $WS^h + \theta TS$ solution will generate the generic node DOF in the form, [2]

$$Q_j(n\Delta t) = G^n q_\kappa^h(j\Delta x, t) \tag{9.38}$$

where the *amplification factor* $G(\kappa, \Delta x, \Delta t, \theta)$ is also a complex function $G = G_{\text{real}} + iG_{\text{imag}}$. Amplification factor magnitude less than unity, that is, $GG^* \equiv |G| < 1$, predicts q_κ^h solution dissipation rate over time quantifying an *artificial diffusion* mechanism.

The second key output from equation (9.38) is the prediction of $WS^h + \theta TS$ algorithm *phase velocity* spectral distribution $U_\kappa(\lambda)$

$$U_\kappa(\lambda) \equiv \frac{\lambda}{\kappa\Delta x} \tan^{-1}\left[G_{\text{real}}/G_{\text{imag}}\right]. \tag{9.39}$$

Fourier mode solutions (9.37)–(9.39) enable a precise theoretical prediction of the wavelength-dependent, that is, *spectral*, performance of *any* WS^h and $WS^h + \theta TS$ algorithm applied to equation (9.35).

Defining *normalized error* in phase velocity as $1 - U_\kappa/u$, then graphed in percent, Figure 9.7 directly compares equation (9.39)-generated data for the discussed FD and FV^h algorithms, plus two higher order FV^h algorithms, and FE Lagrange $1 \leq k \leq 3$ basis GWS^h algorithms for $C = 0.5 = \theta$. The abscissa is logarithmically scaled on wavelength λ, to enhance resolution in the critical short wavelength region, keyed in mesh integer multiplies $n\Delta x$, $n = 2,3,4 \ldots$ As asserted in Chapter 1, the Figure 9.7 data mathematically confirm that *every* $WS^h + \theta TS$ algorithm tested exhibits 100% error (!) in propagation of the $\lambda = 2\Delta x$ wavelength Fourier mode. Present in a vertical column at $\lambda \approx 2.5\Delta x$ are the table labels, for example, FD, UW, GL, $\ldots$, identifying the various algorithms. In rigorous support of the assertion of *optimality*, the $k = 1$ basis GWS^h algorithm (table labeled "GL Gal. linear fe") possesses *much smaller* phase velocity error over the entire λ spectrum than do $O(\Delta x^2)$ FD *and* $O(\Delta x^2, \Delta x^3, \Delta x^5)$ FV^h algorithms.

The data labeled GQ and GC precisely quantify the short wavelength accuracy improvement for $GWS^h + \theta TS$ FE $k = 2,3$ basis implementations, a performance facet that asymptotic error estimation theory cannot predict. Finally, the nondiscussed data labeled TL, TQ, and TC are for the analytically *modified* $mGWS^h + \theta TS$ algorithm, derived in the next section.

The data in Figure 9.7 were generated using non-D time step (Courant Number) $C = 0.5$ and the trapezoidal rule $\theta = 0.5$. The accuracy of each algorithm degrades as C is increased, as computer lab 9.2 will confirm. Definitions $0.5 < \theta \leq 1.0$ yield progressive improvement in U_κ across the entire λ spectrum, at the expense of introducing

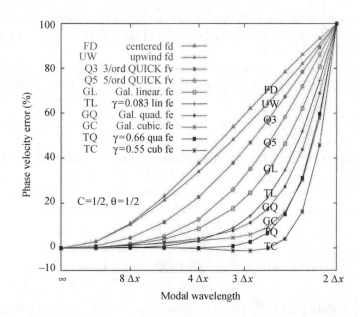

Figure 9.7   Select $WS^h + \theta TS$ algorithm *phase velocity error* spectral distributions, $C = 0.5 = \theta$

artificial diffusion as $|G| < 1$ for all $\theta > 0.5$, clearly illustrated in Figure 9.6f. Definitions $0 < \theta < 0.5$ generate *unstable* time integration algorithms (try them!).

9.8 A Modified Galerkin Weak Statement

Via any number of theories, *artificial diffusion* mechanisms are derived throughout the FD/FV/FE communities to dissipate the dispersive error mechanism intrinsic to discrete algorithms at large Pe. The *modified* Galerkin weak statement $mGWS^h$, referenced in the literature as "*Taylor* weak statement (TWS)," is the *analytical n-D* generalization of the progenitor FD idea, Lax [3]. The $mGWS^h$ theory identifies calculus differential terms that *modify* the textbook conservation principle PDE (9.35). This generates what is termed herein an *mPDE* that facilitates theoretical analyses for *optimal* choices of an embedded parameter set.

Re-expressing equation (9.35) in subscript *kinetic* flux vector notation

$$\mathcal{L}(q) = \frac{\partial q}{\partial t} + \frac{\partial f}{\partial x} \equiv q_t + f_x = 0, \quad \text{with } f \equiv uq \tag{9.40}$$

theory development starts with an explicit time TS to $O(\Delta t^4)$

$$q^{n+1} = q^n + \Delta t \, q_t^n + \frac{1}{2}\Delta t^2 q_{tt}^n + \frac{1}{6}\Delta t^3 q_{ttt}^n + O(\Delta t^4), \tag{9.41}$$

where superscripts n, $n+1$ denote time level. The *mPDE* theory then manipulates equation (9.41) to replace the time derivatives therein using equation (9.40). This process recognizes the kinetic flux vector $f = uq$ is a function of q, not x or t. In derivative subscript notation an exercise will verify that

$$\begin{aligned} q_{tt} &= (q_t)_t = (-f_x)_t = -(f_t)_x = -(f_q q_t)_x = +(f_q f_q q_x)_x \\ q_{ttt} &= \cdots = (f_q f_q q_{tx})_x = -\left(f_q f_q (f_q q_x)_x\right)_x \end{aligned} \tag{9.42}$$

The *jacobian* of the kinetic flux vector f is $\partial f / \partial q \equiv f_q = u$, constant *data* in equation (9.35). Insert this observation into equation (9.42) while introducing coefficients (α, β, γ, δ) to account for terminal expression *arbitrariness*, absorbing the signs therein as well.

Dividing through by Δt this operation generates the homogenous re-expression of equation (9.41)

$$\frac{q^{n+1} - q^n}{\Delta t} + f_x^n - \frac{\Delta t}{2}\left(\alpha u q_t + \beta u^2 q_x\right)_x - \frac{\Delta t^2}{6}\left(\gamma u^2 q_{tx} + \delta u^3 q_{xx}\right)_x + O(\Delta t^3) = 0. \tag{9.43}$$

Forming the limit of equation (9.43) as $\Delta t \Rightarrow \varepsilon > 0$, which retains the *higher order* Δt terms, the first two terms constitute the original PDE (9.40). The resultant TS-modified *mPDE* replacement of equation (9.40) is

$$\mathcal{L}^m(q) \equiv \mathcal{L}(q) - \frac{\Delta t}{2} \frac{\partial}{\partial x}\left(\alpha u \frac{\partial q}{\partial t} + \beta u^2 \frac{\partial q}{\partial x}\right)$$
$$- \frac{\Delta t^2}{6} \frac{\partial}{\partial x}\left(\gamma u^2 \frac{\partial^2 q}{\partial t \partial x} + \delta u^3 \frac{\partial^2 q}{\partial x^2}\right) + \text{TE} = 0 \tag{9.44}$$

Forming GWS^h on equation (9.44), symbolized herein as $mGWS^h$, then implementing θTS generates the analysis framework. The $(\alpha, \beta, \gamma, \delta)$ coefficient set explicitly recognizes the nonuniqueness intrinsic to the TS process (9.42). Thus is generated the opportunity to determine *optimal* $(\alpha, \beta, \gamma, \delta)$, hence also θ, for $mGWS^h$ algorithm moderation of dominant discrete approximation errors intrinsic to GWS^h, or for that matter any WS^h algorithm!

Additionally, and providing truly surprising insight, specific $(\alpha, \beta, \gamma, \delta)$ choices in equation (9.44), also θ, admit recovering a wide range of FD and $WS^h + \theta TS$ algorithms derived using totally distinct discrete theories. Over a dozen such *independently-derived* stabilized algorithms for equation (9.40) are detailed in Table 9.1 to constitute specific parameter selections for FE $k = 1$ basis implementations of $mGWS^h + \theta TS$ [4].

The Fourier phase velocity spectral theory (9.37)–(9.39) enables quantitative prediction of FD, WS^h, GWS^h, $mGWS^h$ algorithm comparative performance for equation (9.35). Of fundamental impact $mGWS^h$ theory admits prediction of *optimal* $\gamma = (-0.5, -0.4, -0.33)$ for Lagrange $k = 1,2,3$ completeness degree basis implementations [2]. The result is

Table 9.1 $mGWS^h + \theta TS$ algorithm definitions recovering select dissipative convection-diffusion algorithms

Algorithm name	θ	$\dfrac{\alpha}{2}$	$\dfrac{\beta}{2}$	$\dfrac{\gamma}{6}$	$\dfrac{\delta}{6}$
$mGWS^h + \theta TS$	All	Arb	Arb	Arb	Arb
$GWS^h + \theta TS$	All	0	0	0	0
Donor cell FD	0	0	1	0	0
Lax–Wendroff FD	0	0	$\text{sgn}(u)$	0	0
Euler Taylor GWS^h	0	0	1	1	0
CN Taylor GWS^h	0.5	0	0.5	1	0
Euler Char. GWS^h	0	0	1	0	1
Swansea Taylor GWS^h	0	0	1	0	0
Wahlbin	0	$\text{sgn}(u)$	$2\text{sgn}(u)$	0	0
Dendy	0	$\Delta x \cdot \text{sgn}(u)$	$\Delta x \cdot \text{sgn}(u)$	0	0
Raymond–Garder	0.5	$\text{sgn}(u)/v_0$	$\text{sgn}(u)/v_0$	0	0
Hughes SUPG	–	0	$\text{sgn}(u)$	0	0
Euler Petrov GWS^h	0	0	0	$(1-v)$	0
CN Petrov GWS^h	0.5	$\text{sgn}(u)$	$v \cdot \text{sgn}(u)$	$-v/2$	0
Warming–Beam FD	0	0	1	0	$-3(1-c)$
VanLeer–MUSCL	1	0	$\text{sgn}(u)$	0	-3
Jiang least squares	all	2θ	2θ	0	0

Note: $\text{sgn}(u)$ is the sign of u; $v_0 = 1/\sqrt{15}$; $c \leq v \leq 1$ where $c = u\Delta t/\Delta x$; arb $\Rightarrow$ arbitrary.

significantly improved real U_κ spectral fidelity compared to GWS^h especially in the *critical* range $2\Delta x < \lambda \le (2+\varepsilon)\Delta x$! The data labeled TL, TQ, TC in Figure 9.7 quantify the $mGWS^h$ $k=1,2,3$ Lagrange basis spectral fidelity improvements for $\alpha=0=\beta=\delta$ and k-dependent *optimal* γ.

The $mGWS^h$ selection $\beta>0$ generates an *artificial diffusion* mechanism. Figure 9.8 compares select FD, FV^h algorithm dissipation generation with that of GWS^h and $mGWS^h(\beta)$ as *amplification factor* spectral distribution magnitude $|G(\gamma)| \equiv GG*$ departure from unity. The abscissa of this theory measure of artificial diffusion is logarithmic in $n\Delta x$ for short wavelength region resolution. The data labeled "centered fd/fe" are standard FD (Crank–Nicolson) and all $k \ge 1$ basis GWS^h algorithms, for which $|G|=1$ on the entire wavelength spectrum.

In Figure 9.8, the $k=1$ basis $mGWS^h(\beta)$ algorithm coefficient is defined to generate a dissipation magnitude at $\gamma=2\Delta x$ identical with the upwind FD and third/fifth order QUICK FV^h algorithms, respectively. These alternative theories possess *no* such control, and upwind FD is *unacceptably* diffusive for all $\gamma < 8\Delta x$. This in fact is genesis of QUICK algorithm derivations, data labeled as Q3, Q5. Definitions $\beta \equiv 0.169, 0.126$ for the $mGWS^h(\beta)$ algorithm spectra are selected to match the Q3, Q5 dissipation magnitude at $\lambda=2\Delta x$, respectively. Algorithm dissipation decreases to $< 20\%$ at $2\Delta x$ for $\beta=0.063$. For all $\theta=0.5$ algorithms using larger Δt (C) increases the departure of $|G|$ from unity, and all $\theta > 0.5$ selections generate added Δt-dependent artificial diffusion.

The output of Fourier *modal analysis* theory, summarized in Figures 9.7 and 9.8, clearly *quantifies* performance expectations for the addressed class of FD, FV^h, GWS^h, and $mGWS^h$ algorithms. For the latter two this extends to the impact of $k=1,2,3$ trial

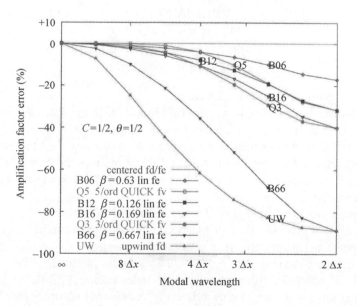

Figure 9.8 Amplification factor magnitude |G| departure from unity

space basis implementations. The Fourier theory functions $G(\gamma)$ and $U_\kappa(\gamma)$ each clearly constitute a solution *norm*.

9.9 Verification Problem Statements Revisited

Computer lab 9.3 suggests replacing the GWS^h template of computer lab 9.2 with the mGWS^h template, then revisiting the detailed verifications. In forming the steady mGWS^h algorithm the Δt multiplier on the β term is replaced, usually with the local timescale $(\ell_e/|u_e|)$. This substitution adds a mesh measure multiplier on the element matrix and $u^2 \Rightarrow |u_e|$. Only characteristic methods employ the δ term, Table 9.1, thus this option is no longer considered.

Recalling template data names PEI $= (\text{Pe})^{-1}$, NU $=$ Nu, TC $= T_c$, and adding U $= u$ and BETA $= \beta$, the suggested exercise will verify the mGWS^h template for linear steady mPDE for equations (9.11) and (9.12) is

$$
\begin{aligned}
\{\text{RES}\}_e = \ &()(){U}(;0)[A3001d]\{Q\} \\
&+(\text{PEI})(){ }(;-1)[A211d]\{Q\} \\
&+(\text{BETA}/2)(){ \text{UMAG}}(;0)[A3011d]\{Q\} \\
&-()(){ }(;1)[A200d]\{\text{SRC}\} \\
&+(\text{PEI},\text{NU})(){ }(;0)[\text{ONE}]\{Q\} \\
&-(\text{PEI},\text{NU},\text{TC})(){ }(;0)[\text{ONE}]\{ \} \\
&+(\text{PEI},\text{FN})(){ }(;0)[\text{ONE}]\{ \}
\end{aligned}
\tag{9.45}
$$

$$
\begin{aligned}
[\text{JAC}]_e = \ &()(){U}(;0)[A3001d][\,] \\
&+(\text{PEI})(){ }(;-1)[A211d][\,] \\
&+(\text{BETA}/2)(){ \text{UMAG}}(;0)[A3011d][\,] \\
&+(\text{PEI},\text{NU})(){ }(;0)[\text{ONE}][\,].
\end{aligned}
$$

Recall $d \Rightarrow L,Q,C$ for Lagrange $k = 1,2,3$ bases.

Returning to the Peclet problem, generating accurate monotone solutions for Pe large has proven a challenge. On a coarse mesh M the mGWS^h algorithm will always generate a *smooth* monotone solution for nominal $\beta > 0$ for any Pe. However, such solutions are *totally incorrect* (!), see Figure 9.9a, which compares the correct wall layer distribution to that produced by the $k = 1$ basis mGWS$^h(\beta = 0.3)$ algorithm on *uniform* $M = 20$ for Pe $= 1000$.

Hence *smooth* does not imply *correct* unless verified via solution-adapted Ω^h refinement. Following this process nominal mesh GWS^h solutions visualize as monotone but can possess *residual* dispersion error oscillations. Figure 9.9b graphs the absolutely monotone, precisely determined $P_r = 0.66$ *solution-adapted non-uniform* $M = 20$, Pe $= 1000$ $k = 1$ basis GWS^h solution. Alternatively, $k = 1$ basis nominal-β mGWS^h will generate nodally exact solutions for any Pe on relatively coarse, not-so-precisely designed non-uniform meshes.

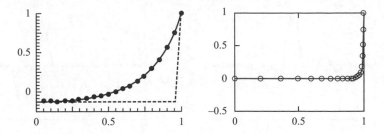

Figure 9.9 $mGWS^h(\beta)$ solutions, steady Peclet problem, Pe = 1000: (a) $M=20$ uniform mesh, (b) $M=20$ solution adapted mesh

Theoretical determination of optimal β for $k=1$ basis steady (only) n-dimensional $mGWS^h$ is derived; details are deferred to the next chapter. Theorization of steady optimal β for $k > 1$ bases has proven intractable, hence algorithm performance using the $d=Q,C$ element libraries requires experimenting. Computing lab experience predicts $0.1 < \beta \leq 0.3$ is practical; for reference the Figure 9.9a uniform M value is $\beta = 1000(0.05)^2/12 = 0.21$, also the value used for the prediction in Figure 9.9b.

A comment is required on assessing $GWS^h/mGWS^h$ algorithm performance, hence convergence, via asymptotic error estimates (9.29) or (9.30). As the energy norm is symmetric quadratic in q^h, dispersion error-induced *modulo* $2\Delta x$ oscillation contributions compute sign insensitive, thereby inject *false energy* into the norm. Therefore, computing $\|q^h\|_E$ makes sense *only* for (essentially) monotone solutions. An exercise suggests proving $\|q^h\|_E \Rightarrow 0.25$ as Pe $\Rightarrow \infty$ for the Peclet problem.

Recalling template unsteady data names DT $=\Delta t$, TDT $=\theta\Delta t$, and adding GAMA $=\gamma$, the linear unsteady $mGWS^h + \theta TS$ algorithm template for $mPDE$ for equations (9.11) and (9.12) is

$$
\begin{aligned}
\Delta t\{RES\}_e = \; &(DT)(\;)\{U\}(;0)[A3001]\{Q\} \\
&+(PEI, DT)(\;)\{\;\}(;-1)[A211]\{Q\} \\
&+(BETA/2, DT^2)(\;)\{U^2\}(;0)[A3011]\{Q\} \\
&-(DT)(\;)\{\;\}(;1)[A200]\{SRC\} \\
&+(PEI, NU, DT)(\;)\{\;\}(;0)[ONE]\{Q\} \\
&-(PEI, NU, DT, TC)(\;)\{\;\}(;0)[ONE]\{\;\} \\
&+(PEI, DT, FN)(\;)\{\;\}(;0)[ONE]\{\;\}
\end{aligned}
\tag{9.46}
$$

$$
\begin{aligned}
[JAC]_e = \; &(\;)(\;)\{\;\}(;1)[A200][\;] \\
&+(TDT)(\;)\{U\}(;0)[A3001][\;] \\
&+(TDT^2, GAMA/6)(\;)\{U^2\}(;0)[A3011][\;] \\
&+(PEI, TDT)(\;)\{\;\}(;-1)[A211][\;] \\
&+(BETA/2, TDT^2)(\;)\{U^2\}(;-1)[A3011][\;] \\
&+(PEI, NU, TDT)(\;)\{\;\}(;0)[ONE][\;]
\end{aligned}
$$

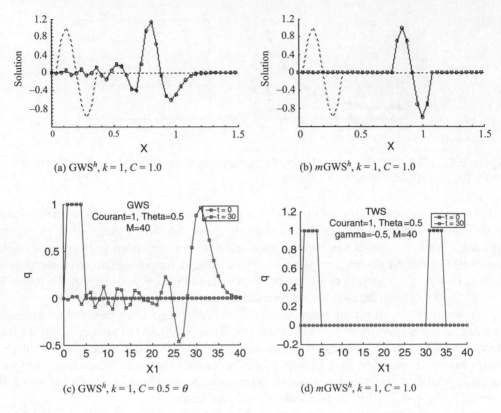

(a) GWS^h, $k = 1$, $C = 1.0$

(b) $mGWS^h$, $k = 1$, $C = 1.0$

(c) GWS^h, $k = 1$, $C = 0.5 = \theta$

(d) $mGWS^h$, $k = 1$, $C = 1.0$

Figure 9.10 Traveling packet, $GWS^h/mGWS^h k = 1$ basis algorithm DOF distributions, $\theta = 0.5$: (a) GWS^h, $C = 1.0$; (b) $mGWS^h$, $\gamma = -0.5$, $C = 1.0$; (c) GWS^h, $C = 0.5$; (d) $mGWS^h$, $\gamma = -0.5$, $C = 1.0$

Returning to translating energy packet verification, the accurate solution is precise preservation of the IC. The *smooth* IC $q_0 = \sin(2\pi x/l)$ possesses full signed Fourier spectral content. Following two IC wavelength translation the $C = 1.0\,k = 1,2$ basis $GWS^h/mGWS^h + \theta TS$ comparative solutions are graphed in Figure 9.10a and b. The GWS^h $\theta = 0.5$ solution preserves waveform essence at best, dispersion error preferentially diminishes the negative peak and a substantial oscillatory trailing wake is generated. The singular specification is $C \equiv 1$ for which $mGWS^h + \theta TS$ dual parameter definitions $\beta = 1 = \gamma$, $\theta = 0$ and $\beta = 0$, $\gamma = -0.5$, $\theta = 0.5$ each generate exact IC preservation, Figure 9.10b.

The nonsmooth IC is a three-element wide square wave for which *any* $WS^h + \theta TS$ algorithm generates total distortion, Figure 9.10c. Re-executing with $\beta > 0$ and $\theta = 0.5$ will confirm a little β generates significant waveform dissipation, further increased using any $\theta > 0.5$. Conversely, $mGWS^h + \theta TS$ with $\beta = 1 = \gamma$, $\theta = 0$, and $\beta = 0$, $\gamma = -0.5$, $\theta = 0.5$ improves solution propagation for all $C < 1$. The selection $C \equiv 1$ remains singular and both $mGWS^h + \theta TS$ parameter options generate *exact* IC translation, Figure 9.10d.

9.10 Unsteady Heat Conduction

Unsteady pure heat conduction is a degenerate form of the $n=1$ DE + BCs + IC system (9.6)–(9.9) as the *kinetic* flux vector that enabled mGWS^h development is absent. The GWS^h + θTS algorithm for this *parabolic* PDE always produces a smooth solution except in the instance of *non-smooth* IC data. Computer lab 9.4 explores the issue and amplifies on the artificial diffusion action of $\theta > 0.5$.

The problem statement is unsteady convective heat transfer in the axisymmetric cylinder, as defined and detailed in Chapter 6.8, minus the fins and radiation. The $n=1$ DE + BCs + IC statement is

$$\mathcal{L}(T) = \frac{\partial T}{\partial t} - \frac{1}{r}\frac{\partial}{\partial r}\left(\kappa r \frac{\partial T}{\partial r}\right) = 0, \qquad \text{on } \Omega \times t \subset \mathfrak{R}^1 \times R^+ \tag{9.47}$$

$$\ell(T) = kdT/dr + h(T - T_c) = 0, \qquad \text{on } \partial\Omega_R \times t \subset \mathfrak{R}^0 \times R^+$$

$$T(r_b) = T_b, \qquad\qquad\qquad \text{on } \partial\Omega_D \times t \subset \mathfrak{R}^0 \times R^+ \tag{9.48}$$

$$T(r, t_0) = T_0, \qquad\qquad\qquad \text{on } \Omega \cup \partial\Omega \times t_0 \tag{9.49}$$

where $\kappa \equiv k/\rho c_p$ is thermal diffusivity. Taking guidance from equation (6.70) an exercise requests confirming the GWS^h + θTS algorithm template is the modest alteration to equation (9.46),

$$\Delta t\{RES\}_e = (DT, KAPA)(\ )\{RAD\}(;-1)[A3011d]\{Q\}$$
$$+(DT, KAPA, RHO, CP, HC, RC)(\ )\{\ \}(;0)[ONE]\{Q\}$$
$$-(DT, KAPA, RHO, CP, HC, RC, TC)(\ )\{\ \}(;0)[ONE]\{\ \}$$
$$\tag{9.50}$$

$$[JAC]_e = (\ )(\ )\{RAD\}(;1)[A3000d][\ ]$$
$$+(TDT, KAPA)(\ )\{RAD\}(;-1)[A3011d][\ ]$$
$$+(TDT, KAPA, RHO, CP, HC, RC)(\ )\{\ \}(;0)[ONE][\ ]$$

For nominal uniform meshes M, implementing the Robin BC (9.48) during the first time step $t_1 = t_0 + \Delta t$ initiates a $2\Delta x$ dispersive error oscillation in the temperature distribution. Dependent on M and time step size Δt, it propagates as a wave throughout the mesh, and/or in time, with continued time integration. Computer lab 9.4 enables comparison of *ad hoc* corrections with the theory-based resolution of regular *solution-adapted* mesh refinement via $\ell_{e+1} \equiv P_r \ell_e$.

The *ad hoc* options are to compromise time accuracy by defining $\theta = 1.0$, which adds artificial diffusion, or replacing the GWS^h theory time term hypermatrix with the FD-derived (Crank-Nicolson) diagonal matrix, [2,4]. The FD construction amounts to *under-interpolation* and is template-implemented in [JAC]$_e$ via the replacement

$$(\ )(\ )\{RAD\}(;1)[A3000d][\ ] \Rightarrow$$
$$(\ )(RADAVG)\{\ \}(;1)[I][\ ] \tag{9.51}$$

where RADAVG is element *average* radius and [A3000d] is replaced by the identity matrix [I].

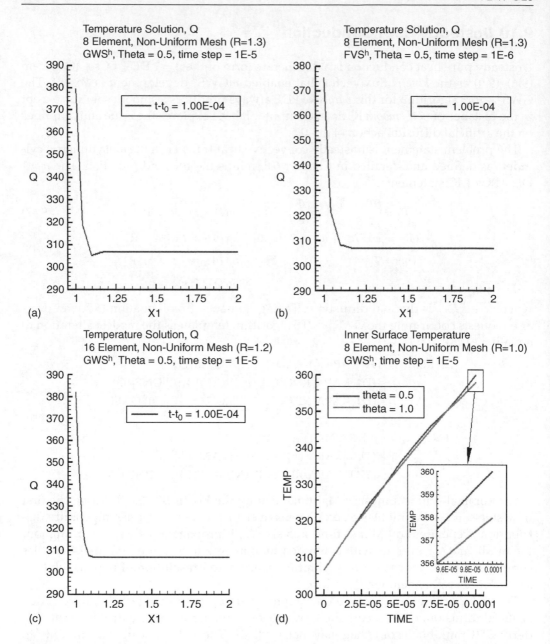

Figure 9.11 Radial unsteady heat conduction, algorithm temperature distributions following 10 time steps at $\Delta t = 0.00001$ hr for meshes: a) $P_r = 1.3$, $M = 8$, GWS^h, b) $P_r = 1.3$, $M = 8$, FD, c) $P_r = 1.2$, $M = 16$, GWS^h, d) $P_r = 1.0$, $M = 8$, GWS^h, $\theta = 0.5$, 1.0

For the substantially solution-adapted $P_r=1.3$, $M=8$ mesh, the GWS^h and FD algorithm temperature distributions following ten time steps at $\Delta t=0.00001$ hr are compared in Figure 9.11a–b. The former is non-monotone with extremum nodal temperature $Q1=380$ K while the FD solution is monotone with extremum $Q1 \sim 4$ degrees lower.

Defining $P_r=1.4$ yields a monotone $M=8$ GWS^h solution, hence the regular refinement mesh is $P_r=1.2$, $M=16$. This mesh supports a monotone GWS^h solution with extremum $Q1=382$ K, Figure 9.1c, confirming the $P_r=1.3$, $M=8$ FD solution is less accurate than GWS^h. For a uniform $M=8$ mesh, replacing $\theta=0.5$ with 1.0 reduces the GWS^h algorithm extremum $Q1$ by ~ 2.5 K, Figure 9.1 d.

Computer lab 9.4 provides ample opportunity to assess impact of the range of implementation factors.

9.11 Chapter Summary

$GWS^h/mGWS^h+\theta TS$ implementation using $k\geq 1$ completeness degree FE trial space bases is thoroughly examined for the $n=1$ unsteady DE + BCs + IC statement (9.6)–(9.9) with fluid convection. Addition of this first spatial derivative term, coupled with nondimensionalizing to introduce the Peclet number, enabled a series of theoretical examinations regarding asymptotic convergence alteration and moderation of the *modulo $2\Delta x$* dispersion error mechanism thus introduced.

In summary, this extension of scope expands the Galerkin weak form algorithm essential ingredients fairly consequentially to:

$$approximation: \quad q(x,t) \approx q^N(x,t) \Rightarrow q^h(x,t) = \cup_e q_e(x,t)$$

$$FE\ basis\ implementation: \quad q_e(x,t) = \{N_k(\zeta_\alpha)\}^T (Q(t))_e$$

$$error\ extremization: \quad mGWS^N = \int_\Omega \Psi_\beta \mathcal{L}^m(q^N)dx \equiv \{0\}, \forall\beta$$

$$matrix\ statement: \quad mGWS^h + \theta TS \Rightarrow [JAC]\{\Delta Q\} = -\Delta t\{RES\}$$

$$element\ formation: \quad [JAC] = S_e([JAC]_e), \{RES\} = S_e(\{RES\}_e)$$

$$\left\| e^h(n\Delta t) \right\|_E \leq C\ell_e^{2\gamma} \|data\|^2_{\Omega,\partial\Omega L2} + C_t\Delta t^{f(\theta)} \|Q_0\|_E$$

$$asymptotic\ convergence: \quad \gamma \equiv \min(k, r-1), f(\theta) = (2,3),\ or$$

$$\left\| e^h(n\Delta t) \right\|_E \leq C\ell_e^2 \int_{t_0}^{n\Delta t} \|T(x,\tau)\|^2_{H^{k+1}}d\tau + C_t\Delta t^{f(\theta)} \|Q_0\|_E$$

$$error\ spectra: \quad U_\kappa(\lambda), G(\lambda) \Rightarrow f(\kappa, k, h, \Delta t, \theta, \alpha, \beta, \gamma). \quad (9.52)$$

The *FEmPSE* toolbox-enabled MATLAB® computer lab environment fully enables detailed examination of all theoretical and code practice issues introduced and discussed.

Exercises

9.1 Verify the nondimensionalization of DE + BCs leading to equations (9.11) and (9.12)

9.2 Derive the truncation error term order for the θ one-step time integration family, (9.19), for $\theta = 0.0, 0.5$, and 1.0.

9.3 Verify the $GWS^h + \theta TS$ algorithm templates (9.25)–(9.28) for $n = 1$ linear, steady, and linear/nonlinear unsteady convection/diffusion problem statements.

9.4 Verify the non-D Peclet problem analytical solution (9.32) for the given Dirichlet BCs.

9.5 Compute the energy norm of the non-D Peclet problem analytical solution, hence verify the norm $\Rightarrow 0.25$ as Pe $\Rightarrow \infty$.

9.6 Verify the Taylor series expressions (9.42).

9.7 Confirm the correctness of the $mPDE$ (9.44).

9.8 Verify the $mGWS^h + \theta TS$ algorithm templates (9.45) and (9.46).

9.9 Develop in completeness the $GWS^h + \theta TS$ algorithm for equations (9.47) and (9.48), hence confirm template (9.50).

Computer Labs

The *FEmPSE* toolbox and MATLAB® .m files required for conducting the computer labs are available for download at www.wiley.com/go/baker/finite.

9.1 Download the .m file $k = 1$ basis template, then execute a convergence study in the energy norm for the steady Peclet problem, (9.31), for Pe $= 1$ and Pe $= 10$. Edit the .m file to define the $k = 2$ basis algorithm and repeat. Then implement the P_r *solution-adapted* regular mesh definition, hence precisely determine the *non-uniform* $M = 20$ mesh that supports generation of $k = 1,2$ basis *monotone* solutions for Pe $= 1000$.

9.2 Edit the computer lab 9.1 .m file to execute a $k = 1$ basis study of the unsteady traveling energy packet problem, (9.33), for Pe$^{-1} \equiv 0$. Interpolate the IC on a uniform $M = 20$ mesh, hence compute the number of time steps $n\Delta t$ required for the exact solution to travel two IC wavelengths at $C = 0.5$. Repeat this exercise for redefined $C = 1.0$. Then alter the $\theta = 0.5$ definition to $\theta = 0$ and/or $\theta > 0.5$ and compare solution accuracy. Edit the .m file to insert the $k = 2$ basis template and repeat.

9.3 Execute a uniform mesh-refinement convergence study in the energy norm for the Pe $= 1000$ steady Peclet problem, equation (9.31). Replace the computer lab 9.1 .m file GWS^h template with that for $k = 1$ basis $mGWS^h$. Confirm that the $k = 1$ basis $mGWS^h$ nominal β solution sequence agrees with the theory asymptotic convergence rate $O(h^2)$. Then implement $k = 2$ basis $mGWS^h$ and repeat the exercise.

9.4 Execute a solution-adapted regular mesh-refinement study for the $DE + BCs + IC$ statement (9.47)–(9.49) implementing $GWS^h + \theta TS$ template (9.50). For $k=1$ basis progressively alter the initial $M=8$ uniform mesh via P_r until this M spatial solution is monotone. Then compute the P_r that renders the Robin BC element of an $M=16$ mesh half the span of this $M=8$ element. Continue this *regular* mesh-refinement process for $M=32$, 64. Decrease the initial $\Delta t = 0.0001$ hr definition appropriately as refined M *a posteriori* data exhibit temporal oscillations.

Compare generated *monotone* solution data with the asymptotic convergence theory (9.52), hence stop the M refinement process when the energy norm confirms the Robin BC surface DOF is accurate to $\approx 0.1°$ at $t - t_0 = 0.001$ hr. Repeat the $M=8$ monotone solution case using $\theta = 1.0$ to quantify time truncation error artificial diffusion on T_{max}. Implement the FD alteration (9.51) and compare computed energy norm data for $M=16$ to the GWS^h solution. Then repeat the solution-adapted regular mesh-refinement study for $k=2$ basis to confirm validity of asymptotic error estimate (9.52)

References

[1] Oden, J.T. and Reddy, J.N. (1976) *An Introduction to the Mathematical Theory of Finite Elements*, Wiley-Interscience, New York.

[2] Chaffin, D.J. and Baker, A.J. (1995) On Taylor weak statement finite element methods for computational fluid dynamics. *J. Num. Methods Fluids*, **21**, 273–294.

[3] Lax, P.D. (1957) Weak solutions of nonlinear hyperbolic equations and their numerical computation. *Commun. Pur. Appl. Math.*, **7**, 537–566.

[4] Baker, A.J. and Kim, J.W. (1987) A Taylor weak statement algorithm for hyperbolic conservation laws. *J. Num. Methods Fluids*, **7**, 489–520.

10

Convection–Diffusion, $n > 1$: $n = 2, 3$ GWS^h/mGWS^h + θTS, stability, error characterization, linear algebra

10.1 The Problem Statement

The FE trial space basis implementation GWS^h/mGWS^h + θTS for the unsteady $n=1$ DE + BCs + IC PDE system, extended to steady, exposed key issues induced by fluid convection and time dependence. This chapter extends the development to unsteady heat/mass transport in $n = 2,3$ dimensions. The theory characterizing dispersion error is generalized leading to mGWS^h n-D optimality prediction. Importantly, as experienced to this juncture, FE completeness degree k bases supporting the process GWS^N $\Rightarrow$ GWS^h/mGWS^h is *totally devoid* of heurism, in sharp distinction to alternative discrete theorization including FD and FV^h methods.

The n-D unsteady convection–diffusion PDE + BCs + IC statement for the *generic scalar* state variable member $q(x,t)$ in non-D tensor indicial notation is

$$\mathcal{L}(q) = \frac{\partial q}{\partial t} + u_j \frac{\partial q}{\partial x_j} - \frac{1}{Pa} \frac{\partial^2 q}{\partial x_j{}^2} - s = 0, \text{ on } \Omega \times t \subset \mathfrak{R}^n \times R^+. \tag{10.1}$$

$$\ell(q) = \nabla q \cdot \hat{\mathbf{n}} + Pb(q - q_r) + \mathbf{f} \cdot \hat{\mathbf{n}} = 0, \text{ on } \partial \Omega_R \times t \subset \mathfrak{R}^{n-1} \times R^+. \tag{10.2}$$

Finite Elements ⇔ Computational Engineering Sciences, First Edition. A. J. Baker.
© 2012 John Wiley & Sons, Ltd. Published 2012 by John Wiley & Sons, Ltd.

$$q(\mathbf{x}_b, t) = q_b(\mathbf{x}_b, t), \qquad \text{on } \partial\Omega_D \times t \subset \Re^{n-1} \times R^+. \tag{10.3}$$

$$q(\mathbf{x}, t_0) = q_0(\mathbf{x}), \qquad \text{on } \Omega \cup \partial\Omega \times t_0. \tag{10.4}$$

In $\mathcal{L}(q)$ and $\ell(q)$, the non-D *parameters* Pa and Pb are the Peclet and Nusselt numbers for DE. Alternatively, for scalar field transport of mass q is identified as mass fraction, Pa is Reynolds × Schmidt number (ReSc), while Pb has no "proper" name but the Robin BC is operative.

Proper specification of BCs is particularly critical for this PDE system. For example, Figure 10.1 graphs an $n = 3$ prediction of the distribution of mass emanating from a point source convected downwind by a diagonal imposed velocity vector field. The proper BCs are (10.3) on all inflow planes and homogeneous (10.2) on all outflow planes. The IC (10.4) translates to DOF $\{Q(t_0)\} \equiv \{1\}$ in the source region and zero everywhere else.

The PDE + BCs + IC domain definitions are precisely expressed in equations (10.1)–(10.4). Specifically, $\Omega \times t$ denotes the outer (tensor) product of n-dimensional Euclidean space $\Re^n$ with time spanning the positive real number axis R^+. The Robin and Dirichlet BCs on boundary segments $\partial\Omega_R$ and $\partial\Omega_D$ are similarly tensor products of $(n-1)$-D space and time. The IC is applied on the entirety of Ω plus $\partial\Omega$ at time $t = t_0$.

(*Note*: The colored version of most figures in this chapter are found and can be downloaded from www.wiley.com/go/baker/finite.)

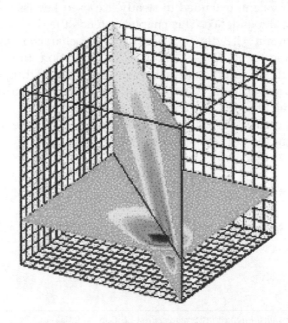

Figure 10.1 Perspective graph of an $n = 3$ mass transport simulation

10.2 GWS^h +θ TS Formulation Reprise

The GWS^N $\Rightarrow$ GWS^h $\Rightarrow$ mGWS^h +θ TS essence for establishing the approximate solution $q^h(\mathbf{x},t)$ and estimating the resultant approximation error $e^h(\mathbf{x},t)$, as coalesced from Chapters 7 and 9, summarizes to:

$$\text{approximation}: \quad q(\mathbf{x},t) \approx q^N(\mathbf{x},t) \equiv q^h(\mathbf{x},t) \equiv \cup_e q_e(\mathbf{x},t). \tag{10.5}$$

$$\text{FE basis}: \quad q_e(\mathbf{x},t) \equiv \left\{ N_k(\zeta_\alpha, \eta_j) \right\}^T \{Q(t)\}_e. \tag{10.6}$$

$$\text{error extremization}: \quad mGWS^N = \int_\Omega \Psi_\beta \mathcal{L}^m(q^N)dx \equiv \{0\}, \forall \beta. \tag{10.7}$$

$$\text{matrix statement}: \quad mGWS^h + \theta TS \Rightarrow [JAC]\{\Delta Q\} = -\Delta t\{RES\}_n. \tag{10.8}$$

$$\text{assembly}: \quad [JAC] = S_e([JAC]_e), \{RES\} = S_e(\{RES\}_e). \tag{10.9}$$

$$\text{asymptotic error}: \quad \begin{array}{c} \left\| e^h(n\Delta t) \right\|_E \leq C\ell_e^{2\gamma}\|data\|_{\Omega,\partial\Omega L2}^2 + C_t \Delta t^{f(\theta)}\|Q_0\|_E \\ \gamma \equiv \min(k, r-1), f(\theta) = (2,3), \text{ or} \\ \left\| e^h(n\Delta t) \right\|_E \leq C\ell_e^2 \int_{t_0}^{n\Delta t} \|T(x,\tau)\|_{H^{k+1}}^2 d\tau + C_t \Delta t^{f(\theta)}\|T_0\|_E \end{array} \tag{10.10}$$

$$\text{error spectra}: \quad U_\kappa(\lambda), G(\lambda) \Rightarrow f(\kappa, k, h, \Delta t, \theta, \beta, \gamma). \tag{10.11}$$

Delaying the mGWS^h $\Rightarrow$ mPDE derivation to later in this chapter, the generalization for GWS^N $\Rightarrow$ GWS^h +θTS to $n=2,3$ lies totally in implementation of the n-D FE trial space bases $\{N_k(\zeta_\alpha)\}$ or $\{N_k^+(\eta_j)\}$, completely detailed in Chapter 7. The process generating the GWS^h matrix statement for $\mathcal{L}(q)$ with $\ell(q)$ is

$$GWS^N \equiv \int_\Omega \Psi_\beta(\mathbf{x})\mathcal{L}(q^N)d\tau \equiv 0 \Rightarrow GWS^h$$

$$GWS^h = S_e \left[\begin{array}{c} \int_{\Omega_e} \{N_k\}\left(\frac{\partial q_e}{\partial t} + u_j\frac{\partial q_e}{\partial x_j} - s\right)d\tau \\ + \int_{\Omega_e} \frac{\partial\{N_k\}}{\partial x_j}\frac{1}{Pa}\frac{\partial q_e}{\partial x_j} + \int_{\partial\Omega_e \cap \partial\Omega_R} \{N_k\}\frac{Pb}{Pa}\frac{\partial q_e}{\partial x_j}\hat{n}_j\,d\sigma \end{array} \right] \tag{10.12}$$

$$= S_e([MASS]_e \frac{d\{Q\}_e}{dt}$$

$$+ ([UVEL]_e + [DIFF]_e + [BC]_e)\{Q\}_e - \{b\}_e)$$

$$= [MASS]\frac{d\{Q\}}{dt} + \{RES\} \equiv \{0\}$$

which differs from $n=1$ GWS^h only in [VEL] becoming n-dimensional [UVEL]. GWS^h constitutes a matrix ODE system describing time evolution of the FE approximation DOF matrix $\{Q(t)\}$. Any ODE algorithm is applicable; selecting the θ-implicit family

GWS^h + θTS for all n generates the algebraic equation system

$$[JAC]\{\Delta Q\} = -\Delta t\{RES\}_n \tag{10.13}$$

the solution of which enables the DOF matrix update

$$\{Q(t_n + \Delta t)\} \equiv \{Q\}_{n+1} = \{Q\}_n + \{\Delta Q\}. \tag{10.14}$$

The global *jacobian* and *residual* matrices [JAC] and {RES} remain formed via *assembly* of element-level operations over all Ω_e of Ω^h. As detailed in Chapter 9, for element matrix generic labels "M" and "N" for Ω_e spanning n and $n-1$ dimensions, and for velocity vector u_j handled as element *nodal* data with DOF {UJ}, the element-level *essence* of GWS^h + θTS formation for any n is

$$[JAC]_e = [M\,200]_e + \theta\Delta t\left(\{UJ\}_e^T[M300J]_e + [M2KK]_e + [N200]_e\right). \tag{10.15}$$

$$\{RES\}_e = \Delta t\left(\{UJ\}_e^T[M300J]_e + [M2KK]_e + [N200]_e\right)\{Q_n\}_e - \{b\}_e. \tag{10.16}$$

$$\{b\}_e = \left([M200]_e\{SRC\}_e + [N200]_e\{QR\}_e - [N200]_e\{F \cdot \hat{\mathbf{n}}\}_e\right)_{n+\theta}. \tag{10.17}$$

By convention M $\Rightarrow$ (A,B,C) for $n = 1,2,3$ while for Robin/Neumann BCs N $\Rightarrow$ (A,B) as $\partial\Omega$ appropriate. Subscript e on a matrix indicates that metric data operations are involved and repeated indices J, K require summation on $1 \leq (J, K) \leq n$.

10.3 Matrix Library Additions, Templates

GWS^h + θTS coding remains clearly visualized via template statements detailing element-dependent operations altering [M . . .]$_e$ and [N . . .]$_e$ to the element-*indepen*dent library matrices. The laplacian, Robin BC, and data-interpolation matrices [M2KK]$_e$, [N200]$_e$, and [M200]$_e$ are fully detailed in Chapter 7.

The sole new element matrix is for multidimensional fluid convection. Velocity data are typically available as nodal DOF requiring the hypermatrix formulation, equations (10.15) and (10.16). For any completeness degree k FE trial space basis implementation the element matrix statement for $[\mathbf{UVEL}]_e\{Q\}_e$ in equation (10.12) is

$$
\begin{aligned}
[\mathbf{UVEL}]_e\{Q\}_e &\equiv \int_{\Omega_e} \{N_k\}\, u_j \frac{\partial q^h}{\partial x_j}\, d\tau \\
&= \int_{\Omega_e} \{N_k\}\{UJ\}_e^T\{N_k\}\frac{\partial\{N_k\}^T}{\partial x_j}\, d\tau\{Q\}_e \\
&= \{UJ\}_e^T\int_{\hat{\Omega}_e} \{N_k\}\{N_k\}\frac{\partial\{N_k\}^T}{\partial\eta_k}\left(\frac{\partial\eta_k}{\partial x_j}\right)_e \det_e d\eta\{Q\}_e \\
&= EKJ_e\{UJ\}_e^T[M300K]\{Q\}_e, 1 \leq (K, J) \leq n
\end{aligned}
\tag{10.18}
$$

In equation (10.18) the coordinate transformation $\partial \eta_k / \partial x_j$ (or $\partial \zeta_\alpha / \partial x_j!$) to $\hat{\Omega}_e$ aligns the velocity vector u_j with the first-order spatial derivative matrix library [M300K], a set of n element-*independent* *hyper*matrices. For equation (10.18) $k > 1$ TP basis implemented, Gauss quadrature is required and the matrix library then contains the element-dependent set $[M300KE]_e$.

The template definition for n-D GWS^h + θTS (10.13)–(10.18) is

$$\{WS\}_e = (\text{const}) \begin{pmatrix} \text{elem} \\ \text{const} \end{pmatrix}_e \left\{ \begin{matrix} \text{distr} \\ \text{data} \end{matrix} \right\}_e^T (\text{metric data; det})[\text{Matrix}] \left\{ \begin{matrix} Q \text{ or} \\ \text{data} \end{matrix} \right\}_e. \tag{10.19}$$

For the $k = 1$ TP basis implementation and introducing compressed notation for equation (10.18) J summation in product $\{UJ\}(EKJ)$, that is, $\{U1,U2,U3\}(10,20,30;0) \equiv \{U1 \times 10 + U2 \times 20 + U3 \times 30\}(;0)$, etc, the $n = 3$ algorithm template is the combination of n-D steady template (7.75) with the n-D expansion on equation (9.50) yielding

$$
\begin{aligned}
\{RES\}_e \;=\; & ([UVEL]_e + [DIFF]_e + [BC]_e)\{Q_n\}_e - \{b(data)\}_e \\
=\; & (\;)(\;)\{U1, U2, U3\}(10, 20, 30; 0)[C3001]\{Q\} \\
& + (\;)(\;)\{U1, U2, U3\}(40, 50, 60; 0)[C3002]\{Q\} \\
& + (\;)(\;)\{U1, U2, U3\}(70, 80, 90; 0)[C3003]\{Q\} \\
& + (1/PA)(\;)\{\;\}(EJI, EKI; -1)[C2JK]\{Q\} \\
& + (PB/PA)(\;)\{\;\}(; 1)[B200](\{Q\} - \{QR\}_{n+\theta}) \\
& - (\;)(\;)\{\;\}(; 1)[C200]\{SRC\}_{n+\theta} \\
& + (\;)(\;)\{\;\}(; 1)[B200]\{FN\}_{n+\theta}
\end{aligned}
\tag{10.20}
$$

$$
\begin{aligned}
[JAC]_e \;=\; & (\;)(\;)\{\;\}(; 1)[C200][\;] \\
& + (TDT)(\;)\{U1, U2, U3\}(10, 20, 30; 0)[C3001][\;] \\
& + (TDT)(\;)\{U1, U2, U3\}(40, 50, 60; 0)[C3002][\;] \\
& + (TDT)(\;)\{U1, U2, U3\}(70, 80, 90; 0)[C3003][\;] \\
& + (TDT/PA)(\;)\{\;\}(EJI, EKI; -1)[C2JK][\;] \\
& + (TDT, PB/PA)(\;)\{\;\}(; 1)[B200][\;]
\end{aligned}
\tag{10.21}
$$

Recall the expansion of the $[C2JK]$ template metric data string contributions in equations (10.20) and (10.21) is detailed in Table 7.3.

Should equations (10.1) and (10.2) define a nonlinear statement, for $p \geq 0$ the iteration index the Newton algorithm for GWS^h + θTS is

$$[JAC]\{\delta Q\}^{p+1} = -\{FQ\}^p \tag{10.22}$$

with {FQ} the placeholder for $GWS^h + \theta TS$, that is,

$$\{FQ\} = [MASS]\{\Delta Q\} + \Delta t(\theta\{RES\}_{n+1} + (1 - \theta)\{RES\}_n). \tag{10.23}$$

The alteration to equation (10.20) to implement equation (10.23) is minor, as is the generation of problem-specific nonlinear contributions to $[JAC]_e$ using calculus. Alteration of template (10.20) and (10.21) for an NC basis implementation requires modification of the metric data string in the $[C2JK]_e$ term *only*, recall Table 7.1, and is a suggested exercise.

10.4 *m*PDE Galerkin Weak Forms, Theoretical Analyses

The theoretical issues of accuracy, asymptotic convergence, and stability are not fundamentally altered in progression to n-dimensional time-dependence with convection. As specified in equation (10.10), for Pa finite and for smooth data the asymptotic convergence rate under solution-adapted *regular* mesh refinement (keeping element aspect ratios constant as Ω^h is refined) depends on FE basis degree k. Conversely, this dependence on k vanishes (as higher order) in the PDE nondiffusive limit $Pa^{-1} = 0$, whereupon convergence rate is quadratic for all k.

The n-D generalization of the mPDE process introduced in Chapter 9 is tensor invariant and generates a unified theoretical framework for $GWS^h \Rightarrow mGWS^h$ performance optimization. As with the $n = 1$ exercise, the starting point is the nondiffusive, source-free flux vector form of equation (10.1)

$$\mathcal{L}(q) = \frac{\partial q}{\partial t} + u_j\frac{\partial q}{\partial x_j} = 0 \equiv q_t + \frac{\partial f_j}{\partial x_j}, \quad f_j \equiv u_j q, \tag{10.24}$$

where subscript t denotes differentiation by time. The fourth-order explicit time Taylor series remains

$$q^{n+1} \equiv q^n + \Delta t q_t^n + \tfrac{1}{2}\Delta t^2 q_{tt}^n + \tfrac{1}{6}\Delta t^3 q_{ttt}^n + O(\Delta t^4) \tag{10.25}$$

and superscripts denote time level. The time derivatives in equation (10.25) can be manipulated using equation (10.24) to

$$
\begin{aligned}
q_t &= -\frac{\partial f_j}{\partial x_j} \\[2mm]
q_{tt} &= -\frac{\partial}{\partial t}\left(\frac{\partial f_j}{\partial x_j}\right) = -\frac{\partial}{\partial x_j}\left(\frac{\partial f_j}{\partial t}\right) \\[2mm]
&= -\frac{\partial}{\partial x_j}\left(\frac{\partial f_j}{\partial q}\frac{\partial q}{\partial t}\right) = +\frac{\partial}{\partial x_j}\left(\frac{\partial f_j}{\partial q}\frac{\partial f_j}{\partial x_k}\right) \\[2mm]
q_{ttt} &= \frac{\partial}{\partial t}\left[-\frac{\partial}{\partial x_j}\left(\frac{\partial f_j}{\partial q}\frac{\partial q}{\partial t}\right) \text{ or } +\frac{\partial}{\partial xy}\left(\frac{\partial f_j}{\partial q}\frac{\partial f_j}{\partial x_k}\right)\right] \\[2mm]
&= +\frac{\partial}{\partial x_j}\left(\frac{\partial f_j}{\partial q}\frac{\partial}{\partial x_k}\left(\frac{\partial f_k}{\partial q}\frac{\partial q}{\partial t}\right)\right) = -\frac{\partial}{\partial x_j}\left(\frac{\partial f_j}{\partial q}\frac{\partial}{\partial x_k}\left(\frac{\partial f_k}{\partial q}\frac{\partial f_l}{\partial x_l}\right)\right).
\end{aligned}
\tag{10.26}
$$

The terminal lines for q_{tt} and q_{ttt} in equation (10.26) represent two completely equivalent expressions. As linear combinations are valid, introduce the parameter set $(\alpha, \beta, \gamma, \delta)$ as multipliers. Embedding the signs in equation (10.26) therein, substitute these terms into equation (10.25), then divide through by Δt. Writing the resultant TS as homogenous yields

$$\frac{q^{n+1} - q^n}{\Delta t} + \frac{\partial f_j}{\partial x_j} - \frac{\Delta t}{2} \frac{\partial}{\partial x_j} \left(\alpha \frac{\partial f_j}{\partial q} \frac{\partial q}{\partial t} + \beta \frac{\partial f_j}{\partial q} \frac{\partial f_k}{\partial x_k} \right)$$

$$- \frac{\Delta t^2}{6} \frac{\partial}{\partial x_j} \left(\gamma \frac{\partial f_j}{\partial q} \frac{\partial}{\partial x_k} \left(\frac{\partial f_k}{\partial q} \frac{\partial q}{\partial t} \right) + \delta \frac{\partial f_j}{\partial q} \frac{\partial}{\partial x_k} \left(\frac{\partial f_k}{\partial q} \frac{\partial f_l}{\partial x_l} \right) \right) + \text{TE} = 0$$

(10.27)

where TE denotes truncation error.

As stated in Chapter 9, the kinetic flux vector f_j is a function of q not x_j. Hence, everywhere in equation (10.27) insert

$$\frac{\partial f_j}{\partial x_j} = \frac{\partial f_j}{\partial q} \frac{\partial q}{\partial x_j} = u_j \frac{\partial q}{\partial x_j}.$$

(10.28)

Then take the limit as $\Delta t \Rightarrow \varepsilon > 0$, which retains the higher order terms. Recognizing these first two terms define equation (10.24), the n-D *modified* PDE (*m*PDE) replacement for equation (10.1) to order Δt^3 is

$$\mathcal{L}^m(q) \equiv \mathcal{L}(q) - \frac{\Delta t}{2} \frac{\partial}{\partial x_j} \left[\alpha u_j \frac{\partial q}{\partial t} + \beta u_j u_k \frac{\partial q}{\partial x_k} \right]$$

$$- \frac{\Delta t^2}{6} \frac{\partial}{\partial x_j} \left[\gamma u_j u_k \frac{\partial^2 q}{\partial x_k \partial t} + \delta u_j u_k \frac{\partial}{\partial x_k} \left(u_l \frac{\partial q}{\partial x_l} \right) \right]$$

(10.29)

Implied in equation (10.29) is the given velocity field u_j is divergence free, recall conservation principle DM for an incompressible flow (2.26). The $\mathcal{L}(q)$ appearing in equation (10.29) is now replaced with equation (10.1) in completeness. Comparing equations (10.29) and (9.44) confirms how elegantly vector calculus generalizes the $n = 1$ formulation while retaining full $(\alpha, \beta, \gamma, \delta)$ arbitrariness.

Table 10.1 extracts from Table 9.1 independently theorized weak form algorithms for equation (10.1) that can be categorized as employing versions of n-D *m*PDE (10.29). Absent the δ term, used only with characteristic methods, the listed $\alpha, \beta, \gamma, \theta$ definitions were individually derived to improve either phase accuracy or stability. Additionally, except for optimal γ *m*GWS^h, only $k = 1$ basis equivalents were derived.

The Fourier stability theory highlighted in Chapter 9.7 can quantify spectral performance alterations intrinsic to weak formulations for *m*PDE (10.29). Recalling

Table 10.1 Select weak form algorithms recoverable in *m*PDE (10.29)

Name	Weak form algorithms on *m*PDE (10.29)			
	α	β	γ	θ
GWS	0	0	0	Arbitrary
Crank Nicolson	0	0	0	0.5
Taylor–Galerkin	0	1	1	0.0
Petrov–Galerkin	Sign($\mathbf{u}$)	Sign($\mathbf{u}$)	0	Arbitrary
Raymond–Garder	$15^{-1/2}$	$15^{-1/2}$	0	0.5
Least Squares	2θ	2θ	0	arbitrary
*m*GWS, optimal γ	0	0	−0.5	arbitrary
*m*GWS, optimal β	0	Pa$h^2/12$		steady

Chapter 8 and dividing through by Δx, a suggested exercise requests confirming the $k=1$ basis GWS^h algorithm *stencil* for equation (10.1) is

$$\frac{1}{6}\frac{d}{dt}\left[Q_{j-1}+4Q_j+Q_{j+1}\right]+\frac{u}{2\Delta x}\left[-Q_{j-1}+Q_{j+1}\right]$$
$$+\frac{1}{\text{Pa}\Delta x^2}\left[-Q_{j-1}+2Q_j-Q_{j+1}\right]=0$$

(10.30)

The discrete Fourier representation $q_\kappa^h(x \Rightarrow j\Delta x, t) = e^{i\kappa(j\Delta x - \tilde{U}t)}$ is shown in (9.37) and the output is *mode velocity* $\tilde{U}(\kappa, \Delta x) \equiv \tilde{U}_{\text{real}} + i\tilde{U}_{\text{imag}}$, a *complex function*. Substituting equation (10.30) into equation (9.37), followed by shifting therein spatial indices $j \pm 1$ to j to extract the mode shape (9.37), generates the matrix statement $[\bullet]q(x, t_0)e^{i\kappa(j\Delta x - \tilde{U}t)} = 0$ [1, Chapter 5].

Since the matrix scalar multiplier does not vanish then [·] must equal zero. The fact that the real and imaginary variables therein are *linearly independent* enables generating the mode velocity complex solution components

$$\tilde{U}_{\text{real}} = u\left[1 - \frac{1}{180}(\kappa\Delta x)^4 + O(\kappa\Delta x)^6\right]$$
$$\tilde{U}_{\text{imag}} = u\left[-\frac{\kappa^2}{\text{Pa}} + O(\kappa\Delta x)^5\right]$$

(10.31)

Thus the $k=1$ basis GWS^h algorithm for equation (10.1) propagates solution Fourier content to $O(\kappa\Delta x)^4$ of the input speed u and will phase-selectively diffuse this content to $O(\kappa^2/\text{Pa})$. Note that $m \equiv \kappa\Delta x = 2\pi\Delta x/\lambda$ defines the mesh *nondimensional wavenumber*.

The Raymond–Garder [2] weak form algorithm for equation (10.1) for $1/\text{Pe} \equiv 0$ is recovered in equation (10.29) for the Table 10.1 definitions *upon* replacing the Δt multiplier with the local timescale $(\ell_e/|u_e|)$. Their theoretical goal was improving the $O(m^4)$ GWS^h Fourier mode velocity, accomplished (in the mGWS^h context) by alteration of equation (10.30) to

$$\frac{d}{dt}\left[\frac{1}{6}\left(Q_{j-1} + 4Q_j + Q_{j+1}\right) + \frac{\alpha \hat{u}}{2}\left(Q_{j-1} - Q_{j+1}\right)\right]$$

$$+ \frac{u}{2\Delta x}\left[-Q_{j-1} + Q_{j+1}\right] + \frac{\beta|u|}{\Delta x}\left[-Q_{j-1} + 2Q_j - Q_{j+1}\right] = 0 \tag{10.32}$$

Comparing equations (10.30) and (10.32), the α term with velocity unit vector $\hat{u}$ augments the time derivative while the β term, obviously diffusive, replaces $1/\text{Pe}$ with $\beta|u|/\Delta x$.

Substituting equation (9.37) into equation (10.32), followed by spatial index shifts $j \pm 1$ to j and defining $\alpha \equiv \beta$, solving the resultant matrix statement generates the Fourier mode velocity complex solution components

$$\tilde{U}_{\text{real}} = u\left[1 + \left(-\frac{1}{180} + \frac{\beta^2}{12}\right)(\kappa\Delta x)^4 + O(\kappa\Delta x)^6\right]$$

$$\tilde{U}_{\text{imag}} = |u|\left[-\frac{\beta(\kappa\Delta x)^3}{12} + O(\kappa\Delta x)^5\right] \tag{10.33}$$

An $O(m^2)$ fidelity improvement in $\tilde{U}_{\text{real}}$ compared to GWS^h results for the *optimal* choice $\beta \equiv 15^{-1/2}$. Attendant is an $O(m^3)$ *artificial* diffusion mechanism replacing $1/\text{Pa}$ in equation (10.31) with $m^3/12(15)^{1/2}$.

The optimal γ mGWS^h algorithm does not replace Δt with the local timescale. For the assumption $1/\text{Pa} \equiv 0$ the stencil is

$$\frac{1}{6}\frac{d}{dt}\left[\left(Q_{j-1} + 4Q_j + Q_{j+1}\right) + \frac{\gamma\Delta t^2 u^2}{\Delta x^2}\left(-Q_{j-1} + 2Q_j - Q_{j+1}\right)\right]$$

$$+ \frac{u}{2\Delta x}\left[-Q_{j-1} + Q_{j+1}\right] = 0 \tag{10.34}$$

and the resulting mode velocity complex solution is

$$\tilde{U}_{\text{real}} = u\left[1 + (\gamma + 0.5)(\kappa\Delta x)^4 + O(\kappa\Delta x)^6\right]$$

$$\tilde{U}_{\text{imag}} = u\left[O(\kappa\Delta x)^5\right] \tag{10.35}$$

The optimal choice is $\gamma \equiv -0.5$, hence mGWS^h accomplishes the $O(m^6)$ goal of Raymond–Garder absent the artificial diffusion.

The terminal comparison is for the classis Crank–Nicolson FD algorithm with stencil

$$\frac{dQ_j}{dt} + \frac{u}{2\Delta x}\left[-Q_{j-1} + Q_{j+1}\right]$$

$$+ \frac{1}{Pa\Delta x^2}\left[-Q_{j-1} + 2Q_j - Q_{j+1}\right] = 0 \qquad (10.36)$$

Proceeding through the Fourier modal solution exercise generates the mode velocity complex solution components

$$\tilde{U}_{real} = u\left[1 - \frac{1}{6}(\kappa\Delta x)^2 + O(\kappa\Delta x)^4\right]$$

$$\tilde{U}_{imag} = u\left[-\frac{\kappa^2}{Pa} + O(\kappa\Delta x)^5\right] \qquad (10.37)$$

which possesses the lowest $O(m^2)$ spectral propagation accuracy of the various detailed algorithms.

The last Table 10.1 entry is the *optimal β mGWS^h* algorithm for generating solutions to steady (10.1). Algorithm derivation requires use of infinite order Taylor series in space rather than time, [3]. Of pertinence to this text, the $β$ differential term in equation (10.29) is identified by the TS theory as the *optimal mPDE* replacement for steady equation (10.1)

$$\mathcal{L}(q) = u_j\frac{\partial q}{\partial x_j} - \frac{1}{Pa}\frac{\partial^2 q}{\partial x_j^2} - \frac{Pah^2}{12}\frac{\partial}{\partial x_j}\left[u_ju_k\frac{\partial q}{\partial x_k}\right] - s = 0,\ on\ \Omega \subset \mathfrak{R}^n. \qquad (10.38)$$

For $k=1$ basis implementation of $mGWS^h$ for equation (10.38), and for *sufficiently large* Pa, the theory predicts *the* significant improvement in the asymptotic error estimate (10.10)

$$\|e^h\|_E \leq C\ell_e^{2\gamma}\|data\|^2_{\Omega,\partial\Omega L2},\ \gamma \equiv min(k(=1)+1 = 2,\ r-1). \qquad (10.39)$$

Hence, for *smooth data k=1* basis $mGWS^h$ solutions are theoretically predicted to exhibit the fourth-order asymptotic convergence rate previously restricted to $k=2$ basis GWS^h! The concomitant bandwidth reduction in the terminal matrix statement is an immense contribution.

10.5 Verification, Benchmarking, and Validation

The FE basis discrete implementation $GWS^h/mGWS^h + \theta TS$ for the n-dimensional unsteady convection–diffusion PDE + BCs + IC system (10.1)–(10.4) with equation (10.29) is complete. At issue is how to quantify performance in comparison to the developed theories for suitable *data* definitions.

Table 10.2 Verification, benchmarking, and validation

Verification	Compare simulation to *analytical* solutions
	Peclet problem
	Traveling wave forms
	Rotating cone
Benchmarking	Compare simulation to *quality* numerical data
	Gaussian plume
	Burgers equation
Validation	Compare simulation to *quality* experimental data
	Genuine geometries
	Realistic diffusion models (turbulence)

The computational fluid dynamics (CFD) community established a well-defined protocol for this in terms of *verification*, *benchmark*, and *validation* problem statements [4]. Table 10.2 describes and gives examples in each category. The intent is to sequentially emulate situations in *genuine*, real-world simulation definitions in a progression of complexity enabling precise performance quantification.

10.6 Mass Transport, the Rotating Cone Verification

A definitive unsteady scalar transport verification that admits precise quantification of dispersion and/or artificial diffusion error mechanisms is the *rotating cone*. It is the $n = 2$ successor to the $n = 1$ traveling wave, Chapter 9.6, using an imposed *rotational* velocity field generating distributed Courant number.

Originally devised by the atmospheric sciences community to assess CFD transport algorithm fidelity, [5], sought is the solution to $n = 2$ non-diffusive ($Pa^{-1} = 0$) PDE (10.1), alternatively various $mPDE$ (10.29). The solution domain Ω is the unit square 80 km on the side. The imposed solid body rotation velocity vector distribution is

$$\mathbf{u}(x, y) \Rightarrow \mathbf{u}(r, \theta) = r\omega\hat{\mathbf{e}}_\theta \tag{10.40}$$

which circulates the axially-rotated gaussian IC around the domain interior, Figure 10.2, with speed of rotation set by ω. The validation statement domain discretization is $M = 32 \times 32$ uniform, which sets h, producing the Courant number distribution $0 < C = r\omega\Delta t/h \leq C_{max} \leq 1$.

From equation (10.40), each cartesian boundary segment $\partial\Omega$ of domain Ω possesses equal partitions of inflow and outflow. Along each inflow segment the mass fraction BC is $q(\mathbf{x}_b) = q_b \equiv 0$, while homogenous Neumann $\hat{\mathbf{n}} \cdot \nabla q = 0$ is the appropriate BC on each outflow segment. The IC is $q(\mathbf{x}, t_0) = q_0(\mathbf{x}) = 0$ throughout $\Omega \cup \partial\Omega$ except for DOF located inside the rotated gaussian IC, Figure 10.2(a).

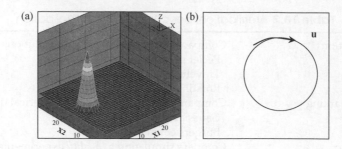

Figure 10.2 Rotating cone verification problem: (a) initial condition (IC), (b) sense of imposed motion (a colored version of this figure is available at www.wiley.com/go/baker/finite)

Recognizing nondiffusive $n = 2$ PDE (10.1) is linear hyperbolic, the analytical (characteristic) solution is the IC distribution $q_0(\mathbf{x})$ undergoes pure rotational translation. Thereby, $q_0(\mathbf{x}) \Rightarrow q(\mathbf{x} - \mathbf{u}t)$ is the *exact solution* at any time $t > t_0$, that is, the gaussian DOF interpolation is *exactly* preserved! The Fourier mode theory is precisely appropriate in accurately predicting comparative performance via algorithm order in non-D wavenumber m.

Figure 10.3 compares the solutions generated by $k = 1$ basis implementations of $WS^h + \theta TS$ algorithms listed in Table 10.1. The solution graphs are generated following the $n\Delta t$ time steps required for the analytical solution to *exactly* complete one circulation about the $x-y$ plane. The Courant number magnitude spans $0 < C \le 0.8$ with $C = 0.4$ at the centroid of q_0. Each graph legend states algorithm FD-equivalent *order of accuracy* in space $O(\Delta x^n)$, spectral $O(m^n)$, and time $O(\Delta t^n)$.

The Fourier theory *accurately predicts* relative performance improvement from finite difference (FD) to FE GWS^h to $mGWS^h$ $(\alpha, \beta, \gamma, \theta)$ parameter selections. The FD Crank–Nicolson solution, Figure 10.3(a), is totally destroyed by dispersion error. The $k = 1$ FE GWS^h solution, (b), is a significant improvement, followed next by $mGWS^h$ least squares $\theta = 0.5$, (c), then $mGWS^h$ Taylor–Galerkin $\alpha = 0$, $\beta = 1 = \gamma$, $\theta = 0$, (e), and finally the optimal $\gamma = -0.5$ $mGWS^h$ $\theta = 0.5$ solution, (f). Note the $mGWS^h$ least squares algorithm solution (d), generated using $\theta = 1$, clearly confirms Δt-induced artificial diffusion.

The .mpg files of these algorithm transient solutions are highly informative regarding visual characterization of dispersion error propagation. They may be accessed for viewing on your PC at the Wiley author website www.wiley.com/go/baker/finite.

Comparative performance in terms of algorithm nodal DOF maximum (IC $= 100$), extremum negative DOF (IC $= 0$) and dispersion error-induced phase lag in mesh units $\Delta x = h$ is summarized in Table 10.3. The $k = 1$ basis $\theta = 0.5$ $mGWS^h$ optimal-γ algorithm solution is clearly superior among those tested, with accuracy surpassing the $k = 2$ basis GWS^h solution! It is also *optimally efficient* in class with minimum bandwidth tridiagonal stencil. The $O(\Delta x^3, \Delta x^5)$ QUICK FV algorithms are not competitive in either efficiency or accuracy. Computer labs 10.1–10.2 support thorough assessment of $WS^h + \theta TS$ algorithm performance.

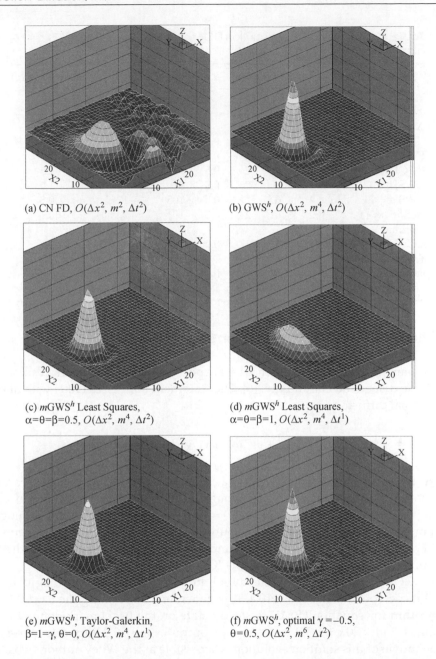

(a) CN FD, $O(\Delta x^2, m^2, \Delta t^2)$

(b) GWS^h, $O(\Delta x^2, m^4, \Delta t^2)$

(c) mGWS^h Least Squares,
$\alpha = \theta = \beta = 0.5$, $O(\Delta x^2, m^4, \Delta t^2)$

(d) mGWS^h Least Squares,
$\alpha = \theta = \beta = 1$, $O(\Delta x^2, m^4, \Delta t^1)$

(e) mGWS^h, Taylor-Galerkin,
$\beta = 1 = \gamma$, $\theta = 0$, $O(\Delta x^2, m^4, \Delta t^1)$

(f) mGWS^h, optimal $\gamma = -0.5$,
$\theta = 0.5$, $O(\Delta x^2, m^6, \Delta t^2)$

Figure 10.3 Select WS^h + θTS algorithm solutions, $k = 1$ basis, rotating cone following one (x, y) plane circulation, $C = 0.4$ at IC centroid: (a) CN FD, $O(\Delta x^2, m^2, \Delta t^2)$, (b) GWS^h, $O(\Delta x^2, m^4, \Delta t^2)$, (c) mGWS^h least squares, $\alpha = \theta = \beta = 0.5$, $O(\Delta x^2, m^4, \Delta t^2)$, (d) mGWS^h least squares, $\alpha = \theta = \beta = 1$, $O(\Delta x^2, m^4, \Delta t^1)$, (e) mGWS^h, Taylor-Galerkin, $\beta = 1 = \gamma$, $\theta = 0$, $O(\Delta x^2, m^4, \Delta t^1)$, (f) mGWS^h, optimal $\gamma = -0.5$, $\theta = 0.5$, $O(\Delta x^2, m^6, \Delta t^2)$ (a colored version of this figure is available at www.wiley.com/go/baker/finite)

Table 10.3 Rotating cone solution DOF extrema, select WS^h + θTS algorithms, $C = 0.4$ at IC centroid

Algorithm	FD order			θ	Q_{max}	Q_{min}	Phase lag ($n\Delta x$)
	Δx	m	Δt				
Analytical	—	—	—		100	0	0.0
CN FD	2	2	2	0.5	44	−28	>4(?)
FV QUICK 3	3	?	2	0.5	45	−8	2.5
FV QUICK 5	5	?	2	0.5	79	−13	1.5
GWS, $k = 1$	2	4	2	0.5	91	−23	1.0
GWS, $k = 2$	4	6	2	0.5	105	−17	0.5
mGWS, $k = 1$	2	6	2	0.5	101	−7	<0.1
Optimal $\gamma = -0.5$							
mGWS, $k = 1$	2	4	1	0.0	78	−6	<0.1
Taylor Galerkin							
mGWS, $k = 1$	2	?	2	0.5	75	−9	2.0
Least Squares			1	1.0	30	−10	<0.1

10.7 The Gaussian Plume Benchmark

The gaussian plume is a steady mass transport $n = 2, 3$ benchmark also from the atmospheric dispersion community, [6]. The statement addresses (10.1) with alteration for directional diffusion

$$\mathcal{L}(q) = \frac{\partial q}{\partial t} + \mathbf{u} \cdot \nabla q - \text{Pa}^{-1}\nabla \cdot \bar{k}\nabla q - s = 0. \tag{10.41}$$

The (non-D) diffusion coefficient is $\bar{k} \equiv \text{diag}[k_x, k_y, k_z]$ and $\text{Pa} = \text{ReSc}$ is the parameter. The domain Ω is rectangular cartesian elongated in the direction of imposed velocity vector $\mathbf{u} - \hat{\mathbf{i}} + 0\hat{\mathbf{j}} + 0\hat{\mathbf{k}}$. The BCs are $q(\mathbf{x}_b, t) = 0$ on the inflow plane and homogenous Neumann on all other $\partial\Omega$ segments. A source $s = s(\mathbf{x}_s)$, symmetrically distributed as a pyramid over the union of four elements and located just downstream from the inflow plane, pumps q into Ω at a steady rate. The IC is $q(\mathbf{x}, t_0) = 0$.

The $n = 2$ GWS^h gaussian plume steady solution for $\text{Pa} = 10$ and $\bar{k} \equiv \text{diag}[0, k_y = 1]$ is graphed in perspective in Figure 10.4. It is generated by running the $\theta = 1.0$ GWS^h + θTS algorithm for equation (10.41) to steady state on the benchmark-defined uniform cartesian $M = 14 \times 20\, k = 1$ TP basis mesh. Computer lab 10.3 supports assessments; the .mpg movie of this solution evolution is accessible at the Wiley author website.

The outflow plane q^h distribution, Figure 10.4, gives credence to the name *gaussian* for this benchmark. The key solution feature is GWS^h *weak enforcement* of the vanishing Neumann BC $\mathbf{n} \cdot \nabla q = 0$ on this boundary. The parent weak form theory clearly admits the solution requisite modestly nonvanishing normal derivative to occur! The lateral span of Ω is sufficient except near the outflow plane whereon weak enforcement of $\mathbf{n} \cdot \nabla q = 0$ is again (!) an accurate implementation.

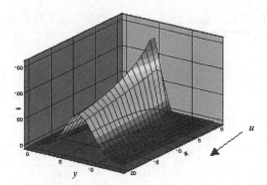

Figure 10.4 Gaussian plume steady GWS^h solution perspective looking upstream, $k = 1$ basis, $n = 2$ (a colored version of this figure is available at www.wiley.com/go/baker/finite)

For the benchmark definition Pa $= 10$ the $k = 1,2$ TP basis GWS^h steady solution non-zero DOF distributions on mirror symmetric half-planes for $\bar{k} \equiv \text{diag}\left[0, k_y = 1\right]_{n=2}$ and $\bar{k} \equiv \text{diag}\left[0, k_y = 1, k_z = 1\right]_{n=3}$ are detailed in Table 10.4. Within the solid line demarcations are the benchmark solution DOF for symmetry plane trajectory and outflow plane lateral ($\sim$ gaussian) distribution. The element DOF containing the source are denoted inside the boxed regions.

The significant feature in the $n = 2$ simulation is symmetry plane improvement of the $k = 2$ basis solution DOF matching the benchmark data. The $k = 1$ basis DOF are essentially displaced one node row downstream in comparison. Conversely, the $n = 3$ solutions are essentially in identical agreement with the benchmark. Both basis solutions for $n = 2$ evidence modest dispersion error-induced negative mass fraction DOF in the near source region. The $mGWS^h(\gamma)$ algorithm will eliminate these; the phenomena is absent in the $n = 3$ solution.

10.8 Steady *n*-D Peclet Problem Verification

The Chapter 9 $n = 1$ steady Peclet problem confirmed that large Pe induces extensive dispersion error into the GWS^h algorithm solution on nominal meshes M. This could be artificially diffused by augmenting equation (9.31) with the $mPDE$ β term, (9.44), or essentially eliminated by a precisely tailored solution-adapted mesh M.

A subsequent publication identified the steady flow replacement of the β coefficient producing the *n*-D $mPDE$ (10.38). The theory requirement is that Pa be *sufficiently large* to admit discarding of higher order terms in Pa^{-1}. For smaller levels, for example, Pa ≤ 100, many additional terms in the spatial TS-generated theory must be retained, [3]. Table 10.5 summarizes the reported uniform mesh-refinement study data for the $n = 2$ Peclet problem at Pa $= 10$ retaining all theory terms. The *a posteriori* data convergence clearly verifies the $O(h^4)$ theoretical prediction, (10.39).

As stated leading to equation (10.39), the steady *n*-D $k = 1$ basis optimal $mGWS^h$ theory is valid only for Pa $\gg 1$. This places no real restriction on problem class

Table 10.4 Gaussian plume GWS^h DOF, $k=1,2$ bases, $n=2,3$

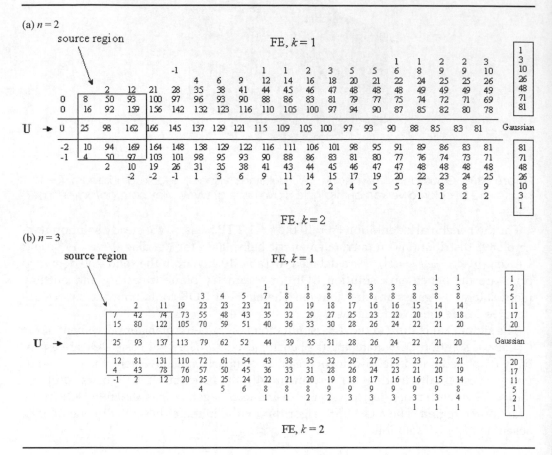

(a) $n=2$

FE, $k=1$

FE, $k=2$

(b) $n=3$

FE, $k=1$

FE, $k=2$

pertinence as Pa always contains Re, typically $O(10^4)$ or larger in practice. Viewing equation (10.38), for theory-induced diffusion to not dominate genuine diffusion requires $Pa(h^2/12) < Pa^{-1}$, preferably $\ll Pa^{-1}$. It is obvious that *solution adapted h*, the measure *distribution* on Ω^h, plays a critical role in theory implementation computing practice.

Table 10.5 Convergence, steady $mGWS^h$ $k=1$ algorithm, $n=2$, Pe $=10$

M	$\|Q\|_E \times 10^{-5}$	$\|\Delta Q\|_E \times 10^{-5}$	Slope
8×8	0.468156		
16×16	0.448714	0.01944	
32×32	0.447985	0.00073	4.74
64×64	0.447945	0.00004	4.19

The definitive verification is the n-D steady, source-free Peclet problem, $n = 1$ defined in Chapter 6. Computer lab 10.4 defines a uniform, then solution-adapted regular mesh-refinement study for the $n = 2$ statement. The imposed non-D velocity vector is $\mathbf{u} \equiv \hat{\mathbf{i}} + \hat{\mathbf{j}}$ and the sole Dirichlet BC is the DOF at the domain corner opposite the confluence of the inflow boundary segments, $Q(x, y = 1) \equiv 1$. The BCs for equation (10.38) are $\{Q\} = \{0\}$ on the two inflow boundary segments and vanishing Neumann on the two outflow segments.

Solution Fourier content, dependent on Pa $\equiv$ Pe, is localized in wall-layers terminating at the Dirichlet DOF, recall Figure 9.3. For the computer lab 10.4 definition Pe $= 1000$, the $k = 1$ basis uniform $mGWS^h$ solution is never monotone, illustrated for $M = 20 \times 20$ in Figure 10.5(a). It can be rendered monotone *only* via a solution adapted

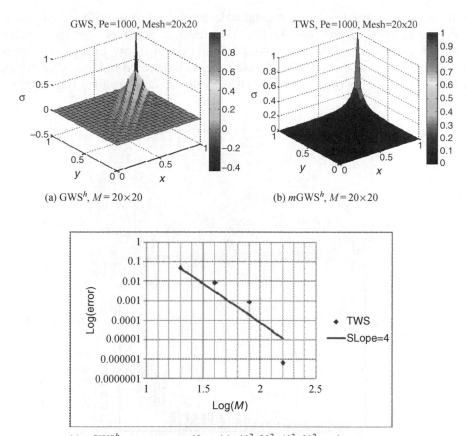

(a) GWS^h, $M = 20 \times 20$ (b) mGWS^h, $M = 20 \times 20$

(c) mGWS^h, convergence, uniform $M = 10^2, 20^2, 40^2, 80^2$ meshe

Figure 10.5 GWS^h/mGWS^h solutions, $n = 2$ Peclet problem, Pe $= 1000$: (a) GWS^h, $M = 20 \times 20$, (b) mGWS^h, $\gamma = -0.5$, $M = 20 \times 20$, (c) mGWS^h, $\gamma = -0.5$, asymptotic convergence, uniform $M = 10^2, 20^2, 40^2, 80^2$ meshes (a colored version of this figure is available at www.wiley.com/go/baker/finite)

mesh; however, determining the dual P_r distributions promoting essential monotonicity is a challenge (try it in the lab!).

Conversely, the optimal steady $mGWS^h$ algorithm solution is always monotone, either on coarse uniform or solution adapted meshes M. The comparative $k=1$ TP basis uniform $M = 20^2$ mesh steady $mGWS^h$ solution is graphed in Figure 10.5(b); it approaches engineering accuracy on uniform $M = 40^2$. One key output of this computer lab is the uniform mesh-refinement exercise will confirm the theory prediction of $O(h^4)$ convergence, Figure 10.5(c) with data label "TWS." The lab exercise will also confirm that uniform $M = 80^2$ solution accuracy is achievable on an $M = 20^2$ solution-adapted mesh.

10.9 Mass Transport, a Validated $n = 3$ Experiment

Validation requires quality experimental data for comparison with a computational engineering sciences prediction. In the fluid-thermal sciences, pertinent velocity-distribution experimental data are reported for buoyant flow in an $n = 3$ full-scale $(9 \times 9 \times 15\,\text{ft})$ HVAC laboratory, [7]. The lab setup is illustrated in Figure 10.6(a) [8]; for a flushing rate of 15 air changes/hour (ACH) velocity data confirm the supply duct

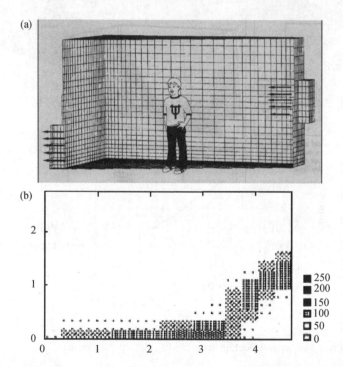

Figure 10.6 Full-scale HVAC laboratory experiment, cold air inflow entering a heated space, ACH = 15: (a) $9 \times 9 \times 15\,\text{ft}$ laboratory geometry, (b) supply flow trajectory and speed distribution (ft/min) experimental data on the supply duct symmetry plane (7,8)

cold inflow entering the initially heated laboratory drops precipitously to the floor, then spreads across it as a wall jet. Figure 10.6(b) is the confirmatory measured speed distribution raster graphic on the symmetry plane of the supply duct [7].

An $n = 3$ CFD prediction of this experiment is reported [8]. Expanding PDE + BCs + IC system (10.1)–(10.4) to state variable $\{q\} = \{u,v,w,\Theta\}^T$ the $\text{GWS}^h/m\text{GWS}^h + \theta\text{TS}$ $n = 3, k = 1$ TP trial space basis algorithm is published [9,10]. The ACH = 15 Reynolds number is Re $\approx 15{,}000$, for length scale the rectangular cross-section supply duct hydraulic diameter, Figure 10.6(a). The flowfield is thus not laminar, but not fully turbulent either, as the supply duct is entirely too short to admit developing a turbulent profile.

Consequently, the simulation employed a scalar parameter Pa^t as a highly simplified "turbulence" model that simply scales the convection–diffusion balance. In concert, the supply inflow velocity assumed a slug profile and boundary layer flow resolution on enclosure walls was discarded in favor of using wall tangency velocity BCs, the Euler flow approximation [9,10]. For ceiling heat flux from [7] and assuming adiabatic side wall and floor BCs, CFD prediction of $n = 3$ steady temperature and velocity vector distributions were generated for equations (10.1)–(10.4) altered to the mPDE form

$$\mathcal{L}(\{q\}) = \frac{\partial\{q\}}{\partial t} + \mathbf{u} \cdot \nabla\{q\} - \frac{1}{\text{Pa}}\nabla \cdot \left((1 + \text{Pa}^t)\nabla\{q\}\right) - s = \{0\}. \tag{10.42}$$

In equation (10.42), Pa $\Rightarrow$ Re, RePr = Pe respectively, for velocity and temperature, and Pr ≈ 0.7 for air. CFD simulations were conducted for $\text{Pa}^t = 14$, 29, and 149 using a cartesian TP basis $M = 26 \times 21 \times 32$ mesh (as limited by 1990's work station compute power). Very good *quantitative* agreement for supply flow speed distribution, trajectory, and floor impact coordinate with the experimental data, Figure 10.6(b), resulted only for $\text{Pa}^t = 14$. Use of the larger Pa^t overdiffused the CFD solutions confirming the supply flow is not "very turbulent."

Figure 10.7(a) graphs in perspective the $\text{Pa}^t = 14$ velocity field as DOF unit vectors, colored by speed, plotted on the supply duct symmetry plane and on planes directly adjacent to the inflow and outflow walls and the floor. Note the supply flow speed increases by a factor of two in accelerating to the floor. The companion temperature DOF distribution plotted on the same planes clearly visualizes the span of heated floor flow, Figure 10.7(b). No comparative experimental data exists for this prediction.

Having this *essentially* validated $n = 3$ velocity vector field enabled a mass transport simulation with the goal to determine if a pathogen, entering with the supply flow, would end up in locally concentrated pockets. Also, since $(1 + \text{Pa}^t)/\text{Pa} = (1 + 14)/{\sim}15{,}000 = {\sim}1000$ initiates a dispersive error mode, the computational experiment probed the $m\text{GWS}^h(\beta)$ algorithm effectiveness in moderating dispersion error for mass transport prediction in a practical $n = 3$ velocity field. (Note: the more appropriate optimal γ $m\text{GWS}^h$ algorithm was not yet derived.)

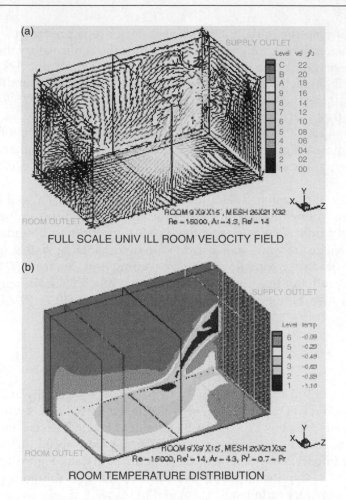

Figure 10.7 Incompressible-thermal Navier–Stokes GWS^h + θTS steady solution, Re = 15,000, Pa^t = 14, supply flow symmetry plane and first off-wall plane graphics: (a) velocity DOF unit vectors colored by speed, (b) temperature distribution (10) (a colored version of this figure is available at www.wiley.com/go/baker/finite)

For mass transport Pa ⇒ ReSc in equation (10.1), and both non-D groups have their "turbulent" counterparts. Hence, the unsteady mPDE(β) form for equation (10.1) is

$$\mathcal{L}(q) = \frac{\partial q}{\partial t} + \mathbf{u} \cdot \nabla q - \frac{1}{\text{ReSc}} \nabla \cdot \left(\left(1 + \frac{Re_t}{Sc^t}\right) \nabla q \right)$$
$$- \frac{\beta \Delta t^2}{2} \frac{\partial}{\partial x_j} \left[u_j u_k \frac{\partial q}{\partial x_k} \right] - s = 0 \tag{10.43}$$

where q is mass fraction of the inert (nonreacting) pathogen. The source s is a continuous emitter in the supply duct with DOF $\{Q(t)\} = \{1.0\}$. The diameter of the nominally circular source s is half the width of the duct and centrally located at the supply duct juncture with the room wall. All BCs are homogeneous Neumann except on the supply duct inflow plane, upstream of the source, wherein the Dirichlet BC is $\{Q(t)\} = \{0\}$. This is also the IC definition except for the DOF in the source.

For the computational experiments equation (10.43) was solved time-accurate ($\theta = 0.5$) over the time $t_0 = n\Delta t$ sufficient for the mass fraction plume to just reach the room plane interface with the exhaust duct. The velocity and temperature fields were held constant at their steady solution, Figure 10.7, for all simulations [10].

As anticipated, the unsteady $k = 1$ basis $GWS^h + \theta TS$ solution is totally polluted by dispersion error. This error is manifested *modulo* $2\Delta x$, hence in the n-D graphical presentation it appears as (color) diamond patterns on the scale of the mesh. This is clearly visible in the perspective graph of mass fraction distributions on the cited DOF planes at $t - t_0 = 25$ s, Figure 10.8(a). The mass fraction range is by definition $0 \le q \le 1.0$, and herein the DOF range is $-1.59 \le \{Q(n\Delta t)\} \le 2.36$! Note, however, there does appear clear indication of a buildup of pathogen local to the floor corner beneath the supply duct.

The unsteady $k = 1$ basis $mGWS^h + \theta TS$ algorithm with $\beta = 0.9$ is significantly less dispersion error polluted, Figure 10.8(b), and retains prediction of the floor corner mass fraction concentration at $t - t_0 = 25$ s. The mass fraction DOF range is $-0.75 \le \{Q(n\Delta t)\} \le 1.45$. Running $mGWS^h + \theta TS$ ($\beta \equiv 0$) and increasing the "turbulent" balance parameter Re^t/Sc^t by an order of magnitude generates an essentially monotone solution with mass fraction DOF range $-0.11 \le \{Q(n\Delta t)\} \le 1.03$, Figure 10.9(a). Due to the added diffusion engendered by $(1 + Pa^t)/Pa = 100$, no pathogen concentration at the floor corner under the supply duct exists.

As a final caveat on dispersion error and its control, a postprocess theory that can reconstruct an n-D dispersion error polluted discrete DOF distribution to absolutely monotone is reported [11]. Termed a *reconstruction algorithm* (RCN), the mathematical theory involves estimating solution Fourier content then clipping offending DOF without addition of diffusion.

The RCN theory alteration of the $mGWS^h + \theta TS$ $\beta = 0.9$ DOF solution for $(1 + Pa^t)/Pa = 1000$ at $t - t_0 = 25$ s is perspective graphed in Figure 10.9(b). The DOF span is precisely $0 \le \{Q(n\Delta t)\} \le 1.0$ and the floor corner extrema evident in Figure 10.8(b), the $mGWS^h(\beta = 0.9)$ solution is clearly preserved, as is the second predicted extrema near floor level at about 80% room span. Note the supply duct location of the pathogen source is clearly visible in the cited solution graphics.

10.10 Numerical Linear Algebra, Matrix Iteration

With the inexpensive compute capability resident in today's PCs, meshes up to $M = 10^6$ nodes, multiplied by algorithm DOF, are common and practical. With the weak form theory exposition complete, numerical linear algebra dominates solution strategies for $n = 3$ $GWS^h/mGWS^h + \theta TS$ algorithms. It is therefore appropriate to complete this chapter with material pertinent to select matrix iteration algorithms.

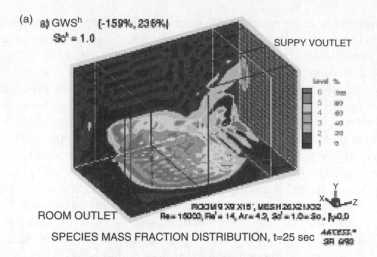

(a) a) GWS^h [-159%, 236%]
Sc^t = 1.0

SUPPY VOUTLET

ROOM OUTLET

SPECIES MASS FRACTION DISTRIBUTION, t=25 sec

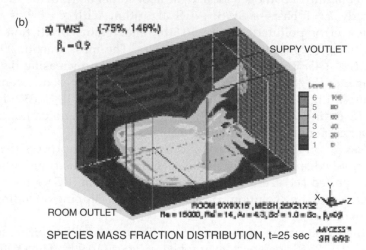

(b) a) TWS^t (-75%, 146%)
β, =0.9

SUPPY VOUTLET

ROOM OUTLET

SPECIES MASS FRACTION DISTRIBUTION, t=25 sec

Figure 10.8 Mass fraction transport predicted distributions, $t - t_0 = 25$ s, Re $= 15,000$, Pa$^t = 14$, supply symmetry plane and first off-wall planes DOF data: (a) GWS$^h + \theta$TS algorithm, (b) mGWS$^h + \theta$TS $\beta = 0.9$ algorithm, from (10) (a colored version of this figure is available at www.wiley.com/go/baker/finite)

The end result of GWS^h/mGWS$^h + \theta$TS algorithms for the linear problem statement (10.1)–(10.4) is the matrix statement

$$[\text{JAC}]\{\Delta Q\} = -\Delta t\{\text{RES}\}_n$$
$$[\text{JAC}] = [\text{MASS}] + \theta \Delta t([\text{UVEL}] + [\text{DIFF}] + [\text{BC}]) \quad . \qquad (10.44)$$
$$\{\text{RES}\}_n = ([\text{UVEL}] + [\text{DIFF}] + [\text{BC}])\{Q\}_n - \{\text{b(data)}\}$$

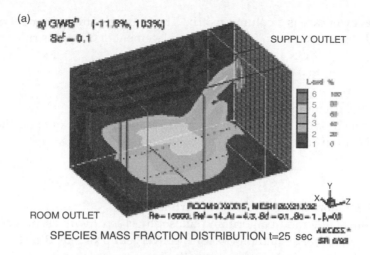

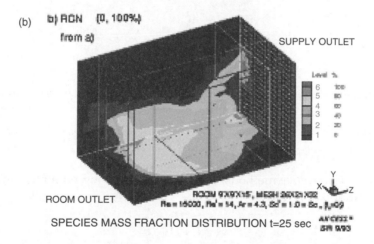

Figure 10.9 Mass fraction transport predicted distributions, $t - t_0 = 25\,\text{s}$, $\text{Re} = 15{,}000$, supply symmetry plane and first off-wall planes DOF data: (a) $\text{GWS}^h = \theta\text{TS}$ algorithm, $\text{Pa}^t = 149$, (b) $m\text{GWS}^h + \theta\text{TS}$ $\beta = 0.9$ algorithm DOF solution reconstruction via RCN theory, from (10) (a colored version of this figure is available at www.wiley.com/go/baker/finite)

Sought are *approximations* to [JAC] and {RES} that compute fast but require many iterations to converge. There exist two distinct iteration strategies; one resolves [JAC] into a sum of matrices with the second a sequence of matrix products [12].

This section's development reverts to standard textbook matrix notation, hence the requirement is to solve $Ax = b$ for A not singular. A capital letter denotes a square matrix while lower case is a column matrix. The Cramer's rule statement of solution is $x = A^{-1}b$ and one general form for iterative replacement of Cramer's rule is

$$x^p = f^p(A, b, x^{p-1}, x^{p-2}, \ldots) \tag{10.45}$$

for p denoting the iteration index.

The iteration (10.45) is called *stationary* if $f^p(\cdot)$ is independent of p. Further, the iteration is called *linear* if equation (10.45) can be written as

$$x^p = N^p x^{p-1} + M^p b \tag{10.46}$$

for N and M iteration matrices related to partitions of A.

The fundamental iteration algorithm philosophy is *speed* with assurance the process *converges*. The proof that equation (10.46) is convergent proceeds directly. The error in any iterate is $e^p \equiv x^p - x$ and via direct substitutions

$$\begin{aligned} e^p &= x^p - A^{-1}b \\ &= N^p x^{p-1} + M^p b - A^{-1}b \\ &= N^p e^{p-1}. \end{aligned} \tag{10.47}$$

Thus, convergence depends strictly on the iteration matrix N, hence ultimately on the initial error e^0

$$e^p = N^p e^{p-1} = N^p N^{p-1} e^{p-2} = \ldots N^p N^{p-1} \ldots N^1 e^0. \tag{10.48}$$

The family of *point iterative* methods, that is, those requiring no matrix solves, centers around the partition of A into

$$A = L + D + U \tag{10.49}$$

where L, D, and U are each square matrices having the same elements as A below the main diagonal (L), on the main diagonal (D), and above it (U). Thus, both L and U are *triangular* matrices while D is *diagonal*.

The familiar iteration algorithms belonging to the point iterative linear-stationary class are *Picard (Jacobi)*, *Gauss–Seidel*, and *successive over-relaxation* (SOR). The *Jacobi* definition for equation (10.46) is

$$x^p = -D^{-1}(L + U)x^{p-1} + D^{-1}b \tag{10.50}$$

hence $N^p \Rightarrow N \equiv -D^{-1}(L + U)$ and $M^p \Rightarrow M \equiv D^{-1}$. Thereby, no matrix solving is required since the elements of the inverse of the diagonal matrix D are simply $1/d_{ii}$, with d_{ii} the diagonal elements of A.

The *Gauss–Seidel* iteration algorithm definition for equation (10.46) is

$$x^p = -(L - D)^{-1}Ux^{p-1} + (L + D)^{-1}b \qquad (10.51)$$

hence $N \equiv -(L + D)^{-1}U$ and $M \equiv (L + D)^{-1}$. This gives the appearance of requiring a (nondiagonal) matrix inverse, but this is clearly not the case upon expressing equation (10.51) in the computational (algebraic) form

$$x^p = -D^{-1}Lx^p - D^{-1}Ux^{p-1} + D^{-1}b. \qquad (10.52)$$

The Gauss–Seidel construction thus amounts to Jacobi iteration, altered by *immediately* using new data x^{p-1} times $D^{-1}L$ as the process scrolls down through the rows of A.

Finally, the successive over-relaxation (SOR) algorithm amounts to Gauss–Seidel with a relaxation factor ω applied in the update step. The optimum value $1 \leq \omega \leq 2$ is problem-dependent, [12], and the algebraic form for SOR is

$$x^p = (D + \omega L)^{-1}\big(((1 - \omega)D - \omega U)x^{p-1} + \omega b\big). \qquad (10.53)$$

The second general category of matrix iteration algorithm applicable to equation (10.44) is named *approximate factorization* (AF). Constructions employ a matrix product approximation to [JAC] as

$$[JAC] \cong [J1][J2][J3]. \qquad (10.54)$$

Substitution of equation (10.54) into equation (10.44) yields

$$[J1][J2][J3]\{\Delta Q\} = -\Delta t\{RES\}_n \qquad (10.55)$$

and defining the intermediate matrix products in equation (10.55) as $\{P\}$ and $\{S\}$, an AF matrix iteration solution sequence is

$$[J1]\{P\} \equiv -\Delta t\{RES\}_n$$
$$[J2]\{S\} \equiv \{P_s\} \qquad \qquad (10.56)$$
$$[J3]\{\Delta \tilde{Q}\} \equiv \{S_s\}$$

In equation (10.56) the superscript tilde denotes the solution is but an approximation to $\{\Delta Q\}$. The column matrices $\{P\}$ and $\{S\}$ are *data* generated in the solution sequence, the matrix orderings of which must be row–column *shuffled* (subscripts s) as the sequences in equation (10.56) are processed.

The key to AF efficiency, hence speed, is identifying matrices [J1], [J2], and [J3] of much *lower order* than is apparent in the definition (10.54). The rationale is a matrix *outer product*, which in the literature is called a matrix *tensor product* symbolized as $\otimes$. The tensor product (TP) of two matrices is an *outer* multiplication whereby, for example, two square matrices of order m when multiplied together form a square matrix of order $2m$.

For $GWS^h/mGWS^h + \theta TS$ algorithms the theory generated element matrix $[A200d]_e$, which assembles to $[MASS]$ in equation (10.44), provides the matrix TP strategy for constructing the AF factors in equation (10.56). Selecting $k = 1$ for illustration of the matrix TP of $[A200L]_e$ with itself, for alignments with coordinate directions x and y, is

$$[A200L]_e \otimes [A200L]_e = \frac{l_x}{6}\begin{bmatrix} 2 & 1 \\ 1 & 2 \end{bmatrix} \otimes \frac{l_y}{6}\begin{bmatrix} 2 & 1 \\ 1 & 2 \end{bmatrix}$$

$$= \frac{l_x l_y}{36}\begin{bmatrix} 4 & 2 & 1 & 2 \\ 2 & 4 & 2 & 1 \\ 1 & 2 & 4 & 2 \\ 2 & 1 & 2 & 4 \end{bmatrix} = [B200L]_e \tag{10.57}$$

In the second line of equation (10.57) the row–column entries in $[B200L]_e$ reflect the FE basis counter-clockwise DOF convention in direct forming of the $n = 2$, $k = 1$ TP basis matrix, recall Table 7.3.

An exercise will verify that $[A200L]_e \otimes [A200L]_e \otimes [A200L]_e = [C200L]_e$. One can further easily verify that for the fluid convection term in equation (10.1) the $n = 1 = k$ basis implementation generates

$$[A200L]_e \otimes [A201L]_e = \frac{l_x}{6}\begin{bmatrix} 2 & 1 \\ 1 & 2 \end{bmatrix} \otimes \frac{1}{2}\begin{bmatrix} -1 & 1 \\ -1 & 1 \end{bmatrix}$$

$$= \frac{l_x}{12}\begin{bmatrix} -2 & 2 & 1 & -1 \\ -2 & 2 & 1 & -1 \\ -1 & 1 & 2 & -2 \\ -1 & 1 & 2 & -2 \end{bmatrix} = [B201L]_e \tag{10.58}$$

the $n = 2$ TP basis-generated matrix, and similarly for the diffusion term

$$[A200L]_e \otimes [A211L]_e = [B211L]_e. \tag{10.59}$$

In summary, element level $n = 1 = k$ FE basis $GWS^h + \theta TS$ algorithm matrix TPs generate element matrices for $n > 1$ exactly reproducing those formed directly using the multidimensional TP bases $\{N_1^+(\eta)\}$. (Note: that both the element and matrix operations are named TP is not a coincidence!)

Thereby, the generic AF matrix $[JK]$, $K = 1,2,3$, equation (10.54), is defined at the Ω_e level using the $k = 1$ basis on $n = 1$. Now suppressing the L notation

$$[JK]_e \equiv [A200]_e + \theta \Delta t \left(u_K [A20K]_e + Pa^{-1}[A2KK]_e + [BCK]_e \right). \tag{10.60}$$

The index K (*not* italics) in equation (10.60) is the *indicator* for the velocity component and Ω_e measure l_e appropriate for the η_k coordinate direction alignment of $[JK]_e$. In practice, the η_k directions K $\Rightarrow$ 1,2,3 for the resolution $[JAC] \cong [J1]\,[J2]\,[J3]$ are identical with the curvilinear coordinate system lines spanning (blocks of) a body-fitted mesh Ω^h constituted of quadrilateral or hexahedron Ω_e [13].

The development is not restricted to the $k = 1$ basis. The AF factors $[JK]_e$ can also be formulated using the $k = 2$, $n = 1$ basis, which produces element matrices comparable to those generated by the n-dimensional Lagrange TP basis $\{N_2^+(\eta)\}$. This basis differs from the *serendipity* basis, equation (7.13), in adding a geometric centroid DOF to meet the biquadratic completeness degree requirement.

The element-dependence in matrices forming $[JK]_e$, equation (10.60), is removed by extracting the η_k-direction element measure l_e

$$[JK]_e = \ell_{e,K}[A200] + \theta\Delta t\left(u_K[A20K] + (\ell_e Pa)^{-1}[A2KK] + [BCK]\right) \tag{10.61}$$

formed using $n = 1$, $k = 1,2$ bases. While the TP of $[A200]_e$ with any other $n = 1$ GWS^h algorithm matrix produces the next-dimensional directional component of the TP k basis element matrix, the TP of the $n = 1$ element matrix sum (10.61) generates error matrices. Using X and Y as K-directional indices, $[J1]_e \otimes [J2]_e$ on $n = 2$ generates, in intermediate shorthand notation

$$
\begin{aligned}
[J1]_e \otimes [J2]_e &\equiv ([MX] + \theta\Delta t[CX + DX + BCX])\otimes \\
&\quad ([MY] + \theta\Delta t[CY + DY + BCY]) \\
&= [MX] \otimes [MY] + [MX] \otimes \theta\Delta t[CY + DY + BCY] \\
&\quad + [MY] \otimes \theta\Delta t[CX + DX + BCX] \quad . \\
&\quad +(\theta\Delta t)^2[CX + DX + BCX] \otimes [CY + DY + BCY] \\
&= [B200] + \theta\Delta t[B201 + B202 + B211 + B222 + BC] \\
&\quad + (\theta\Delta t)^2[\text{Error matrix}]
\end{aligned}
\tag{10.62}
$$

Clearly, an error of order $(\theta\Delta t)^2$ is introduced in the $n = 2$ matrix TP process, which is also the significant order error for $n = 3$. This TP jacobian approximation error acts to limit usable Δt for an AF algorithm. Note however that *no approximation* has been made to the $GWS^h + \theta TS$ matrix $\{RES\}$ in equation (10.44). Thereby, if the AF construction involves an iterative process solving for $\{\delta Q\}^{p+1}$ rather than $\{\Delta\tilde{Q}\}$ in equation (10.56), then at convergence the precision of $\{Q\}^{n+1}$ is not compromised by the matrix TP error identified in equation (10.62).

Alternatively, there exist two *legacy* noniterative AF algorithms that make approximations to $\{RES\}$ in equation (10.55). The *time-split* algorithm, [14], decomposes $\{RES\}_n$ into directional factors parallel to each AF sweep direction. Using $n = 2$ for exposition, and X

210   FE ⇔ CES

and Y as the direction indicators, the time-split AF noniterative solution sequence is

$$
\begin{aligned}
[JX]\{\Delta Q\}^{n+1/2} &= -\Delta t\{RESX(Q^n)\} \\
\{Q\}^{n+1/2} &= \{Q\}^n + \{\Delta Q\}^{n+1/2} \\
[JY]\{\Delta Q\}^{n+1} &= -\Delta t\{RESY(Q^{n+1/2})\} \\
\{Q\}^{n+1} &= \{Q\}^{n+1/2} + \{\Delta Q\}^{n+1}
\end{aligned}
\tag{10.63}
$$

where $\{RESX\}$ and $\{RESY\}$ denote $n=1$ basis evaluations of $\{RES\}$, equation (10.16), with η_1 and η_2 coordinate direction data respectively. The $n+1/2$ superscript in equation (10.63) denotes data generated conceptually at the "half" time interval $\Delta t/2$. The $n=3$ implementation adds another sweep sequence, replaces superscript $1/2$ with $1/3$ and adds one at $2/3$.

The famous *alternating direction implicit* (ADI) AF algorithm, [5], interchanges the directional resolutions of $\{RES\}$ expressed in equation (10.63). The noniterative ADI matrix solution sequence on $n=2$ is

$$
\begin{aligned}
[JX]\{\Delta Q\}^{n+1/2} &= -\Delta t\{RESY(Q^n)\} \\
\{Q\}^{n+1/2} &= \{Q\}^n + \{\Delta Q\}^{n+1/2} \\
[JY]\{\Delta Q\}^{n+1} &= -\Delta t\{RESX(Q^{n+1/2})\} \\
\{Q\}^{n+1} &= \{Q\}^n + \{\Delta Q\}^{n+1/2} + \{\Delta Q\}^{n+1}
\end{aligned}
\tag{10.64}
$$

10.11 Newton and AF TP Jacobian Templates

A *quasi*-Newton method is *any* iterative procedure employing an *approximation* to the exact matrix statement jacobian. For the linear $GWS^h + \theta TS$ convection–diffusion problem statement, equation (10.44) defines the Newton jacobian. Recalling the stated notations the $n=3$, TP $k=1$ basis template for the Newton jacobian $[JAC]_e$ is

$$
\begin{aligned}
[JAC]_e = (\ )(\ )\{\ \}(0;1)[C200][\ ] \\
+(TDT)(\ )\{U1+U2+U3\}(102030;0)[C3001][\ ] \\
+(TDT)(\ )\{U1+U2+U3\}(405060;0)[C3002][\ ] \\
+(TDT)(\ )\{U1+U2+U3\}(708090;0)[C3003][\ ] \\
+(PAI,TDT)(\ )\{\}(112233;-1)[C211][\ ] \\
+(PAI,TDT)(\ )\{\ \}(445566;-1)[C222][\ ] \\
+(PAI,TDT)(\ )\{\ \}(778899;-1)[C233][\ ] \\
+(PAI,TDT)(\ )\{\ \}(142536;-1)[C221][\ ] \\
+(PAI,TDT)(\ )\{\ \}(142536;-1)[C212][\ ] \\
+(PAI,TDT)(\ )\{\ \}(475869;-1)[C223][\ ] \\
+(PAI,TDT)(\ )\{\ \}(475869;-1)[C232][\ ] \\
+(PAI,TDT)(\ )\{\ \}(172839;-1)[C213][\ ] \\
+(PAI,TDT)(\ )\{\ \}(172839;-1)[C231][\ ] \\
+(PB,PAI,TDT)(\ )\{\ \}(0;1)[B200][\ ]
\end{aligned}
\tag{10.65}
$$

For the *generic* AF TP jacobian (10.60), and metric data from equation (10.65), the template for the $n = 3$ AF TP quasi-Newton jacobian algorithm is

$$[J1]_e = (\;)(\;)\{\;\}(\;)[A200]\text{diag}[(det)^{1/n}]$$
$$+(\text{TDT})(\;)\{U1 + U2 + U3\}(102030; 0)[A3001][\;]$$
$$+(\text{PAI, TDT})(\;)\{\;\}(114477; -1)[A211][\;]$$
$$+(\text{PB, PAI, TDT})(\;)\{\;\}(0; 1)[A200][\;]$$

$$[J2]_e = (\;)(\;)\{\;\}(\;)[A200]\text{diag}[(det)^{1/n}]$$
$$+(\text{TDT})(\;)\{U1 + U2 + U3\}(405060; -1)[A3001][\;]$$
$$+(\text{PAI, TDT})(\;)\{\;\}(225588; -1)[A211][\;] \qquad (10.66)$$
$$+(\text{PB, PAI, TDT})(\;)\{\;\}(0; 1)[A200][\;]$$

$$[J3]_e = (\;)(\;)\{\;\}(\;)[A200]\text{diag}[(det)^{1/n}]$$
$$+(\text{TDT})(\;)\{U1 + U2 + U3\}(708090; 0)[A3001][\;]$$
$$+(\text{PAI, TDT})(\;)\{\;\}(336699; -1)[A211][\;]$$
$$+(\text{PB, PAI, TDT})(\;)\{\;\}(0; 1)[A200][\;]$$

In the template lines with [A3001] the $n = 3$ metric data strings align the velocity vector resolution appropriate for TP jacobian sweeps along boundary-fitted mesh lines. Computational experience proves placing the one-third power of the $n = 3$ det on the diagonal in the sixth bracket enhances accuracy. Finally, note that only diagonal members of the coordinate transformation matrix appear in [A211], an added TP matrix approximation error limiting useful Δt for $\text{Pa}^{-1} > 0$.

10.12 Chapter Summary

Mathematical underpinnings and computing practice *substance* of $\text{GWS}^N \Rightarrow \text{GWS}^h \Rightarrow m\text{GWS}^h + \theta\,\text{TS}$ for scalar transport applications in n-dimensions is complete. The n-D unsteady PDE + BCs + IC statement for *generic scalar* state variable member $q(\mathbf{x},t)$ in non-D indicial notation is

$$\mathcal{L}(q) = \frac{\partial q}{\partial t} + u_j \frac{\partial q}{\partial x_j} - \frac{1}{\text{Pa}} \frac{\partial^2 q}{\partial x_j{}^2} - s = 0, \text{ on } \Omega \times t \subset \Re^n \times R^+. \qquad (10.1)$$

$$\ell(q) = \nabla q \cdot \hat{\mathbf{n}} + \text{Pb}(q - q_r) + \mathbf{f} \cdot \hat{\mathbf{n}} = 0, \text{ on } \partial\Omega_R \times t \subset \Re^{n-1} \times R^+. \qquad (10.2)$$

$$q(\mathbf{x}_b, t) = q_b(\mathbf{x}_b, t), \text{ on } \partial\Omega_D \times t \subset \Re^{n-1} \times R^+. \qquad (10.3)$$

$$q(\mathbf{x}, t_0) = q_0(\mathbf{x}), \text{ on } \Omega \cup \partial\Omega \times t_0. \qquad (10.4)$$

For any basis degree k the $\text{GWS}^h + \theta\text{TS}$ process generates the matrix statement for DOF determination with component definitions

$$[\text{JAC}]\{\Delta Q\} = -\Delta t\{\text{RES}\}_n. \qquad (10.13)$$

$$[\text{JAC}]_e = [\text{M200}]_e + \theta \Delta t ((UJ)_e [\text{M20}J]_e + [\text{M2}KK]_e + [\text{N200}]_e). \tag{10.15}$$

$$\{\text{RES}\}_e = \Delta t ((UJ)_e [\text{M20}J]_e + [\text{M2}KK]_e + [\text{N200}]_e) \{Q_n\}_e - \{b\}_e. \tag{10.16}$$

$$\{b\}_e = ([\text{M200}]_e \{\text{SRC}\}_e + [\text{N200}]_e \{\text{QR}\}_e - [\text{N200}]_e \{\text{FN}\}_e)_{n+\theta}. \tag{10.17}$$

The asymptotic error estimates for solutions generated by equations (10.13)–(10.17) for $\text{Pa}^{-1} > 0$ and $\text{Pa}^{-1} = 0$ are, respectively

$$\left\| e^h(n\Delta t) \right\|_E \leq C \ell_e^{2\gamma} \|\text{data}\|_{\Omega,\partial\Omega L2}^2 + C_t \Delta t^{f(\theta)} \|Q_0\|_E$$
$$\gamma \equiv \min(k, r-1), f(\theta) = (2,3), \text{ or} \tag{10.10}$$
$$\left\| e^h(n\Delta t) \right\|_E \leq C \ell_e^2 \int_{t_0}^{n\Delta t} \|T(x,\tau)\|_{H^{k+1}}^2 d\tau + C_t \Delta t^{f(\theta)} \|T_0\|_E$$

Thereafter, a variety of *m*PDE modifications to equation (10.1) were derived with the purpose to moderate and/or annihilate significant-order discrete approximation error mechanisms. Generalizing to *n*-D the prior chapter Taylor series manipulation, the limit process $\Delta t \Rightarrow \varepsilon > 0$ identifies the parent *m*PDE as

$$\mathcal{L}^m(q) \equiv \mathcal{L}(q) - \frac{\Delta t}{2} \frac{\partial}{\partial x_j} \left[\alpha u_j \frac{\partial q}{\partial t} + \beta u_j u_k \frac{\partial q}{\partial x_k} \right]$$
$$- \frac{\Delta t^2}{6} \frac{\partial}{\partial x_j} \left[\gamma u_j u_k \frac{\partial^2 q}{\partial x_k \partial t} + \delta u_j u_k \frac{\partial}{\partial x_k} \left(u_l \frac{\partial q}{\partial x_l} \right) \right] \tag{10.29}$$

which proved capable of recovering, via appropriate definitions for the parameter set $(\alpha, \beta, \gamma, \delta)$ plus θ, a wide variety of algorithms derived by totally alternative strictly discrete theories.

Fourier spectral theory enabled detailed exposition of algorithm distinctions in terms of approximation error order in nondimensional wave number $m = \kappa/\Delta x$. Two optimally performing $m\text{GWS}^h$ algorithm implementations using the *efficient* (tridiagonal stencil) $k = 1$ trial space basis were identified. Unsteady $m\text{GWS}^h + \theta\text{TS}$ solutions exhibiting $O(m^6)$ spectral fidelity without added artificial diffusion resulted on identification of the *optimal* γ *m*PDE

$$\mathcal{L}^m(q) = \left(1 - \frac{\gamma \Delta t^2}{6} \frac{\partial}{\partial x_j} \left(u_j u_k \frac{\partial}{\partial x_k} \right) \right) \frac{\partial q}{\partial t} + u_j \frac{\partial q}{\partial x_j} - \frac{1}{\text{Pa}} \frac{\partial^2 q}{\partial x_j^2} - s = 0 \tag{10.67}$$

for $\gamma \equiv -0.5$ for the $k = 1$ basis implementation. Theory-determined optimal γ are derived for $k = 2,3$ basis implementations as well [15].

The $m\text{GWS}^h$ steady solution algorithm results upon retaining the β term in equation (10.29), but with a theoretically derived alternative parameter generating the *optimal* β *m*PDE

$$\mathcal{L}^m(q) = u_j \frac{\partial q}{\partial x_j} - \frac{1}{\text{Pa}} \frac{\partial^2 q}{\partial x_j^2} - \frac{\text{Pa}h^2}{12} \frac{\partial}{\partial x_j} \left[u_j u_k \frac{\partial q}{\partial x_k} \right] - s = 0. \tag{10.38}$$

Solutions to equation (10.38) are theoretically predicted to adhere to the *modified $k=1$* basis (only!) asymptotic error estimate

$$\|e^h\|_E \leq C\ell_e^{2\gamma}\|data\|^2_{\Omega,\partial\Omega L2}, \; \gamma \equiv \min(k(=1)+1 = 2, \, r-1). \tag{10.39}$$

Then follows ample computational experiment-generated *a posteriori* data in the verification, benchmark, and validation categories. Algorithm templates converting theory in a visually transparent manner to code practice are stated. The final topic is a brief exposure to numerical linear algebra, identifying matrix iterative methods possessing n-dimensional speed advantages over full Newton methodology.

Exercises

10.1 Confirm the n-D $GWS^h + \theta TS$ FE implementation template essence (10.20) and (10.21).

10.2 From the answer to 10.1, generate the template for the nonlinear $GWS^h + \theta TS$ matrix statement (10.22) and (10.23).

10.3 Develop the template replacement for 10.2 for the natural coordinate $k=1$ basis implementation.

10.4 Verify the TS expansion manipulations leading to $\mathcal{L}^m(q)$, (10.29).

10.5 Confirm that equation (10.31) is the Fourier spectral theory solution for $k=1$ basis GWS^h algorithm complex convection velocity.

10.6 Repeat the exercise 10.5 analysis selectively for the other $mGWS^h$ algorithms identified in Section 10.4.

Computer Labs

The *FEmPSE* toolbox and all MATLAB® .m files required for conducting the chapter computer labs are available for download from www.wiley.com/go/baker/finite.

10.1 Execute the MATLAB® .m file for the $n=2$ unsteady rotating cone validation statement. The file data specification is $GWS^h + \theta TS$ for centroid $C=0.4$ on a uniform $M=33 \times 33 \, k=1$ TP basis mesh. Repeat this experiment using the $k=1$ NC basis on the identical mesh, hence note the modest accuracy degradation. Then distort either the TP or NC basis mesh via P_r definitions and re-execute to quantify solution accuracy loss due to mesh nonuniformity.

10.2 Update the computer lab 1 template implementing one or more of the Chapter 10.4-identified $mGWS^h + \theta TS$ algorithms. Execute simulations on uniform and nonuniform TP and NC $k=1$ basis meshes and compare results.

10.3 Adapt the computer lab 1 MATLAB® .m file template for the $n=2$ gaussian plume benchmark definition. Evaluate specification of $10 < Pa \leq 100$ on solution accuracy using both TP and NC bases for GWS^h and optimal steady mGWS$^h(\beta)$ algorithms.

10.4   Adapt computer lab 3. MATLAB® .m file template for the $n = 2$ steady Peclet problem definition. Confirm the optimal steady mGWS$^h(\beta)$ algorithm asymptotic error estimate (10.39) for Pe = 500, 1000 for both TP and NC $k = 1$ basis meshes.

References

[1] Baker, A.J. (2013) *Optimal Modified Continuous Galerkin CFD*, Wiley, London.
[2] Raymond, W.H. and Garder, A. (1976) Selective damping in a Galerkin method for solving wave problems on a variable grid. *Mon. Weather Rev.*, **104**, 1583–1590.
[3] Kolesnikov, A. and Baker, A.J. (2001) An efficient high order Taylor weak statement formulation for the Navier-Stokes equations. *J. Comput. Phys.*, **22**, 45–53.
[4] American Institute Aeronautics & Astronautics (1998) AIAA Guide G-077–1998.
[5] Dendy, J.E. and Fairweather, G. (1975) Alternating-direction Galerkin methods for parabolic and hyperbolic problems on rectangular polygons. *SIAM J. Numerical Analysis*, **12**, 144–163.
[6] Fabrick, A., Sklarew, R., and Wilson, J. (1977) "Point Source Model Evaluation and Development Study," Science Applications Inc. Tech. Report A5-058-87.
[7] Spittler, J.D. (1990) PhD dissertation, Univ. Illinois/Urbana.
[8] Williams, P.T., Baker, A.J., and Kelso, R.M. (1994) Numerical calculation of room air motion, III. 3-D CFD simulation of a full scale room air experiment. *Trans. ASHRAE*, **100**, 549–564.
[9] Williams, P.T. and Baker, A.J. (1996) Incompressible computational fluid dynamics and the continuity constraint method for the 3-D Navier–Stokes equations. *J. Numerical Heat Transfer, Part B*, **29**, 137–273.
[10] Baker, A.J., Roy, S., and Kelso, R.M. (1994) CFD experiment characterization of airborne contaminant transport for two practical 3-D room air flowfields. *J. Building Environ.*, **29**, 253–273.
[11] Roy, S. and Baker, A.J. (1995) A post-processing algorithm for CFD dispersion error annihilation. in Proceedings CFD '95 International Symposium, P. Thibault and D. Bergeron (Eds.), CFD Society of Canada, pp. 333–339.
[12] Varga, R.S. (1962) *Matrix Iterative Analysis*, Prentice-Hall, Englewood Cliffs, NJ.
[13] Thompson, J.F., Thames, F., and Mastin, W. (1977) TOMCAT — A code for numerical generation of boundary-fitted curvilinear coordinate systems on fields containing any number of arbitrary two-dimensional bodies. *J. Comput. Phys.*, **24**, 274–302.
[14] Yanenko, N.N. (1971) *The Method of Fractional Steps*, Springer-Verlag, Berlin.
[15] Chaffin, D.J. and Baker, A.J. (1995) On Taylor weak statement finite element methods for computational fluid dynamics. *J. Num. Methods Fluids*, **21**, 273–294.

11

Engineering Sciences, $n > 1$: $\text{GWS}^h \{N_k(\zeta_\alpha, \eta_j)\}$ implementations in the n-D computational engineering sciences

11.1 Introduction

The $\text{WF} \Rightarrow \text{WS}^N \Rightarrow \text{GWS}^N \Rightarrow \text{GWS}^h/m\text{GWS}^h + \theta\text{TS}$ sequence for formulating approximate solutions to PDEs + BCs + IC statements is now thoroughly exposed. The methodology is broadly applicable across the *computational engineering sciences* spectrum, for example, heat transfer, structural mechanics, heat/mass transport, fluid mechanics, mechanical vibrations, electromagnetics. Deriving a specific formulation addresses coupled conservation principles DM, DP, and DE, Chapter 2, augmented with defining closure models pertinent to "the physics" with well-posed boundary conditions guarantee and an initial condition.

This chapter details $\text{GWS}^h/m\text{GWS}^h + \theta\text{TS}$ algorithms with *template* essence for n-D statements in select computational engineering sciences disciplines. The formulation sequence is thoroughly developed for each addressed problem class with key computer lab-generated *a posteriori* solution data critically discussed.

The MATLAB® *FEmPSE* toolbox compute lab environment utilized to this juncture is replaced herein with COMSOL, a GUI-enabled, *multiphysics* GWS^h-implemented *problem solving environment* (PSE) [1]. Organization of these compute lab .mph files is selectively presented, with goal to provide guidance for COMSOL multiphysics

Finite Elements ⇔ Computational Engineering Sciences, First Edition. A. J. Baker.
© 2012 John Wiley & Sons, Ltd. Published 2012 by John Wiley & Sons, Ltd.

PDEs + BCs + IC setup. Alternatively, all computer lab COMSOL .mph files are available at www.comsol.com/community/exchange/?page=2.

11.2 Structural Mechanics

Structural systems analysis is the engineering science discipline where FE methods were first reduced to compute practice. The mass and thermal energy principles DM and DE are usually identically satisfied, leaving Newton's law DP. Recalling equation (2.19), assuming stationary media with stress tensor σ_{ij} and body force B_i

$$\text{DP}: \quad \frac{\partial \sigma_{ij}}{\partial x_j} + B_i = 0, 1 \le (i,j) \le n. \tag{11.1}$$

Rather than forming GWS^h for DP in PDE form (11.1), the traditional structural mechanics community approach is via a *mechanical energy* principle DE called the *Principle of Virtual Work* [2]. This principle expresses the balance of mechanical energy stored in a structure with the work done by external forces in its deformation. Recalling equation (2.23) the mechanical DE principle in tensor index notation is

$$\text{DE}: \Pi \equiv \int_\Omega de = \frac{1}{2} \int_\Omega dvol \int_0^\varepsilon \sigma_{ij} d\varepsilon_{ij} - \int_\Omega u_j B_j dvol - \int_{\partial\Omega} u_j T_j d\,\text{surf}. \tag{11.2}$$

The differential elements are labeled dvol and dsurf to avoid confusion with τ and σ, the classic variables in elasticity. For *linear* elasticity, stress and strain are correlated via Hooke's law, equation (2.20), and displacement and strain are assumed kinematically linear. Substituting Hooke's law and replacing tensor index with matrix notation, equation (11.2) for a linear elastic medium is

$$\text{DE}: \Pi = \int_\Omega \left(\frac{1}{2} \{\varepsilon\}^T [E] \{\varepsilon\} - \{u\}^T \{B\} \right) dvol - \int_{\partial\Omega} \{u\}^T \{T\} d\,\text{surf}. \tag{11.3}$$

The variables in equation (11.3) are introduced in Chapter 2.4. Equation (11.3) is the *variational calculus* expression which early in FE structural mechanics theorization replaced the original classic engineering mechanics foundation. The engineers' finite element constructions then became expressed via *extremization* of a discrete implementation of equation (11.3), a process restricted to structural mechanics. Lacking this theoretical background, engineers/analysts outside the structures field experience great difficulty comprehending this FE formulation strategy.

The process $\text{GWS}^N \Rightarrow \text{GWS}^h$ for the structural mechanics DP PDE + BCs statement (11.1), replacing the tensor form with a matrix expression, will *exactly reproduce* extremization of the FE-discretized evaluation of Π, equation (11.3). Therefore, herein is developed the parallel processes of GWS^h formulation on DP and FE implemented DE discrete extremization, to enable bridging the theory and *lingo* gaps.

The first requirement is to determine that the DP and Π principles describe an elliptic boundary value (EBV) problem to identify well-posed BCs. The exposition vehicle is plate theory leading to *plane stress* and/or *plane strain* formulations. The GWS^h starting point is expanding equation (11.1) for $n=2$

$$\frac{\partial \sigma_x}{\partial x} + \frac{\partial \tau_{xy}}{\partial y} + B_x = 0. \tag{11.4}$$

$$\frac{\partial \tau_{xy}}{\partial x} + \frac{\partial \sigma_y}{\partial y} + B_y = 0. \tag{11.5}$$

The ultimate GWS^h formulation dependent variable is the displacement vector $\mathbf{u}(x,y) = u(x,y)\hat{\mathbf{i}} + v(x,y)\hat{\mathbf{j}}$. The algorithm degrees-of-freedom (DOF) are the vector scalar resolution at mesh geometric nodes (at least). Displacement is related kinematically to strain (displacement/unit length) $\{\varepsilon\}^T \equiv \{\varepsilon_x, \varepsilon_y, \gamma_{xy}\}$ with definitions

$$\varepsilon_x = \frac{\partial u}{\partial x}, \quad \varepsilon_y = \frac{\partial v}{\partial y}, \quad \gamma_{xy} = \frac{1}{2}\left(\frac{\partial u}{\partial y} + \frac{\partial v}{\partial x}\right). \tag{11.6}$$

Strain is correlated with stress via a *constitutive* closure model, which brings in properties of the material present. For $n=2$ linear *plane stress*, and a homogenous medium, the field definitions are

$$\begin{aligned} \{\boldsymbol{\sigma}\} &\equiv \{\sigma_x, \sigma_y, \tau_{xy}\}^T \\ \{\boldsymbol{\varepsilon}\} &\equiv \{\varepsilon_x, \varepsilon_y, \tau_{xy}\}^T \\ \varepsilon_z &= -v(\varepsilon_x + \varepsilon_y)/(1-v) \end{aligned}, \tag{11.7}$$

where v is the Poisson ratio and Hooke's law is given in equation (2.20)

$$[\mathbf{E}] \equiv \frac{E}{1-v^2}\begin{bmatrix} 1 & v & 0 \\ v & 1 & 0 \\ 0 & 0 & (1-v)/2 \end{bmatrix}. \tag{11.8}$$

For linear *plane strain* $\{\boldsymbol{\sigma}\}$ and $\{\varepsilon\}$ remain as in equation (11.7), a dilatation stress $\sigma_z = v(\sigma_x + \sigma_y)$ is admitted and Hooke's law is

$$[\mathbf{E}] \equiv \frac{E}{(1+v)(1-2v)}\begin{bmatrix} 1-v & v & 0 \\ v & 1-v & 0 \\ 0 & 0 & (1-v)/2 \end{bmatrix}. \tag{11.9}$$

The result of combining the plane stress definitions was introduced as equation (2.21). Completing the suggested exercise will confirm that combining equations (11.4)–(11.9) produces PDE systems on the scalar resolution of $\mathbf{u}(x,y) = u(x,y)\hat{\mathbf{i}} + v(x,y)\hat{\mathbf{j}}$ in the forms:
Plane stress

$$\begin{aligned} \mathcal{L}(u) &= \nabla^2 u + \frac{1+v}{1-v}\frac{\partial}{\partial x}\left(\frac{\partial u}{\partial x} + \frac{\partial v}{\partial y}\right) + \frac{B_x}{G} = 0 \\ \mathcal{L}(v) &= \nabla^2 v + \frac{1+v}{1-v}\frac{\partial}{\partial y}\left(\frac{\partial u}{\partial x} + \frac{\partial v}{\partial y}\right) + \frac{B_y}{G} = 0 \end{aligned}. \tag{11.10}$$

Plane strain

$$\mathcal{L}(u) = \nabla^2 u + \frac{1}{1 - 2\upsilon}\frac{\partial}{\partial x}\left(\frac{\partial u}{\partial x} + \frac{\partial v}{\partial y}\right) + \frac{B_x}{G} = 0$$

$$\mathcal{L}(v) = \nabla^2 v + \frac{1}{1 - 2\upsilon}\frac{\partial}{\partial y}\left(\frac{\partial u}{\partial x} + \frac{\partial v}{\partial y}\right) + \frac{B_y}{G} = 0$$

(11.11)

The constitutive closure data in equations (11.10)–(11.11) are Poisson ratio υ and elastic shear modulus $G = E/2(1 + \upsilon)$ where E is the Youngs modulus.

Since weak forms employ vector calculus, the GWS^N starting point is to coalesce (11.10)–(11.11) in vector form. This simple operation leads to

$$\mathcal{L}(\mathbf{u}) = -\nabla^2 \mathbf{u} - g(\upsilon)\nabla(\nabla \bullet \mathbf{u}) - \mathbf{b} = 0$$

(11.12)

for definitions $g(\upsilon) \equiv (1 + \upsilon)/(1 - \upsilon)$ or $1/(1 - 2\upsilon)$ for plane stress or strain, and $\mathbf{b} \equiv \mathbf{B}/G$. This vector PDE is indeed EBV via the laplacian, hence $\mathbf{u}$ and/or its normal derivative must be constrained on the entirety of the domain boundary $\partial\Omega$.

Traditional formation of GWS^N on equation (11.12) encounters an impass with the divergence constraint term $\nabla(\nabla\cdot\mathbf{u})$, as the Green–Gauss theorem is not applicable to symmetrize differentiability support. This issue is resolved by expressing DP, equation (11.1), as a *vector matrix* differential equation, [3, Chapter 6]. Returning to $n = 3$ dimensionality and denoting vector matrices by *boldface* type, equation (11.1) is equivalently

$$\text{DP}: \quad [\mathbf{D}]\{\boldsymbol{\sigma}\} + \{\mathbf{B}\} = \{\mathbf{0}\}.$$

(11.13)

For stress tensor matrix order $\{\boldsymbol{\sigma}\} = \{\sigma_x, \sigma_y, \sigma_z, \tau_{xy}, \tau_{xz}, \tau_{yz}\}^T$ the matrix derivative definition in equation (11.13) is

$$[\mathbf{D}] \equiv \begin{bmatrix} \partial/\partial x & 0 & 0 & \partial/\partial y & \partial/\partial z & 0 \\ 0 & \partial/\partial y & 0 & \partial/\partial x & 0 & \partial/\partial z \\ 0 & 0 & \partial/\partial z & 0 & \partial/\partial x & \partial/\partial y \end{bmatrix}.$$

(11.14)

Since $\{\boldsymbol{\sigma}\} = [\mathbf{E}]\{\boldsymbol{\varepsilon}\}$ generalizes to n-D, and noting that equation (11.6) for the definition $\{\boldsymbol{\varepsilon}\} \equiv \{\varepsilon_x, \varepsilon_y, \varepsilon_z, 2\gamma_{xy}, 2\gamma_{xz}, 2\gamma_{yz}\}^T$ possesses the n-D vector matrix form $\{\boldsymbol{\varepsilon}\} = [\mathbf{D}]\{\mathbf{u}\}$, then DP matrix PDE (11.13) for the linear elastic closure assumption is

$$\text{DP}: \quad [\mathbf{D}]^T[\mathbf{E}][\mathbf{D}]\{\mathbf{u}\} + \{\mathbf{B}\} = \{\mathbf{0}\}.$$

(11.15)

The Galerkin WS^N written on equation (11.15) for any approximation $\mathbf{u}^N(\mathbf{x})$, supported by the trial space function set $\Psi_\beta(\mathbf{x})$, is

$$\text{GWS}^N \equiv \int_\Omega \Psi_\beta(\mathbf{x})\left([\mathbf{D}]^T[\mathbf{E}][\mathbf{D}]\{\mathbf{u}^N\} + \{\mathbf{B}\}\right)d\tau \equiv 0, \forall \beta.$$

(11.16)

Now applying the *matrix* Green–Gauss divergence theorem, which symmetrizes the trial and test function differentiability support, and inserting a sign change generates

$$
\begin{aligned}
\mathrm{GWS}^N = &\int_\Omega [\mathbf{D}\,\Psi_\beta]^T [\mathbf{E}][\mathbf{D}]\{\mathbf{u}^N\}\mathrm{d}\tau - \int_\Omega \Psi_\beta\{\mathbf{B}\}\mathrm{d}\tau \\
&- \int_{\partial\Omega} \Psi_\beta[\mathbf{N}]^T [\mathbf{E}][\mathbf{D}]\{\mathbf{u}^N\}\mathrm{d}\sigma \equiv 0
\end{aligned}
\tag{11.17}
$$

In equation (11.17) the Green–Gauss theorem-generated flux (Neumann) BC surface integral operator $[\mathbf{N}]^T \equiv \hat{\mathbf{n}}\cdot[\mathbf{D}]^T$ is the matrix gradient derivative dot product. This term becomes the placeholder for surface tractions $\mathbf{T}$ loading the structure. The companion mechanical energy principle DE, equation (11.3), when FE discretely implemented and extremized, will identify its functional form.

11.3 Structural Mechanics, Virtual Work FE Implementation

In addition to identifying the surface integral integrand in equation (11.17), the mechanical energy principle (11.3) readily extends the analysis statement for initial stress and strain states and point loads. For the definitions $\{\sigma_0\}$ and $\{\varepsilon_0\}$, and replacing the differential elements $\mathrm{d}\tau$ and $\mathrm{d}\sigma$ with dvol and dsurf for notational clarity, the augmented replacement for DE (11.3) is [4]

$$
\begin{aligned}
\mathrm{DE} : \Pi = &\int_\Omega \left(\frac{1}{2}\{\varepsilon\}^T[\mathbf{E}]\{\varepsilon\} + \{\varepsilon\}^T[\mathbf{E}]\,\varepsilon_0 + \{\varepsilon\}^T\{\sigma_0\} - \{\mathbf{u}\}^T\{\mathbf{B}\} \right)\mathrm{d\,vol} \\
&- \int_\Omega \{\mathbf{u}\}^T\{\mathbf{B}\}\mathrm{d\,vol} - \int_{\partial\Omega}\{\mathbf{u}\}^T\{\mathbf{T}\}\mathrm{d\,surf} - \{\mathbf{U}\}^T\{\mathbf{P}\}
\end{aligned}
\tag{11.18}
$$

In equation (11.18) $[\mathbf{E}]$ remains matrix Hooke's law and $\{\mathbf{U}\}$ contains the FE displacement nodal DOF where point loads $\{\mathbf{P}\}$ are admitted applied.

The FE implementation of equation (11.18) is simply the sum of integrals formed on the mesh M, the union of finite element domains Ω_e with boundaries $\partial\Omega_e$. The FE *discrete approximation* statement essence is

$$
\Pi \approx \Pi^h \equiv \sum_{\Omega^h} \Pi_e = \sum_{e=1}^{M} \left[\int_{\Omega_e} (\bullet)_e \mathrm{d\,vol} - \int_{\partial\Omega\cap\partial\Omega_e} (\bullet)_e \,\mathrm{d\,surf} - (\bullet)_e \right].
\tag{11.19}
$$

The formulation requirement is to *extremize* (render *stationary*) Π^h with respect to all *admissible* FE approximation DOF, that is, *only* those DOF not constrained by Dirichlet BCs. This operation transforms the *scalar* (11.19) into the matrix equation system

$$
\partial\Pi^h/\partial\{\mathrm{DOF}\}_{\mathrm{admissible}} \equiv \{0\} \quad \Rightarrow \quad [\mathbf{K}]\{\mathbf{U}\} - \{\mathbf{R}\} = \{0\}.
\tag{11.20}
$$

The notation in equation (11.20), typical for structural mechanics FE algorithms, denotes the global *stiffness matrix* as $[\mathbf{K}]$. The global *load matrix* $\{\mathbf{R}\}$ contains *all* contributions defined in equation (11.18) except that from the lead term [4].

Evaluating contributions to the element energy functional $\prod_e$ utilizes (of course!) FE basis functions. For exposition clarity, this detailing reverts to the $n=2$ plane stress/strain class. The integrals are fully matrix coupled which requires the *matrix statement* of approximation to be $\mathbf{u}_e \equiv [\{N_k\}^T] \{\mathbf{U}\}_e$ with DOF resolution

$$\{\mathbf{u}\}_e \equiv \left\{ \begin{matrix} u \\ v \end{matrix} \right\}_e = \begin{bmatrix} \{N_k\}^T & \{0\} \\ \{0\} & \{N_k\}^T \end{bmatrix} \left\{ \begin{matrix} U \\ V \end{matrix} \right\}_e. \tag{11.21}$$

Reducing the rank of $[\mathbf{D}]$ in equation (11.14) to $n=2$, the kinematic strain-displacement matrix definition with equation (11.21) inserted is

$$\left\{ \begin{matrix} \varepsilon_x \\ \varepsilon_x \\ \gamma_{xy} \end{matrix} \right\}_e = [\mathbf{D}]\{\mathbf{u}\}_e \equiv \begin{bmatrix} \partial/\partial x & 0 \\ 0 & \partial/\partial y \\ \partial/2\partial y & \partial/2\partial x \end{bmatrix} \begin{bmatrix} \{N_k\}^T & \{0\} \\ \{0\} & \{N_k\}^T \end{bmatrix} \left\{ \begin{matrix} U \\ V \end{matrix} \right\}_e \equiv [\mathbf{B}]\{\mathbf{U}\}_e. \tag{11.22}$$

The generic element level *stiffness matrix* contribution to Π^h is then

$$\Pi_e \equiv \int_{\Omega_e} \left(\frac{1}{2}\{\boldsymbol{\varepsilon}\}_e^T[\mathbf{E}]\{\boldsymbol{\varepsilon}\}_e \right) \mathrm{d}x\mathrm{d}y$$

$$= \frac{1}{2}\{\mathbf{U}\}_e^T \int_{\Omega_e} \left([\mathbf{B}]_e^T[\mathbf{E}][\mathbf{B}]_e \right) \mathrm{d}x\mathrm{d}y\{\mathbf{U}\}_e \equiv \frac{1}{2}\{\mathbf{U}\}_e^T[\mathbf{k}]_e\{\mathbf{U}\}_e \tag{11.23}$$

with $[\mathbf{k}]_e$ again typical notation for *the* element stiffness matrix.

Via the chain rule the extremization of equation (11.23) with respect to *admissible* DOF in $\{\mathbf{U}\}_e$ produces $\frac{1}{2}[\mathbf{k}]_e\{\mathbf{U}\}_e + \frac{1}{2}[\mathbf{k}]_e^T\{\mathbf{U}\}_e = [\mathbf{k}]_e\{\mathbf{U}\}_e$ since $[\mathbf{k}]_e$ is symmetric. The assembly S_e $([\mathbf{k}]_e)$ over all Ω_e generates the global stiffness matrix $[\mathbf{K}]$ defined in equation (11.20).

The load matrix $\{\mathbf{R}\}$ is also formed via assembly of all Ω_e contributions $\{\mathbf{r}\}_e$. The *discrete extremization* of all load terms contained in equation (11.18) generates the element load matrix

$$\{\mathbf{r}\}_e = \int_{\Omega_e} \left([\mathbf{B}]^T[\mathbf{E}]\{\boldsymbol{\varepsilon}_0\}_e - [\mathbf{B}]^T\{\boldsymbol{\sigma}_0\}_e + [\{N_k\}]\left[\{N_k\}^T\right]\{\rho\mathbf{g}\}_e \right)\mathrm{d\ vol}$$

$$+ \int_{\partial\Omega_e \cap \partial\Omega} [\{N_k\}]\left[\{N_k\}^T\right]\{\mathbf{T}\}_e \,\mathrm{d\ surf} + \{\mathbf{P}\}_e \tag{11.24}$$

The first two terms in $\{\mathbf{r}\}_e$ result from initial strain and stress distributions, and the body force term is familiar. The surface traction term in equation (11.24) is the Neumann BC required for completing GWS^h, the FE implementation of GWS^N, equation (11.17). Hence, in the discrete implementation the surface integral determination is

$$\int_{\partial\Omega} \Psi_\beta(\cdot)\mathrm{d}\sigma = S_e\left(\int_{\partial\Omega \cap \partial\Omega_e} \begin{bmatrix} \{N_k\} & \{0\} \\ \{0\} & \{N_k\} \end{bmatrix} \begin{bmatrix} \{N_k\}^T & \{0\} \\ \{0\} & \{N_k\}^T \end{bmatrix} \left\{ \begin{matrix} TX \\ TY \end{matrix} \right\}_e \mathrm{d}\sigma \right). \tag{11.25}$$

11.4 Plane Stress/Strain GWS^h Implementation

Structural mechanics FE texts rarely detail extremization of $\prod^h$ as commercial code availability makes individual implementation rarely required. In support of absolute clarity, the approach herein remains detailing the matrix expressions in GWS$^h =$ $S_e\{WS\}_e \equiv \{0\}$, first in general form then followed by NC $\{N_1(\zeta_\alpha)\}$ basis implementation.

The GWS$^N \Rightarrow$ GWS^h process in equation (11.17) after inserting BC (11.25) generates

$$\{WS\}_e = [STIFF]_e\{U\}_e - \{b\}_e \tag{11.26}$$

for $\{U\}_e = \{U, V\}_e^T$ the DOF of the $n = 2$ cartesian resolution of the displacement vector (11.21). The major expenditure of effort remains in forming $[STIFF]_e$, the structural mechanics equivalent of $[DIFF]_e$, in Chapter 7. From equation (11.23) and recalling Chapter 7 nomenclature conventions, the formation *essence* is

$$\begin{aligned}
[STIFF]_e &\equiv \int_{\Omega_e} \left([\mathbf{B}]_e^T[\mathbf{E}][\mathbf{B}]_e \right) dxdy \\
&= CONST\, DET_e^{-1}[ETLI]_e^T[B2LKE][ETKI]_e
\end{aligned} \tag{11.27}$$

The coordinate transformation arrays $[ETLI]_e$ and $[ETKI]_e$ are now matrices of order element DOF $\times (n+1)$, and in $[B2LKE]$ boldface $\mathbf{E}$ emphasizes the Hooke's law matrix expression, for example, equations (11.8), (11.9).

In the same manner, the load matrix formation essence is

$$\begin{aligned}
\{\mathbf{b}\}_e &\equiv \int_{\Omega_e} \left(\begin{array}{l} [\mathbf{B}]_e^T[\mathbf{E}]\{\varepsilon_0\} - [\mathbf{B}]_e^T\{\boldsymbol{\sigma}_0\} \\ +[\{N_k\}][\{N_k\}^T]\{\rho\mathbf{g}\}_e \end{array} \right) dxdy \\
&\quad + \int_{\partial\Omega_e \cap \partial\Omega_R} [\{N_k\}][\{N_k\}^T]\{\mathbf{T}\}_e d\sigma + \mathbf{P}\{\delta\}_e \\
&= CONST[ETLI]_e^T[B2L0E]\{EPS0\}_e \\
&\quad - CONST[ETLI]_e^T[B1L]\{SIG0\}_e \\
&\quad + DET_e[[B200]]\{RHOG\}_e + DET_e[[A200]]\{\mathbf{T}\}_e + \mathbf{P}\{\delta\}_e
\end{aligned} \tag{11.28}$$

In equations (11.27)–(11.28) all element matrix orders are twice that developed for scalar DOF, as $\{U\}_e \Rightarrow \{UJ\}_e$ contains two DOF/node, $1 \leq J \leq 2 = n$.

The detailed GWS^h completion assumes plane stress and FE NC $\{N_1(\zeta_\alpha)\}$ linear basis triangular Ω_e. The transformation matrix $[\mathbf{B}]_e$ thus contains constant data. The starting point is equation (11.22) and moving to index notation

$$[\mathbf{B}]_e \equiv \begin{bmatrix} \partial/\partial x_1 & 0 \\ 0 & \partial/\partial x_2 \\ (1/2)\partial/\partial x_2 & (1/2)\partial/\partial x_1 \end{bmatrix} \begin{bmatrix} \{N_1\}^T & \{0\} \\ \{0\} & \{N_1\}^T \end{bmatrix}. \tag{11.29}$$

The coordinate transformation chain rule is $\partial/\partial x_i = (\partial\zeta_\alpha/\partial x_i)_e\partial/\partial\zeta_\alpha$; recalling $\partial\{N_1\}/\partial\zeta_\alpha \Rightarrow \{\delta_\alpha\}$ generates Kronecker delta matrices, Section 7.4, the suggested exercise will confirm equation (11.29) becomes

$$[\mathbf{B}]_e = \frac{1}{\det_e}\begin{bmatrix} \zeta_{\alpha 1} & 0 \\ 0 & \zeta_{\alpha 2} \\ (1/2)\zeta_{\alpha 2} & (1/2)\zeta_{\alpha 1} \end{bmatrix}_e \begin{bmatrix} \{\delta_\alpha\}^T & \{0\} \\ \{0\} & \{\delta_\alpha\}^T \end{bmatrix}$$

$$\equiv \frac{1}{\det_e}\begin{bmatrix} \zeta_{11} & \zeta_{21} & \zeta_{31} & 0 & 0 & 0 \\ 0 & 0 & 0 & \zeta_{12} & \zeta_{22} & \zeta_{32} \\ \frac{1}{2}\zeta_{12} & \frac{1}{2}\zeta_{22} & \frac{1}{2}\zeta_{32} & \frac{1}{2}\zeta_{11} & \frac{1}{2}\zeta_{21} & \frac{1}{2}\zeta_{31} \end{bmatrix}_e = \det_e^{-1}[ETKI]_e$$ (11.30)

for the metric data definition $\det_e(\partial\zeta_\alpha/\partial x_i)_e \equiv \zeta_{\alpha i}$. Thereby, the $[\mathbf{B}]_e$ solution (11.30) defines $[ETKI]_e$ in equation (11.27) as a 3×6 matrix with elements various linear NC basis triangle node coordinate data, (7.33).

With $[ETKI]_e$ thus identified, hence also $[ETLI]_e$, the element *stiffness matrix* (11.27) for plane stress becomes

$$[STIFF]_e \equiv \int_{\Omega_e} [\mathbf{B}]^T[\mathbf{E}][\mathbf{B}]_e \, dxdy$$

$$= \frac{E(\det_e)^{-2}}{4(1-v^2)}\begin{bmatrix} \zeta_{11} & 0 & \frac{1}{2}\zeta_{12} \\ \zeta_{21} & 0 & \frac{1}{2}\zeta_{22} \\ \zeta_{31} & 0 & \frac{1}{2}\zeta_{32} \\ 0 & \zeta_{12} & \frac{1}{2}\zeta_{11} \\ 0 & \zeta_{22} & \frac{1}{2}\zeta_{21} \\ 0 & \zeta_{32} & \frac{1}{2}\zeta_{31} \end{bmatrix}_e \begin{bmatrix} 1 & v & 0 \\ v & 1 & 0 \\ 0 & 0 & (1-v)/2 \end{bmatrix}[ETKI]_e \int_{\Omega_e} dxdy$$

(11.31)

$$= \frac{\det_e^{-1} E}{2(1-v^2)}\begin{bmatrix} \zeta_{11} & 0 & \frac{1}{2}\zeta_{12} \\ \zeta_{21} & 0 & \frac{1}{2}\zeta_{22} \\ \zeta_{31} & 0 & \frac{1}{2}\zeta_{32} \\ 0 & \zeta_{12} & \frac{1}{2}\zeta_{11} \\ 0 & \zeta_{22} & \frac{1}{2}\zeta_{21} \\ 0 & \zeta_{32} & \frac{1}{2}\zeta_{31} \end{bmatrix}_e \begin{bmatrix} \zeta_{11} & \zeta_{21} & \zeta_{31} & v\zeta_{12} & v\zeta_{22} & v\zeta_{32} \\ v\zeta_{11} & v\zeta_{21} & v\zeta_{31} & \zeta_{12} & \zeta_{22} & \zeta_{32} \\ (\zeta_{12} & \zeta_{22} & \zeta_{32} & \zeta_{11} & \zeta_{21} & \zeta_{31})\frac{1-v}{4} \end{bmatrix}_e.$$

A decomposition of [STIFF]$_e$ into normal and shear stress components organizes interpretation of its constitution. For the resolution

$$[STIFF]_e = \frac{E}{2det_e(1-v^2)}[NORML]_e + \frac{E}{8det_e(1+v)}[SHEAR]_e \tag{11.32}$$

the [NORML]$_e$ and [SHEAR]$_e$ compliments are

$$[NORML]_e = \begin{bmatrix} \zeta_{11}^2 & & & & & & \\ \zeta_{11}\zeta_{21} & \zeta_{21}^2 & & & (sym) & & \\ \zeta_{11}\zeta_{31} & \zeta_{21}\zeta_{31} & \zeta_{31}^2 & & & & \\ v\zeta_{11}\zeta_{12} & v\zeta_{12}\zeta_{21} & v\zeta_{12}\zeta_{31} & \zeta_{12}^2 & & & \\ v\zeta_{11}\zeta_{22} & v\zeta_{22}\zeta_{21} & v\zeta_{22}\zeta_{31} & \zeta_{12}\zeta_{22} & \zeta_{22}^2 & & \\ v\zeta_{11}\zeta_{32} & v\zeta_{32}\zeta_{21} & v\zeta_{32}\zeta_{31} & \zeta_{12}\zeta_{32} & \zeta_{22}\zeta_{32} & \zeta_{32}^2 \end{bmatrix}_e \tag{11.33}$$

$$[SHEAR]_e = \begin{bmatrix} \zeta_{12}^2 & & & & & & \\ \zeta_{12}\zeta_{22} & \zeta_{22}^2 & & & (sym) & & \\ \zeta_{12}\zeta_{32} & \zeta_{32}\zeta_{22} & \zeta_{32}^2 & & & & \\ v\zeta_{11}\zeta_{12} & v\zeta_{11}\zeta_{22} & v\zeta_{11}\zeta_{32} & \zeta_{11}^2 & & & \\ v\zeta_{21}\zeta_{12} & v\zeta_{21}\zeta_{22} & v\zeta_{21}\zeta_{32} & \zeta_{11}\zeta_{21} & \zeta_{21}^2 & & \\ v\zeta_{31}\zeta_{12} & v\zeta_{31}\zeta_{22} & v\zeta_{31}\zeta_{32} & \zeta_{11}\zeta_{31} & \zeta_{21}\zeta_{31} & \zeta_{31}^2 \end{bmatrix}_e \tag{11.34}$$

which indeed are geometric node metric data difference products with select insertion of Poisson ratio. DOF order in equations (11.32)–(11.34) is

$$\{U\}_e \equiv \{U1, U2, U3, V1, V2, V3\}_e^T. \tag{11.35}$$

The $\{N_1(\zeta_\alpha)\}$ basis *stiffness matrix* for plane stress is fully identified. Now extracting a common multiplier the DOF coupling in [STIFF]$_e$ can be symbolized as

$$\begin{aligned} [STIFF]_e &= \frac{E}{2det_e(1+v)}\left(\frac{1}{1-v}[NORML]_e + \frac{1}{4}[SHEAR]_e\right) \\ &\equiv \begin{bmatrix} [DIFFUU]_e, & [DIFFUV]_e \\ [DIFFVU]_e, & [DIFFVV]_e \end{bmatrix} \end{aligned} \tag{11.36}$$

The matrix partitions in equation (11.36) enable the observation that the self-coupling matrices [DIFFUU] and [DIFFVV] contain the identical metric data products, compare the upper left triad in [NORML]$_e$ to the lower right triad in [SHEAR]$_e$.

A highly informative comparison is now possible with the scalar heat transfer $n = 2$ [DIFF]$_e$ matrix. From Section 7.4

$$[DIFF]_e = \frac{\bar{k}_e}{2det_e}\begin{bmatrix} \zeta_{11}^2+\zeta_{12}^2, & \zeta_{11}\zeta_{21}+\zeta_{12}\zeta_{22}, & \zeta_{11}\zeta_{31}+\zeta_{12}\zeta_{32} \\ \zeta_{21}\zeta_{11}+\zeta_{22}\zeta_{12}, & \zeta_{21}^2+\zeta_{22}^2, & \zeta_{21}\zeta_{31}+\zeta_{22}\zeta_{32} \\ \zeta_{31}\zeta_{11}+\zeta_{32}\zeta_{12}, & \zeta_{31}\zeta_{21}+\zeta_{32}\zeta_{22}, & \zeta_{31}^2+\zeta_{32}^2 \end{bmatrix}_e . \tag{7.37}$$

The divisor common to equations (11.36) and (7.37) is $2\det_e$, hence the diffusion coefficient correlation is $\bar{k}_e \Leftrightarrow E/(1+v)$. Thereby, a problem statement in plane stress is *thoroughly* diffusive, as $v < 1$ and $E \approx O(10^6)$ while thermal conductivity average ranges $\bar{k}_e < 10^3$.

It remains to form the NC $k = 1$ basis load vector $\{b\}_e$ for the plane stress GWS^h formulation. The initial strain and stress contributions, $\{\varepsilon_0\}_e$ and $\{\sigma_0\}_e$ in equation (11.28), are but special cases of $[STIFF]_e$ and are left as an exercise. The body force, surface traction, and point load contributions involve already established element matrices

$$\{b\}_e = \frac{\det_e}{6} \begin{bmatrix} [B\,200L] & [0] \\ [0] & [B\,200L] \end{bmatrix} \begin{Bmatrix} \text{RHOGX} \\ \text{RHOGY} \end{Bmatrix}_e$$
$$+ \frac{\det_e}{3} \begin{bmatrix} [A\,200L] & [0] \\ [0] & [A\,200L] \end{bmatrix} \begin{Bmatrix} \text{TX} \\ \text{TY} \end{Bmatrix}_e + P \begin{Bmatrix} \delta_{\text{DOF,X}} \\ \delta_{\text{DOF,Y}} \end{Bmatrix}_e . \tag{11.37}$$

11.5 Plane Elasticity Computer Lab

Computer lab 11.1 specifies the design *optimization* requirement of minimizing the stress concentration in a plate loaded in tension by geometry alteration of the initially circular hole. This elasticity statement is plane stress; the associated GWS^h COMSOL . mph file is available at www.comsol.com/community/exchange/?page=2.

The definition assumes the plate lies in the x–y plane, Figure 11.1 The first requirement is the identification of well-posed BCs. The plate is assumed rigidly attached to the wall on the left leading to the displacement vector BC $\mathbf{u}(0, y) = \mathbf{0}$. No constraints

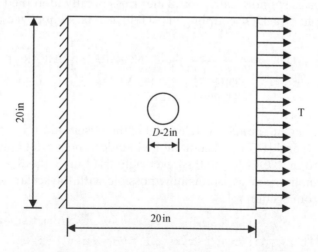

Figure 11.1 Elastic plate with a hole in tension

exist on the upper and lower edges, hence homogeneous Neumann BCs are appropriate. The surface traction $\mathbf{T} = T\,\hat{\mathbf{i}}$ is applied parallel to the x axis and is assumed uniform with magnitude $T = 100$ psi.

The plate is square of size (w,w,t) with dimensions $(20a,20a,1)$ for a = one inch. The initial circular hole radius is $r = a$. The plate is steel with (COMSOL notation) material properties thickness density (rho0) $= 0.35405$ psi, Poisson ratio (nu0) $= 0.3$ and, Youngs modulus (E0) $= 30.E06$ psi.

The design's goal is to alter the original circular hole shape to minimize the associated local stress concentration. The *von Mises stress* is an accepted scalar metric of an n-D state of stress for assessing design margins in structural mechanics. Its definition corresponds to the stress state during failure of ductile materials in the uniaxial tensile test. COMSOL outputs the *effective* von Mises stress with definition

$$\sigma_{\text{eff}} \equiv \sqrt{\sigma_x^2 + \sigma_y^2 + \sigma_z^2 - \sigma_x\sigma_y - \sigma_y\sigma_z - \sigma_z\sigma_x + 3\tau_{xy}^2 + 3\tau_{yz}^2 + 3\tau_{zx}^2}. \tag{11.38}$$

Since the plane stress/strain PDE system is linear second order EBV, the GWS^h theory (9.5) is pertinent for predicting asymptotic convergence in the energy norm under regular solution-adapted mesh refinement. COMSOL does not output the energy norm but it can be readily postprocessed. In the discussion leading to equation (11.26), the global matrix statement is $[\text{STIFF}]\{U\} = \{b\}$. An exercise suggests verifying the energy norm definition is $\frac{1}{2}\{U\}^T[\text{STIFF}]\{U\}$. The COMSOL solve process generates $\{U\}$ hence norm computation is the postprocess operation $\frac{1}{2}\{U\}^T\{b\}$. The lab suggests adding this computation to the .mph file.

The computer lab 11.1 protocol is:

1. access the COMSOL .mph file set up for plane stress analysis of a plate with hole in tension
2. execute the .mph file base specification, hence generate the $\{N_1(\zeta_\alpha)\}$ solution on the initial self-generated mesh
3. add the energy norm postprocess computation to 2.
4. execute a uniform mesh-refinement study hence observe solution divergence (!) at plate corner wall attachments
5. the cause for divergence is the over constraining wall BC, hence change it to $u(0, y) = 0$ only
6. retaining the area of the hole constant alter its geometry to generate the *minimum* local stress concentration with *norm* the effective von Mises stress
7. verify the accuracy of the step 6 design objective via solution-adapted mesh refinement.

(*Note*: The colored versions of most figures in this chapter are found and can be downloaded from www.wiley.com/go/baker/finite.)

A summary of the circular hole specification computer lab 11.1 results follows. Figure 11.2a graphs the step 2 COMSOL self-generated linear NC basis mesh, which possesses enhanced resolution about the hole. This mesh GWS^h algorithm solution for displacement resolution $\mathbf{u}^h = u^h(x,y)\hat{\mathbf{i}} + v^h(x,y)\hat{\mathbf{j}}$ is graphed in Figure 11.2b and c. The black circle is the original location of the hole; plate lateral shrinking distribution is consistent with the Neumann BC and axial stretching is clearly illustrated. That the initial mesh provides adequate resolution for an *engineering accurate* solution of the circular hole case is confirmed by the data in Figure 11.3.

The von Mises effective stress distribution graphic, Figure 11.4, gives indication of wall BC specification incorrectness. The local extrema about the hole appear well-resolved; however the solution lighter shade local to both corners of the wall attachment indicates the problem. Locally refining this mesh region leads to progressively unrealistic results, that is, the solution diverges! The cause is violation of the structural mechanics Poisson ratio effect, clearly illustrated by the divergence term in equation (11.10). The resolution is the step 4 BC alteration.

Classic textbook guidance on circular hole stress concentration reduction is elongation parallel to the applied traction. Keeping the hole area constant, Table 11.1 summarizes the alteration of von Mises stress extremum for the cited meshes. These data are required validated via the solution adapted mesh-refinement lab component.

11.6 Fluid Mechanics, Incompressible-Thermal Flow

The incompressible Navier–Stokes (INS) DM + DP conservation principle PDE system is shown in equations (2.26)–(2.28), Chapter 2. The *fundamental* mathematical issue in formulating a discrete approximation algorithm is enforcing the DM *differential constraint* that the velocity field be divergence free

$$DM: \quad \nabla \cdot \mathbf{u} = 0. \tag{2.26}$$

Numerous mathematically exact and computationally inexact approaches exist to account for equation (2.26); a detailed comparative summary is given in [5]. Mathematically exact enforcement involves vector field theory with state variable transformation to a combination of vorticity, stream, and velocity vector fields. The explicit appearance of pressure in the DP PDE is removed in this process. Alternatively, a uniquely weak form theory implementing constraint (2.26) requires the velocity vector trial space reside in a Hilbert space H^m containing functions one differentiability order higher than that for pressure [6].

Computationally inexact theories universally involve an iterative process wherein equation (2.26) is enforced *only* approximately via a pressure accumulation sequence within the DP algorithm matrix iteration. Under this *iteration* genre exist a multitude of approaches, for example, MAC, SMAC, SIMPLE, SIMPLER, SIMPLEC, PISO, . . . A theoretical unification of this family of algorithms under the umbrella of *pressure projection* theory is detailed in [7].

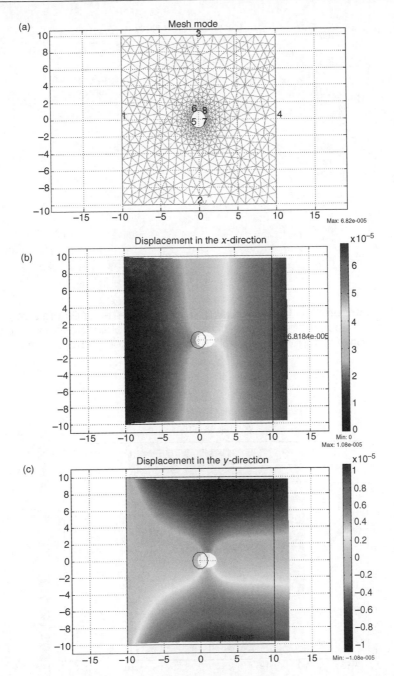

Figure 11.2 Computer lab 11.1 GWS^h $\{N_1(\zeta_\alpha)\}$ base solution, plate with a hole plane stress formulation: (a) COMSOL self-generated mesh, (b) $u^h(x,y)$ displacement field, (c) $v^h(x,y)$ displacement field (a colored version of this figure is available at www.wiley. com/go/baker/finite)

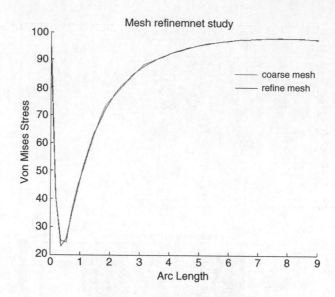

Figure 11.3 Computer lab 11.1 GWS^h {$N_1(\zeta_\alpha)$} solution mesh-refinement study, von Mises effective stress distribution on a vertical traverse, circular hole edge to top of plate (a colored version of this figure is available at www.wiley.com/go/baker/finite)

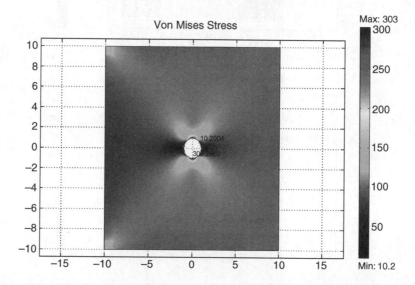

Figure 11.4 Computer lab 11.1 GWS^h {$N_1(\zeta_\alpha)$} base solution, plate with a hole plane stress formulation, von Mises stress distribution (a colored version of this figure is available at www.wiley.com/go/baker/finite)

Table 11.1   von Mises stress extremum with hole shape

Shape	Area	x semi-axis	y semi-axis	Elements	Max stress
Circle	π	1	1	16,896	309.142
Ellipse	π	0.5	2	22,432	954.041
Ellipse	π	2	0.5	22,816	257.289
Ellipse	π	4	0.25	32,576	255.468
Ellipse	π	8	0.125	32,672	1273.695

This chapter section uses vector field theory to exactly enforce equation (2.26) for the $n = 2$ restriction. With addition of the DE conservation principle PDE, the subject thermal-incompressible $n = 2$ PDE system in non-dimensional (non-D) form is equation (2.26) with

$$\mathcal{L}(u_i) = \frac{\partial u_i}{\partial t} + u_j \frac{\partial u_i}{\partial x_j} + \mathrm{Eu}\frac{\partial P}{\partial x_i} - \frac{1}{\mathrm{Re}}\frac{\partial^2 u_i}{\partial x_j^2} + \frac{\mathrm{Gr}}{\mathrm{Re}^2}\Theta\hat{g}_i = 0 \qquad (11.39)$$

$$\mathcal{L}(\Theta) = \frac{\partial\Theta}{\partial t} + u_j \frac{\partial\Theta}{\partial x_j} - \frac{1}{\mathrm{Pe}}\frac{\partial^2\Theta}{\partial x_j^2} - s_\Theta = 0. \qquad (11.40)$$

The PDE system *state variable* is $q(\mathbf{x},t) = \{u,v,P,\Theta\}$, the cartesian velocity vector resolution, kinematic pressure $P \equiv p/\rho_0$, and potential temperature $\Theta \equiv (T - T_{\min})/(T_{\max} - T_{\min})$. The reference scales for the identified non-D groups are $(L, U, P, \rho_0, \upsilon, \Delta T, c_\mathrm{p}, k)_\mathrm{ref}$ with

$\mathrm{Re} = UL/\upsilon$	Reynolds number
$\mathrm{Eu} = P/U^2$	Euler number
$\mathrm{Gr} = \Delta T \beta L^3/\upsilon^2$	Grashoff number ($\beta \equiv 1/T_{\mathrm{absolute}}$)
$\mathrm{Pe} = \mathrm{RePr}$	Peclet number
$\mathrm{Pr} = c_\mathrm{p}\upsilon/k$	Prandtl number

Exact satisfaction of equation (2.26) results by defining the $n = 2$ velocity vector as the out-of-plane ($\hat{\mathbf{k}}$) component of the stream vector, familiarly called *streamfunction* ψ. The operation is *curl*

$$\mathbf{u} \equiv \nabla \times \psi\hat{\mathbf{k}} \qquad (11.41)$$

and $\nabla \cdot \mathbf{u} = \nabla \cdot \nabla \times \psi\hat{\mathbf{k}} = 0$ is obvious. Substitution of equation (11.41) into equation (11.39) introduces ψ as the replacement for $\mathbf{u}$ but pressure remains present in the state variable.

Its removal accrues to forming the $\hat{\mathbf{k}}$ component of the curl of DP, equation (11.39), then substitute therein the definition of the $\hat{\mathbf{k}}$ component of vorticity

$$\omega \equiv \nabla \times \mathbf{u} \cdot \hat{\mathbf{k}}. \qquad (11.42)$$

This removes pressure from the state variable, as $\nabla \times \nabla P \cdot \hat{\mathbf{k}} = 0$ identically via vector calculus.

The net result is generation of the $n=2$ thermal *vorticity* form of the DP conservation principle (11.39)

$$\text{DP}: \quad \mathcal{L}(\omega) = \frac{\partial \omega}{\partial t} + \nabla \times \psi \hat{\mathbf{k}} \cdot \nabla \omega - \frac{1}{\text{Re}} \nabla^2 \omega + \frac{\text{Gr}}{\text{Re}^2} \nabla \times \Theta \hat{g} \cdot \hat{\mathbf{k}} = 0. \tag{11.43}$$

The DP *vorticity–streamfunction* PDE system replacement for equations (11.39),(11.40) is established by substituting equation (11.41) into equation (11.42). This generates the linear laplacian PDE

$$\mathcal{L}(\psi) = -\nabla^2 \psi - \omega = 0 \tag{11.44}$$

a statement of *kinematic compatibility* valid *everywhere* on the INS domain Ω *and its* enclosure $\partial\Omega$.

The system (11.43)–(11.44), along with equation (11.40) with **u** replaced by equation (11.41), is the sought DP+DE restatement with pressure removed and rigorous enforcement of DM. As Re and Pe are non-negative, this PDE system is EBV, hence BC specifications are required for state variable $\{q(\mathbf{x}, t)\} = \{\omega, \Theta, \psi\}^T$ on the *entirety* of $\partial\Omega$.

The distribution of velocity and temperature on $\partial\Omega$ segments defined as inflow are typically given *data*. From these data, inflow Dirichlet BC distributions for state variable $\{q\}$ are readily established via the calculus operations

$$\omega_{\text{in}} = \nabla \times u(s)_{\text{in}} \hat{\mathbf{n}} \cdot \hat{\mathbf{k}} = \frac{du(s)_{\text{in}}}{ds}, \quad \Delta\psi_{\text{in}} = \int_0^s u(s)_{\text{in}} ds \tag{11.45}$$

for s the coordinate spanning the inflow $\partial\Omega$ segment.

For boundary segments $\partial\Omega$ with outflow generally no *a priori* knowledge exists on $\{q\}$. Thereby, the requirement is to ensure that an outflow boundary is located sufficiently downstream from flowfield action such that the homogeneous Neumann BC $\nabla\{q\}\cdot\hat{\mathbf{n}} = 0$ is applicable. The remaining $\partial\Omega$ segments are impervious to flow, hence $\psi = $ constant by definition, with the actual constant(s) determined by solution of the second expression in equation (11.45). Thereon, a fixed temperature Θ and/or Robin BCs are admissible.

The remaining BC required identified is for ω on impervious boundary segments. As stated, the kinematic relation (11.44) is valid *everywhere* on Ω *and* $\partial\Omega$, and since $\psi = $ constant on any such segment the defining BC is determinable from dimensionally *reduced* equation (11.44)

$$\frac{d^2\psi}{dn^2} + \omega = 0 \tag{11.46}$$

with n the coordinate spanning the boundary segment normal direction.

Writing a Taylor series expansion on ψ from the wall into the near-field flow a distance Δn and inserting equation (11.46) generates

$$\psi(\Delta n) = \psi_w + \Delta n \frac{d\psi}{dn}\bigg|_w + \frac{1}{2}(\Delta n)^2 \frac{d^2\psi}{dn^2}\bigg|_w + \frac{1}{3!}(\Delta n)^3 \frac{d^3\psi}{dn^3}\bigg|_w + O(\Delta n^4)$$

$$= \psi_w - U_w \Delta n - \frac{(\Delta n)^2}{2}\omega_w - \frac{(\Delta n)^3}{6}\frac{d\omega}{dn}\bigg|_w + O(\Delta n^4). \qquad (11.47)$$

In equation (11.47) U_w is the tangential velocity of the impervious segment, ω_w is the vorticity at the wall, and $\psi(\Delta n)$ is the nearest wall-adjacent ψ.

Rearrangement of equation (11.47) to homogeneous form generates the Robin BC for vorticity

$$\ell(\omega) = \nabla\omega \cdot \hat{\mathbf{n}} + \frac{3}{\Delta n}\omega_w + \frac{6}{\Delta n^2}U_w + \frac{6}{\Delta n^3}(\psi_{w+1} - \psi_w) + O(\Delta n^4) \qquad (11.48)$$

readily implemented in $\mathrm{GWS}^h + \theta\mathrm{TS}$. However, traditional practice in the CFD world is to consider equation (11.47) a strong-form *discrete* Dirichlet BC statement for $\omega_w \Rightarrow \Omega_w$, the algorithm DOF. The resultant first and second order-accurate discrete DOF Dirichlet BC specifications are

$$\Omega_w + \frac{2}{(\Delta n)^2}(\Psi_{w+1} - \Psi_w) + \frac{2}{\Delta n}U_w + O(\Delta n) = 0$$

$$\Omega_w + \frac{1}{2}\Omega_{w+1} + \frac{3}{(\Delta n)^2}(\Psi_{w+1} - \Psi_w) + \frac{3}{\Delta n}U_w + O(\Delta n^2) = 0 \qquad (11.49)$$

Subscripts w and w + 1 denote DOF on the wall and the first off-wall DOF in the direction of positive n.

11.7 Vorticity–Streamfunction $\mathrm{GWS}^h + \theta\mathrm{TS}$ Algorithm

The $\mathrm{GWS}^N + \theta\mathrm{TS}$ for the vorticity–streamfunction–thermal PDE system follows the developed process. For the three-member state variable $\{q(\mathbf{x}, t)\} = \{\omega, \Theta, \psi\}^T$, the approximate solution is

$$q(\mathbf{x}, t) \approx q^N(\mathbf{x}, t) \equiv \sum_{\alpha=1}^{N}\mathrm{diag}[\Psi_\alpha(\mathbf{x})]\{Q(t)\}_\alpha \Rightarrow q^h(\mathbf{x}, t)$$

$$q^h(\mathbf{x}, t) = \cup_e q_e(\mathbf{x}, t), \quad q_e = \{N_k(\zeta_\alpha, \boldsymbol{\eta})\}^T\{Q(t)\}_e \qquad (11.50)$$

$$\{Q(t)\}_e \equiv \{\Omega, \Theta, \Psi\}_e^T$$

anticipating the terminal discrete implementation. For the vorticity–temperature PDE system, equations (11.43) and (11.40), GWS^N produces the coupled global matrix ODE system for $\{Q\} = \{\Omega, \Theta\}^T$

$$\mathrm{GWS}^N = [\mathrm{MASS}]\frac{d\{Q\}}{dt} + \{\mathrm{RES}\{Q\}\} = \{0\} \tag{11.51}$$

while GWS^N for the kinematic compatibility PDE (11.44) directly produces the algebraic statement for $\{Q\} = \{\Psi\}$

$$\{F(\{Q\})\} = [\mathrm{DIFF}]\{\Psi\} - [\mathrm{MASS}]\{\Omega\} = \{0\}. \tag{11.52}$$

The GWS^N (11.51) constitutes data for the θ-implicit time TS. From the Chapter 9.3 process, upon clearing $[\mathrm{MASS}]^{-1}$ the $\mathrm{GWS}^N + \theta\mathrm{TS}$ generates the $\{Q\} = \{\Omega, \Theta\}^T$ coupled algebraic system

$$\{F(\{Q\})\} = [\mathrm{MASS}]\{Q_{n+1} - Q_n\}$$
$$+ \Delta t\left(\theta\{\mathrm{RES}(\{Q\})\}_{n+1} + (1-\theta)\{\mathrm{RES}(\{Q\})\}_n\right) = \{0\}. \tag{11.53}$$

Solution of the fully coupled algebraic system (11.52)–(11.53)is formulated symbolically as the Newton algorithm

$$[\mathrm{JAC}(\{Q\})]\{\delta Q\}^{p+1} = -\{F(\{Q\})\}^p$$

$$[\mathrm{JAC}(\{Q\})] \equiv \frac{\partial\{F(\{Q\})\}}{\partial\{Q\}} = \begin{bmatrix} [\mathrm{J}\Omega\Omega], & [\mathrm{J}\Omega\Psi], & [\mathrm{J}\Omega\Theta] \\ [\mathrm{J}\Psi\Omega], & [\mathrm{J}\Psi\Psi], & [0] \\ [0], & [\mathrm{J}\Theta\Psi], & [\mathrm{J}\Theta\Theta] \end{bmatrix}. \tag{11.54}$$

In equation (11.54), $[\mathrm{JAC}(\{Q\})]$ is block banded by state variable member coupling with the recognition that $[\mathrm{J}\Psi\Theta]$ and $[\mathrm{J}\Theta\Omega]$ identically vanish.

The $\mathrm{GWS}^N + \theta\mathrm{TS}$ theory now complete, discrete implementation $\mathrm{GWS}^h + \theta\mathrm{TS} = S_e\{WS\}_e$ involves selecting an FE trial space basis $\{N_k(\cdot)\}$ and moving the integrals to the generic $n = 2$ domain Ω_e. In equations (11.52)–(11.53), $[\mathrm{MASS}] \Rightarrow [\mathrm{B}200d]_e$ and $[\mathrm{DIFF}] \Rightarrow [\mathrm{B}2KKd]_e$. Switching DOF notation to $\{\mathrm{OMG}, \mathrm{TMP}, \mathrm{PSI}\}_e^T$ from $\{\Omega, \Theta, \Psi\}_e^T$ in anticipation of template statements, an exercise will verify

$$\{\mathrm{RES}(\{\mathrm{OMG}\})\}_e = [\mathrm{VEL}(\mathrm{PSI})]_e\{\mathrm{OMG}\}_e + \mathrm{Re}^{-1}[\mathrm{B}2KKd]_e\{\mathrm{OMG}\}_e$$
$$- \mathrm{Gr}\,\mathrm{Re}^{-2}[\mathrm{B}20Xd]_e\{\mathrm{TMP}\}_e \tag{11.55}$$

$$\{\mathrm{RES}(\{\mathrm{TMP}\})\}_e = [\mathrm{VEL}(\mathrm{PSI})]_e\{\mathrm{TMP}\}_e + \mathrm{Pe}^{-1}[\mathrm{B}2KKd]_e\{\mathrm{TMP}\}_e$$

The element matrix $[\mathrm{VEL}(\mathrm{PSI})]$ is entirely new, resulting from the convection nonlinearity in equation (11.43). The continuum calculus operation is $\nabla \times \psi_e\hat{\mathbf{k}} \cdot \nabla\omega$; another exercise suggests direct substitution of the FE basis matrix definition therein which *precisely* (!) establishes the element *hypermatrix* as

$$[\mathrm{VEL}(\mathrm{PSI})]_e = \det_e^{-1}(\mathrm{ET}\,J1_e\,\mathrm{ET}\,K2_e)\{\mathrm{PSI}\}_e^T([\mathrm{B}3K0Jd] - [\mathrm{B}3J0Kd]) \tag{11.56}$$

valid for any basis completeness degree k implementation.

For $k=1$ NC/TP bases the element-independent *skew-symmetric* hypermatrices [B3K0JL] are readily evaluated. Recalling that products of metric data ETJI_e are denoted by their single subscript locations in the metric data array, equation (7.71), the cited exercise will verify the $k=1$ TP basis implementation of equation (11.56) simplifies to

$$[\text{VEL(PSI)}]_e = \det_e^{-1}\{\text{PSI}\}_e^T \begin{pmatrix} (14-23)[\text{B3201L}] \\ -(32-14)[\text{B3102L}] \end{pmatrix} . \tag{11.57}$$

$$= \det_e^{-1}\{\text{PSI}\}_e^T (14-23)([\text{B3201L}] + [\text{B3102L}])$$

The TP hypermatrix sum, [B3201L] + [B3102L], in equation (11.57) is

$$[B3K0KTPL] = \frac{1}{12} \left[\begin{Bmatrix} 0 \\ 2 \\ 0 \\ -2 \\ 0 \\ 2 \\ -1 \\ -1 \\ 0 \\ 1 \\ 0 \\ -1 \\ 0 \\ 1 \\ 1 \\ -2 \end{Bmatrix} \begin{Bmatrix} -2 \\ 0 \\ 1 \\ 1 \\ -2 \\ 0 \\ 2 \\ 0 \\ -1 \\ 0 \\ 2 \\ -1 \\ -1 \\ 0 \\ 1 \\ 0 \end{Bmatrix} \begin{Bmatrix} 0 \\ -1 \\ 0 \\ 1 \\ 1 \\ -2 \\ 0 \\ 1 \\ 0 \\ -2 \\ 0 \\ 2 \\ -1 \\ -1 \\ 0 \\ 2 \end{Bmatrix} \begin{Bmatrix} 2 \\ -1 \\ -1 \\ 0 \\ 1 \\ 0 \\ -1 \\ 0 \\ 1 \\ 1 \\ -2 \\ 0 \\ 2 \\ 0 \\ -2 \\ 0 \end{Bmatrix} \right] . \tag{11.58}$$

The companion NC $k=1$ basis hypermatrix matrix sum is

$$[B3K0KNCL] = \frac{1}{6} \left[\begin{Bmatrix} 0 \\ 1 \\ -1 \\ 0 \\ 1 \\ -1 \\ 0 \\ 1 \\ -1 \end{Bmatrix} \begin{Bmatrix} -1 \\ 0 \\ 1 \\ -1 \\ 0 \\ 1 \\ -1 \\ 0 \\ 1 \end{Bmatrix} \begin{Bmatrix} 1 \\ -1 \\ 0 \\ 1 \\ -1 \\ 0 \\ 1 \\ -1 \\ 0 \end{Bmatrix} \right] . \tag{11.59}$$

Recalling definition (7.40), the template statement for equation (11.57) inserted into equation (11.55) is

$$\begin{aligned} \{WS\}_e &= [\text{VEL(PSI)}]_e\{Q\}_e \\ &= (\ )(\ )\{\text{PSI}\}(14; -1)[B3K0KTPL]\{Q\} \\ &\quad +(-)(\ )\{\text{PSI}\}(23; -1)[B3K0KTPL]\{Q\} \end{aligned} \tag{11.60}$$

The curl operation on the gravity body force term redirects its impact to "horizontal" in equation (11.43), hence the matrix name $[B20Xd]_e$. The matrix statement and template for $k = 1$ TP basis implementation are

$$[B\,20XL]_e = \det_e^0 \text{ET } J1_e[B\,20JL]$$
$$= (1)_e[B\,201L] + (3)_e[B\,202L] \qquad (11.61)$$

$$\{WS\}_e = \frac{-Gr}{Re^2}[B\,20XL]_e\{TMP\}_e$$
$$= (-GR, RE^{-2})(\ )\{\ \}(1;0)[B\,201L]\{TMP\} \qquad (11.62)$$
$$+(-GR, RE^{-2})(\ )\{\ \}(3;0)[B\,202L]\{TMP\}$$

The templates for all remaining terms in equations (11.52)–(11.53) are available, Tables 7.1 and 7.3, as are their contributions to $[JAC]_e$. The new nonlinear contribution results from the product of equation (11.57) with $\{Q\}_e = \{OMG, TMP\}_e$. Differentiating equation (11.57) by $\{Q\}_e$ is elementary, and following interchange of the $\{Q\}$ and $\{PSI\}^T$ template positions, differentiation by $\{PSI\}$ simply generates a sign change on the terms in equation (11.57)!

The final jacobian formation topic is accounting for the vorticity impervious boundary BCs. The Robin BC (11.48) ends up in the algorithm statement (11.53), hence differentiation is analytical. Conversely, for the discrete DOF Dirichlet option (11.49), the results of the scalar DOF differentiation must be properly (manually) inserted into $[J\Omega\Omega]_e$ and $[J\Omega\Psi]_e$.

11.8 An Isothermal INS Validation Experiment

The classic validation problem for incompressible Navier–Stokes is prediction of flow in a close-coupled stepwall diffuser. Figure 11.5 graphs the classic 1:2 duct cross-section area ratio of the diffuser with boundary surface notation consistent with COMSOL execution directives. The description for creating a COMSOL executable for the derived vorticity–streamfunction $GWS^h + \theta TS$ $k = 1$ NC basis algorithm, equations (11.52)–(11.54), follows. Alternatively, the .mph file for this lab is available at www.comsol.com/community/exchange/?page=2.

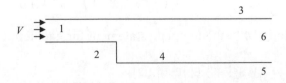

Figure 11.5 Close-coupled step wall diffuser, $n = 2$

To set up the computer lab .mph file, open the 2D general, nonlinear stationary PDE mode from the COMSOL model navigator. This mode is selected when the built-in mode is not available, hence one must define the PDE system. Set the number of dependent variables to 2 and name them psi (ψ) and omega (ω). In subdomain settings define

Subdomain	1
$\Gamma(1)$	-omegax –omegay
$\Gamma(2)$	-psix –psiy
F(1)	Re*(psix*omegay - psiy*omegax)
F(2)	omega

The $O(\Delta n)$ vorticity BC (11.49) is the COMSOL default. BCs for streamfunction ψ are applied on the inflow boundary 1 and on all impervious segments 2, 3, 4, 5, Figure 11.5. The homogeneous Neumann BC exists on segment 6; the COMSOL input definitions are

Boundary	1	2,4,5	3	6
R(1)	$-psi + (3*s^2/2-s^3)$	$-psi$	$-psi + \frac{1}{2}$	0
R(2)	0	0	0	0

The astute reader will recognize the formula for psi on boundary segment R(1) defines the inflow velocity as fully developed, a parabola in s.

In the COMSOL solver option, choose the parametric mode to enable executing the lab for $100 \le Re \le 600$ in increments of 100. As usual, the code self-generates the initial FE NC linear basis mesh for the first specification $Re = 100$. Via the restart function, each converged Re solution for {Q} is used as the IC for the next larger Re execution. (Termed Re-*continuation*, this IC-generator compensates for the INS diffusion mechanism diminishing with increasing Re, which makes *cold-starting* a solution at large Re very compute intensive.)

The computer lab 11.2 computational experiment protocol is:

1. secure a COMSOL .mph file for the $GWS^h + \theta TS$ linear NC basis algorithm for the step-wall diffuser.
2. determine the step-induced primary recirculation region reattachment coordinate on the lower wall for Re specifications $100 \le Re \le 600$.
3. for $Re = 600$, note the upper recirculation region intersects the outflow plane which violates the outflow BC, hence corrupts the solution.

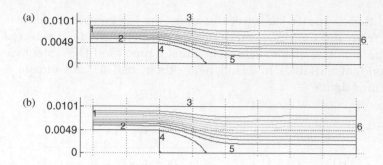

Figure 11.6 Step wall diffuser, GWS^h $k=1$ NC basis steady ψ^h solution distributions, Re $=100$: (a) $M=348$, (b) $M=2573$ (a colored version of this figure is available at www. wiley.com/go/baker/finite)

4. therefore, elongate the base solution domain to ensure the outflow Neumann BC specification is appropriate.
5. via a regular mesh-refinement study confirm achieving solution engineering accuracy for Re $=100$ and 600.
6. validate the GWS^h steady solution prediction of primary recirculation region reattachment coordinate using the available experimental data [8].

Following is a brief summary of computer lab 11.2 generated *a posteriori* data. The COMSOL self-generated mesh supports an engineering accurate solution for Re $=100$. Figure 11.6 graphs steady solution streamfunction ψ^h distributions for, (a) the initial $M=348$ mesh, and (b) the next automated mesh refinement, $M=2573$. The ψ^h distributions are visually indistinguishable.

The steady ψ^h solution gives the first indication of a second, upper-wall recirculation bubble at Re $=400$, Figure 11.7a. This bubble extends to the outflow boundary at Re $=500$, Figure 11.7b, which violates the Neumann BC application. The steady ψ^h

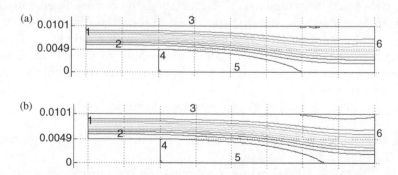

Figure 11.7 Step wall diffuser, GWS^h $k=1$ NC basis steady ψ^h solution distributions, $M=2573$: (a) Re $=400$, (b) Re $=500$ (a colored version of this figure is available at www. wiley.com/go/baker/finite)

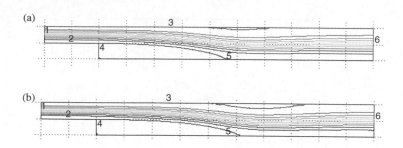

Figure 11.8 Step wall diffuser, GWS^h $k=1$ NC basis steady ψ^h solution distributions, $M=3807$: (a) Re $=500$, (b) Re $=600$ (a colored version of this figure is available at www.wiley.com/go/baker/finite)

solution on the elongated domain $M=3807$ mesh solves the BC issue, Figure 11.8a, which also proves adequate for the Re $=600$ solution, Figure 11.8b.

Having assessed these solutions are compatible with the applied BCs, the validation exercise is to compare the lower wall intercept of the primary recirculation bubble with the experimental data. For any Re this coordinate is mathematically identified by a sign change in the wall vorticity DOF distribution, which indicates a local stagnation point. The comparison of GWS^h $k=1$ NC basis steady solution prediction to published experimental data, [8], confirms excellent quantitative agreement for all Re <400 which degrades at the larger Re, Table 11.2.

The solution intercept data graph confirms the disagreement is progressive with increasing Re >400, Figure 11.9. The departure sequence is not a flaw in the GWS^h algorithm, but rather the assumption that larger Re stepwall diffuser flowfields are two-dimensional. The CFD literature confirms this progression for dozens of algorithms for Re >400. An $n=3$ $mGWS^h(\beta)$ algorithm prediction, [9], generates firm quantitative validation, Figure 11.10.

Table 11.2 Step wall diffuser primary recirculation bubble lower wall x intercept, $100 \le Re \le 600$, comparison normalized on step height s

Re	x	x/s	expt x/s, [8]
100	0.0160	3.27	3.10
200	0.0240	4.90	4.95
300	0.0330	6.7	6.8
400	0.0405	8.3	8.6
500	0.0460	9.4	10.4
600	0.0520	10.6	11.6

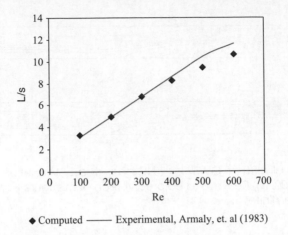

◆ Computed ———— Experimental, Armaly, et. al (1983)

Figure 11.9 Step wall diffuser primary recirculation bubble lower wall intercept, $n = 2$ GWS^h-predicted and experiment, $100 \leq Re \leq 600$ (a colored version of this figure is available at www.wiley.com/go/baker/finite)

11.9 Multimode Convection Heat Transfer

Predicting heat transfer in tube bank cross-flow heat exchangers is a classic mechanical engineering design requirement. Consider the array of heated tubes submerged in a chamber with fluid cross-flow, Figure 11.11a. Neglecting end effects admits assuming the flowfield uniform in horizontal planes with normal parallel to tube axes. For modest onset velocity U_∞, corresponding to small Re, the solution domain can be reduced to 2D executing a steady solution process. For this restriction, due to multiple planes of

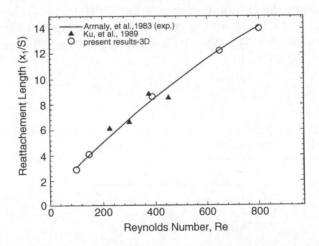

Figure 11.10 Step wall diffuser symmetry plane primary recirculation bubble lower wall intercept, $n = 3$ mGWS^h algorithm comparison with experiment, $100 \leq Re \leq 800$ (9)

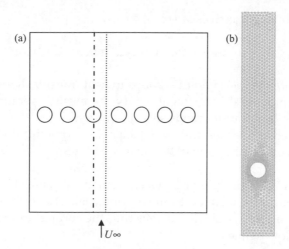

Figure 11.11 Tube bank cross-flow heat exchanger (a) geometric essential with symmetry planes, (b) NC $k=1$ basis $M=2226$ mesh Ω^h

symmetry, the minimum span solution domain Ω is the narrow region bounded by dashed and solid lines in Figure 11.11a. Preferable, as will become verified, is to place the tube central in the domain; the resultant COMSOL self-generated NC $k=1$ basis mesh is graphed in Figure 11.11b.

Thermal fluid systems are characterized by their nondimensional groups Re, Pr, and Gr. The three modes of convective heat transfer are termed *forced, natural,* and *mixed* as determined by the buoyancy term multiplier $Gr/Re^2 \ll 1$, $Gr/Re^2 \gg 1$, and $Gr/Re^2 \approx 1$ in equation (11.39). Literature correlations for convective heat transfer in tubes employ the non-D Nusselt number, $Nu \equiv hL/k$, and the Rayleigh number, $Ra \equiv GrPr$. In all non-D groups $L \equiv D$, the tube hydraulic diameter, and published correlations for natural and forced convection in terms of average Nu are [10]

$$\overline{Nu}_N = \left(0.6 + \frac{0.387 * (Ra)^{1/6}}{\left(1 + \left(\frac{0.559}{Pr} \right)^{9/16} \right)^{8/27}} \right)^2, \quad \text{for} \quad Ra \leq 10^{12}$$

$$(11.63)$$

$$\overline{Nu}_F = \left(0.3 + \frac{0.62(Re)^{1/2}(Pr)^{1/3}}{\left(1 + \left(\frac{0.4}{Pr} \right)^{2/3} \right)^{1/4}} \left(1 + \left(\frac{Re}{282,000} \right)^{5/8} \right)^{4/5} \right), \quad Re\,Pr > 0.2$$

For mixed convection $\left(\overline{Nu}_M \right)^n \equiv \left(\overline{Nu}_F \right)^n \pm \left(\overline{Nu}_N \right)^n$ with $n=4$ for a cylinder. The plus sign is used when the tube mean temperature exceeds that of the cross-flow fluid and vice versa.

The computer lab 11.3 objectives are to:

1. generate $GWS^h + \theta TS$ solutions for natural, mixed, and forced convection heat transfer
2. thereby appreciate the role of non-D groups in heat transfer characterization
3. via solution-adapted mesh refinement, verify the mesh M required for engineering accuracy for each heat transfer mode steady solution
4. perturb the forced convection inflow velocity data to nonuniform, hence critically assess the erroneous assumption of flowfield steadiness.

The INS conservation principle PDE system remains (11.40), (11.43), (11.44). The lab COMSOL .mph file can be adapted from computer lab 11.2, or downloaded from the COMSOL user community web site. If selecting the former, in the COMSOL subdomain settings window define:

$$F_y = \text{rho } 0^* \text{g} 0^* \text{alpha } 0^* (T - T0), \text{ and}$$
$$Q = -\text{rho } 0^* \text{cp}^* (Tx^*u + Ty^*v)$$

The sought steady solutions are generated running the unsteady solver $GWS^h + \theta TS$ to time-invariant solutions. The thermal data definitions are uniform for each heat-transfer mode, hence alteration of the onset velocity U_∞ generates each heat-transfer mode according to computed Re, Gr, and Pr. Thereafter, the computation of average Nu will quantify the convection mode level.

Data input to COMSOL is dimensional in any consistent set of units and Re, Pr, Gr, and Nu are internally computed. The computer lab 11.3 base case data specification in COMSOL nomenclature is:

T0 (inlet temperature of the fluid)	$= 293\,\text{K}$
T1 (outside temperature of the tube)	$= 303\,\text{K}$
D (outside diameter of the tube)	$= 0.005\,\text{m}$
rho0 (density of the fluid)	$= 1\text{e}3\,\text{kg/m}^3$
mu0 (dynamic viscosity of the fluid)	$= 1\text{e-}3\,\text{kg/ms}$
c_p (heat capacity of fluid)	$= 4.2\text{e}3\,\text{J/kg·K}$
k_c (thermal conductivity of fluid)	$= 0.6\,\text{W/m·K}$
beta0 (volume expansion coefficient)	$= 0.18\text{e-}3\,\text{K}^{-1}$
G0 (acceleration due to gravity)	$= 9.8\,\text{m/s}^2$
V0 (inlet velocity)	$= 5\text{e-}3\,\text{m/s}$

yielding thermal diffusivity $\kappa = k/\rho c_p = 1.14\text{E-}07\,\text{m}^2/\text{s}$, Pr $= 7.0$, kinematic viscosity nu $= 1\text{e-}6\,\text{m/s}^2$, hence Gr $= 1.10\text{E}03$ for all cases. The base case onset fluid speed $U_\infty = 5.0\text{E-}3\,\text{m/s}$ generates Re $= 25$ and Gr/Re$^2 = 1.76$, which corresponds to mixed convection.

The following summarizes the *a posteriori* data generated in computer lab 11.3. The base $M = 2226$ mesh steady temperature solutions with velocity vector overlay for natural, mixed, and forced convection are compared, left to right, in Figure 11.12. Included is the COMSOL temperature color bar and $Gr/Re^2 = 11.0$, 1.76, 0.0176 for the three modes respectively. Key solution distinction is the velocity distribution in the cylinder wake thermal plume. The wake velocity extrema occurs within the plume for natural convection becoming rather uniform for mixed convection. For these modes average $Nu = 1.49$, 1.28 respectively, fluid significant heating is local to the cylinder and central within the plume.

For forced convection with average $Nu = 18.6$ the onset velocity field essentially bypasses the cylinder inducing a wake recirculation bubble extending to the solution domain outflow boundary. The vanishing Neumann BC applied thereon is therefore inappropriate as fluid of unspecified temperature is injected into the solution domain. This is a moot issue however, as a $Re = 250$ steady forced convection solution is unlikely to occur in nature. Instead, a slight perturbation to perfect BC symmetry will cause the flowfield to remain unsteady shedding a periodic, mirror-symmetric vortex

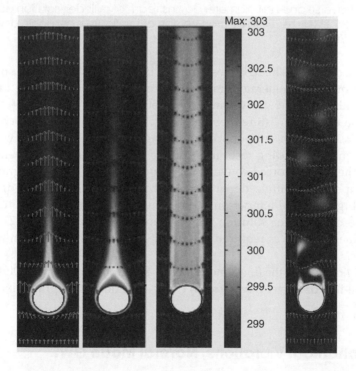

Figure 11.12 $GWS^h + \theta TS$ NC $k = 1$ basis solutions for cross-flow heat exchanger, $M = 2226$ mesh, steady temperature distributions with velocity vector overlay, left to right, natural, mixed, forced convection heat transfer; temperature color bar; unsteady solution snapshot for forced convection heat transfer, far right (a colored version of this figure is available at www.wiley.com/go/baker/finite)

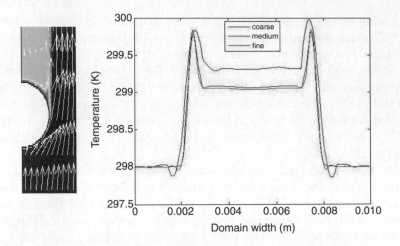

Figure 11.13 GWS^h + θTS NC $k=1$ steady solution, cross-flow heat exchanger, forced convection, Re = 250: left, high-temperature plume departing cylinder; right, temperature DOF channel traverse one-half diameter downstream of cylinder, solution adapted meshes M = 2226, 8904, 18,204 (a colored version of this figure is available at www.wiley. com/go/baker/finite)

street from the cylinder. The rightmost graphic in Figure 11.12 is a time snapshot of this solution; computer lab-generated .mv files for all heat-transfer mode unsteady solutions are available at www.wiley.com/go/baker/finite.

The solution adapted mesh-refinement study for natural and mixed convection confirms the $M=2226$ mesh supports engineering accuracy. Conversely, this mesh is totally inadequate regarding resolution of the temperature plume being shed from each side of the tube in forced convection. Figure 11.13a is a close in graphic showing the high temperature plume departing the tube right side at mid-diameter while Figure 11.13b graphs the temperature distributions through the plumes just downstream of the cylinder for solution adapted mesh refinements $M=2226$, 8904, and 18,204. The finest mesh was required to generate a monotone temperature distribution in the plumes.

Figure 11.14 compares the finest mesh steady solution temperature DOF channel distributions one-half diameter downstream of the cylinder for natural, mixed, and forced convection.

11.10 Mechanical Vibrations, Normal Mode GWS^h

Analyses for n-D mechanical vibrations center on considerations for DP, as $DM = 0 = DE$ are identically satisfied in general. DP for a specific problem is typically formed using the *variational calculus* to construct the *Lagrangian* [11]. Briefly, the Lagrangian L of a mechanical continuum is defined as the balance between kinetic (T) and

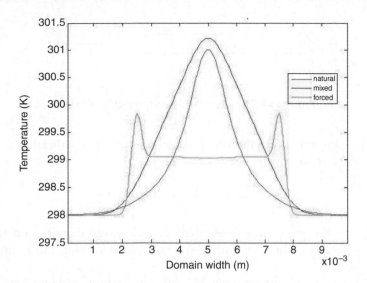

Figure 11.14 Tube bank cross-flow heat exchanger, comparison of natural, mixed, forced convection temperature distributions in channel (a colored version of this figure is available at www.wiley.com/go/baker/finite)

potential (V) energies in the form

$$L \equiv T - V \equiv f(\dot{\mathbf{u}}, \nabla\mathbf{u}, \text{data}), \tag{11.64}$$

where $\mathbf{u}$ is the displacement state, superscript *dot* denotes time derivative, and *data* include pertinent material properties. Via variational principles, the extremum of the Lagrangian, with respect to all admissible states, produces the *Euler–Lagrange equation* [11]

$$E\text{-}L: \quad \frac{\partial}{\partial t}\left(\frac{\partial L}{\partial \dot{\mathbf{u}}}\right) - \nabla \cdot \frac{\partial L}{\partial(\nabla\mathbf{u})} = 0 \tag{11.65}$$

which is the mathematical equivalent of equilibrium DP.

A simple $n = 1$ example illustrates the process. Consider a bar into which longitudinal vibrations are induced by striking the free end, Figure 11.15. The resultant displacement state expressions for kinetic and potential energy in the bar are readily formed

$$L \equiv T - V = \frac{1}{2}\rho\dot{u}^2 - \frac{1}{2}Eu_x^2 \tag{11.66}$$

where the *data* are mass density ρ and the Youngs modulus E. Substituting equation (11.66) into equation (11.65) (an exercise) yields

$$E\text{-}L \Rightarrow \mathbf{DP}: \quad \ddot{u} - c^2 u_{xx} = 0, \tag{11.67}$$

where $c \equiv \sqrt{E/\rho}$ is the elastic wave propagation speed in the bar.

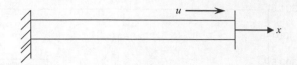

Figure 11.15 Longitudinal vibrations of a bar

A practical n-D mechanical vibration problem is transverse vibration of a plate. The corresponding E-L equation is

$$E\text{-}L \Rightarrow DP: \quad \frac{\partial^2 y}{\partial t^2} - \nabla \cdot f(E, v)\nabla y = 0, \tag{11.68}$$

where $y(x, z, t)$ is the plate time-dependent displacement distribution from the x–z plane and data $f(E, v)$ is the constitutive relationship involving the Youngs modulus and Poisson ratio.

Recalling Chapter 6, for homogenous BCs the usual engineering goal is to seek a normal mode solution. Hence, assuming

$$y(x, z, t) \equiv Y(x, z)e^{i\omega t} \tag{11.69}$$

and substituting into equation (11.68) produces the DP normal mode form

$$DP(n\,\text{mode}): \quad -\nabla \cdot f(E, v)\nabla Y - \omega^2 Y = 0. \tag{11.70}$$

The solution of equation (11.70) then describes the distribution of admissible vibration normal mode displacement states $Y(x, z)$.

The GWS^N for any approximation $Y^N(x,z)$ to the solution of equation (11.70) is

$$\begin{aligned}
GWS^N &= \int_\Omega \Psi_\beta(x)\left[-\nabla \cdot f(E, v)\nabla Y^N - \omega^2 Y^N\right]d\tau = 0, \forall \beta \\
&= \int_\Omega \left[\nabla\Psi_\beta \bullet f(E, v)\nabla Y^N - \omega^2\Psi_\beta Y^N\right]d\tau \\
&\quad - \int_{\partial\Omega} f(E, v)\nabla Y^N \bullet \mathbf{n}\, d\sigma = 0
\end{aligned} \tag{11.71}$$

Implementing the process $GWS^N \Rightarrow GWS^h$ via any FE basis with DOF array $\{Q\}$ generates the algebraic statement

$$GWS^h = [STIFF(E, v)]\{Q\} - \omega^2[MASS]\{Q\} = \{0\} \tag{11.72}$$

which for homogenous BCs defines an *eigenvalue* problem.

Thereby the GWS^h prediction of natural frequencies of vibration are the discrete solution approximations ω_i^h, $1 \le i \le n$, generated via n solutions to the determinantal statement associated with equation (11.72)

$$\det\left([\text{MASS}]^{-1}[\text{STIFF}(E, \upsilon)] - (\omega_i^h)^2[\text{I}]\right) = \{0\}. \tag{11.73}$$

The corresponding n normal mode eigenvector solutions are the DOF arrays $\{Q_i\}$, the GWS^h approximate solution displacement distributions

$$Y^N \Rightarrow Y^h(\omega_i^h) = \cup_e \{N_k(\zeta_\alpha, \boldsymbol{\eta})\}^T \{Q_i\}_e \tag{11.74}$$

as generated by solving a nonsingular rank reduction of equation (11.72) for the first n discrete frequencies ω_i^h, $1 \leq i \leq n$, as input data.

11.11 Normal Modes of a Vibrating Membrane

Computer lab 11.4 is a design study with the goal to adjust the first n natural frequencies of vibration of an L-shaped membrane subject to an impulsively applied transverse load (think of a drum head). Figure 11.16 graphs the membrane initial geometry the BCs for which are zero displacement on the entire $\partial\Omega$. The design component is to assess the effect of filleting the L-shape convex corner to reduce the high stresses that occur at sharp interior corners. The fillet is created by fitting a circle of radius r tangent to both legs of the L-shape.

The computer lab .mph file may be downloaded from www.comsol.com/community/exchange/?page=2 or be self-generated. The following sequence of COMSOL commands creates the required geometry, generates the .mph file to solve for the

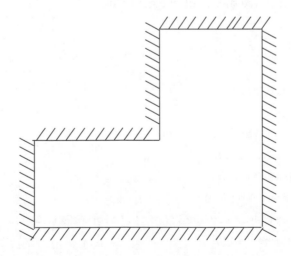

Figure 11.16 L-shaped membrane undergoing transverse vibration

eigenvalues, and thereafter plots the variation of the first eigenvalue as a function of membrane fillet radius.

```
a = 0.5;
sq1 = square2(0,0,1);
sq2 = move(sq1,0,-1);
sq3 = move(sq1,-1,-1);
sq4 = square2(-a,0,a);
c1 = circ2(-a,a,a);
geom. = sq1+sq2+sq3+(sq4-c1);
  geomplot(geom)
```

To solve the eigenvalue problem on this geometry, start by defining the COMSOL data string

```
clear fem
fem.shape = 2;
fem.sshape = 2;
fem.dim = 1;
fem.bnd.h = 1;
fem.equ.c = 1;
fem.equ.da = 1;
```

Next, solve the GWS^h normal mode DOF matrix statement for each r from 0.1 to 1 with a step of 0.05 and save the smallest eigenvalue

```
r = 0.1:0.05:1;
l = [];
for a = r
sq4 = square2(-a,0,a);
c1 = circ2(-a,a,a);
fem.geom. = sq1+sq2+sq3+(sq4-c1);
fem.mesh = meshinit(fem);
fem.mesh = meshrefine(fem);
fem.xmesh = meshextend(fem);
fem.sol = femeig(fem,'eigfun','fleig','eigpar',[0 10]);
l = [l fem.sol.lambda(1)];
end
plot(r,l) ]]>
```

The computer lab 11.4 objectives are to:

1. set up the COMSOL analysis for mechanical vibration of an L-shaped membrane
2. generate the first ten FE solution eigenvalues and eigenmodes on the self-generated base mesh
3. refine the base mesh, hence estimate the mesh necessary to predict the highest natural frequency to engineering accuracy, then
4. modify the plate geometry by introducing a fillet in the concave corner and characterize the effect of fillet radius on select natural frequencies.

The GWS^h normal mode solution displacement content increases remarkably with each increase in natural frequency. The self-generated $M = 2304$ base mesh solutions are graphed in Figures 11.17–11.20. The displacement distributions alternate between symmetric and skew-symmetric about the diagonal bisecting the L-shape into mirror halves. Figures 11.20–11.21 confirm the original $M = 2304$ and uniformly refined $M = 9216$ mesh solutions are indistinguishable. Thereby, the self-generated initial mesh supports engineering accuracy with tenth natural frequency altered by less than an 0.01%.

The smallest eigenfrequency normal mode solution for L-shape fillet of maximum radius $r = 0.5$ is graphed in Figure 11.22. The effect of the fillet is to greatly reduce the GWS^h displacement solution gradient near the concave corner, compare to Figure 11.17b. The effect of fillet radius on the first, third, sixth, and tenth eigenfrequency is graphed in Figure 11.23. The third frequency is most responsive and all magnitudes decrease with increasing fillet radius.

Table 11.3 details the smallest natural frequency dependence on fillet radius. The frequency decrease is monotone with a net change of 7.5% from the original L-shape to the full fillet.

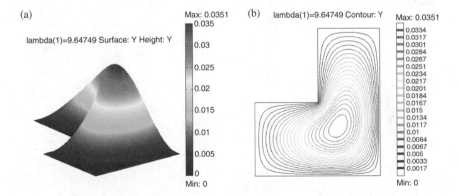

Figure 11.17 GWS^h normal mode solution, $\omega_1^h = 3.10604$ rad/s (a colored version of this figure is available at www.wiley.com/go/baker/finite)

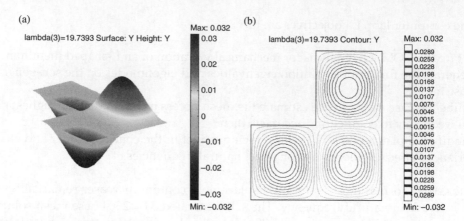

Figure 11.18 GWS^h normal mode solution, $\omega_3^h = 4.44$ rad/s (a colored version of this figure is available at www.wiley.com/go/baker/finite)

11.12 Multiphysics Solid–Fluid Mass Transport

The vast majority of chemical engineering processes involves the transport of mass within a fluid mechanics system creating a new substance. This process is typically not isothermal, hence the pertinent INS conservation principle PDE system is (2.26), (11.39)–(11.40) augmented with a PDE describing chemical species mass transport DM_c.

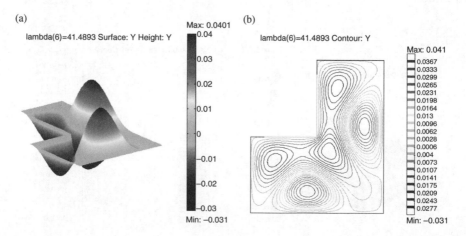

Figure 11.19 GWS^h normal mode solution, $\omega_6^h = 6.44$ rad/s (a colored version of this figure is available at www.wiley.com/go/baker/finite)

(a)
lambda(10)=56.7269 Surface: Y Height: Y

(b)
lambda(10)=56.7269 Contour: Y

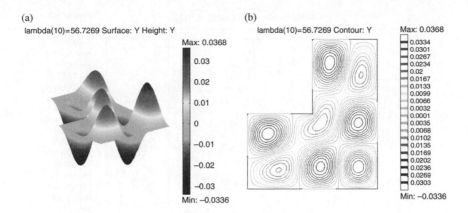

Figure 11.20 GWS^h normal mode solution, $\omega_{10}^h = 7.53$ rad/s (a colored version of this figure is available at www.wiley.com/go/baker/finite)

In non-D tensor form for *mass fraction* definition $0 \leq \Upsilon_\alpha \leq 1.0$, the added state variable member PDE for DM_c is

$$\mathcal{L}(\Upsilon_\alpha) = \frac{\partial \Upsilon_\alpha}{\partial t} + u_j \frac{\partial \Upsilon_\alpha}{\partial x_j} - \frac{1}{\text{ReSc}} \frac{\partial^2 \Upsilon_\alpha}{\partial x_j^2} - s_{\Upsilon_\alpha} = 0. \tag{11.75}$$

The new non-D parameter in equation (11.75) is the *Schmidt number*, $\text{Sc} \equiv \mu/\rho D$, the ratio of diffusion of momentum to mass fraction with coefficient D.

Anticipating the $n=2$ computer lab the INS conservation principle PDE system (11.43)–(11.44) is augmented with equation (11.75). The extension on GWS^N $\Rightarrow$ GWS^h + θTS $= S_e\{WS\}_e$ is the minor augmentation to equations (11.52)–(11.54) with block

(a)
lambda(10)=56.716 Surface: Y Height: Y

(b)
lambda(10)=56.716 Contour: Y

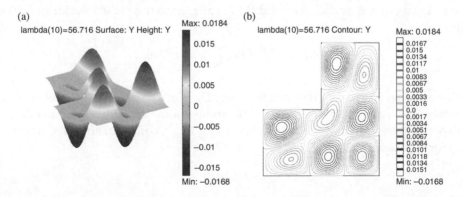

Figure 11.21 GWS^h normal mode solution, $\omega_{10}^h = 7.53$ rad/s, $M = 9216$ (a colored version of this figure is available at www.wiley.com/go/baker/finite)

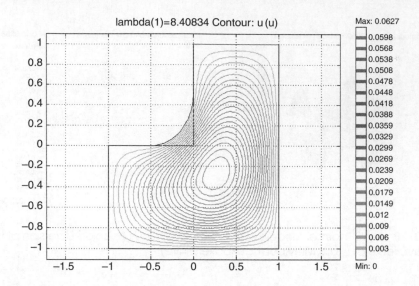

Figure 11.22 GWS^h normal mode solution with fillet, $r=0.5$, $\omega_1^h = 2.89$, $M=3004$ (a colored version of this figure is available at www.wiley.com/go/baker/finite)

banded jacobian (11.54) re-definition

$$[\text{JAC}(\{Q\})] \equiv \frac{\partial\{F(\{Q\})\}}{\partial\{Q\}} = \begin{bmatrix} [J\Omega\Omega], & [J\Omega\Psi], & [J\Omega\Theta], & [0] \\ [J\Psi\Omega], & [J\Psi\Psi], & [0], & [0] \\ [0], & [J\Theta\Psi], & [J\Theta\Theta], & [0] \\ [0], & [J\Psi\Upsilon], & [0], & [J\Upsilon\Upsilon] \end{bmatrix}. \tag{11.76}$$

The COMSOL *pellet problem* .mph file supports prediction of the coupled steady solid–fluid mass transport interaction PDE system, computer lab 11.5. The pellet is immersed in a cross-flow; being a solid (11.75) altered to the pellet interior in COMSOL notation is

$$\mathcal{L}(c) = -\frac{\partial}{\partial x_j}\left(D_{\text{eff}}\frac{\partial c}{\partial x_j}\right) - kc^2 = 0. \tag{11.77}$$

The dimensional *effective* diffusion coefficient D_{eff} (m^2/s) is related to pellet porosity and tortuosity, with k the rate constant (m^3/mol·s) of the second-order reaction generating species mass c within the pellet. Assuming D_{eff} a constant, specified as data, the non-D form of equation (11.77) is

$$-D^*\nabla^2 c - k^* c^2 = 0 \tag{11.78}$$

where superscript * denotes the variable nondimensionalized by the reference value for D_{eff}.

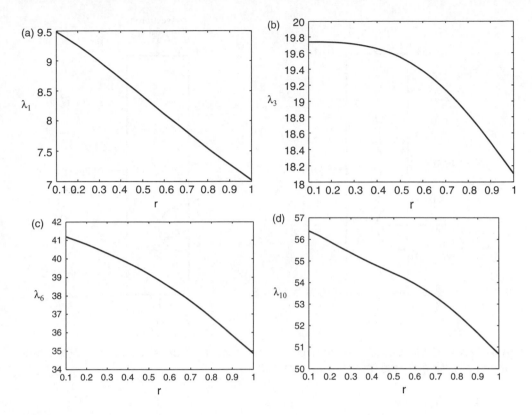

Figure 11.23 Variation of (a) first, (b) third, (c) sixth, and (d) tenth eigenfrequency with convex corner fillet radius r (a colored version of this figure is available at www.wiley.com/go/baker/finite)

Table 11.3 First normal mode frequency dependence on fillet radius

Fillet radius (r)	Frequency (ω_1)
0.1	3.0799
0.2	3.0621
0.3	3.0422
0.4	3.0208
0.5	2.9981
0.6	2.9744
0.7	2.9500
0.8	2.9251
0.9	2.8997
1.0	2.8741

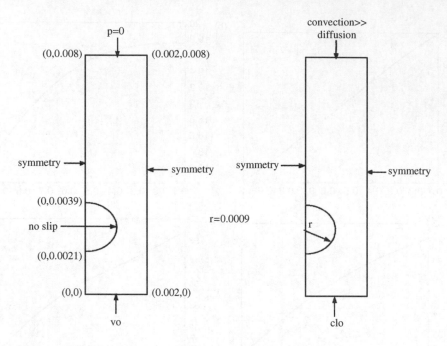

Figure 11.24 COMSOL BC specifications for the pellet multiphysics simulation, (a) cross-flow velocity domain, (b) mass transport domain

Multiphysics computer lab 11.5 requires coupling INS with convection–conduction in the COMSOL Chemical Engineering module. The symmetric half-domain is selected; in COMSOL nomenclature the BCs for the INS and mass transport domains are defined in Figure 11.24a and b. For the INS domain *symmetry* specifies $\mathbf{u} \cdot \hat{\mathbf{n}} = 0$ while *no-slip* corresponds to $\mathbf{u} = \mathbf{0}$. For the mass transport domain *symmetry* is the Robin BC $(-D^*\nabla c + c\mathbf{u}) \cdot \hat{\mathbf{n}} = 0$ while *convection* ≫ *diffusion* defines $(-D^*\nabla c + c\mathbf{u}) \cdot \hat{\mathbf{n}} = c\mathbf{u} \cdot \hat{\mathbf{n}}$ as the BC, COMSOL internally-evaluated during the matrix solution process.

Setup of the pellet geometry, solution domains and grid in COMSOL terminology is

Axis		Grid	
X min	−0.001	X spacing	0.001
X max	0.003	Extra X	0.0009
Y min	−0.001	Y spacing	0.001
Y max	0.007	Extra Y	0.0021 0.0039

The base case data specification is

Parameter	Value
ro (density)	$0.66\,\text{kg/m}^3$
mu (dynamic viscosity)	$2.6\text{e-}5\,\text{kg/ms}$
D (diffusion coefficient)	$1\text{e-}5\,\text{m}^2/\text{s}$
D_{eff} (effective diffusion coefficient)	$9\,\text{e-}6\,\text{m}^2/\text{s}$
k (rate constant)	$100\,\text{m}^3/\text{mol·s}$
v_{o} (inflow velocity)	$0.1\,\text{m/s}$
c1o (inflow concentration)	$1.3\,\text{mol/m}^3$

The computer lab 11.5 objectives are:

1. access the COMSOL .mph file for the pellet problem
2. for the base data, which generates Re ≈ 5 and Sc ≈ 4, compute the steady velocity and mass fraction DOF distributions on the self-generated mesh
3. conduct a regular mesh-refinement study, hence confirm this Re–Sc solution is engineering accurate
4. the base data specification is very diffusive. Alter inflow velocity to produce $50 \leq \text{Re} \leq 100$, also decrease effective diffusion coefficient to generate $15 \leq \text{Sc} \leq 35$
5. compute the velocity and mass fraction DOF distributions for various data combinations in 4
6. for solutions exhibiting dispersion error oscillation, perform solution-adapted mesh refinement to determine M necessary to generate monotone mass fraction distributions.

The following summarizes the essence of *a posteriori* data generated in computer lab 11.5. The base specification self-generated $M = 1961$ mesh, and Re $= 4$ solution mass fraction distributions, color coded in decile (0.1) increments for $D = 0.9$, 0.25, 0.125 E-05 m^2/s, yielding Sc $= 4$ (base), 16, 32 are graphed left to right in Figure 11.25. The influence of decreasing diffusion narrows the plume in the fluid trailing the pellet and sharpens gradients in the pellet upstream flowfield interface. The mass fraction in the pellet is essentially uniform.

Viewing these data the self-generated mesh is poorly designed for mass fraction plume resolution. As can be anticipated, the mass fraction distribution computed on this mesh for Re $= 75$, Sc $= 32$ is totally dispersion-error polluted, left graphic in Figure 11.26. Uniform mesh refinement is obviously impractical hence solution-adapted mesh refinement is required.

The cogent approach is decreasing mesh measure keeping COMSOL scaling factor constant. For this factor fixed at 1.3, the mesh measure sequential reduction 1.0 (base), 0.5, 0.25, 0.1 generated meshes containing $M = 1961$, 2206, 3533, 15,009 triangular

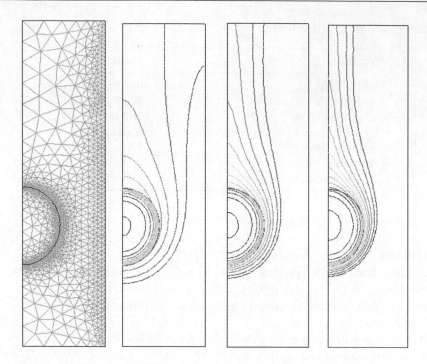

Figure 11.25 GWS^h $k = 1$ NC basis algorithm steady solutions, pellet problem mass fraction distributions, Re $= 4$, left to right: COMSOL base mesh, $D = 0.9$, 0.25, 0.125 E-05 m^2/s, Sc $= 4$, 16, 32 (a colored version of this figure is available at www.wiley.com/go/baker/finite)

elements with DOF ranging 11,346 to 92,437. These meshes supported the Re $= 75$, Sc $= 32$ steady mass fraction distributions graphed left to right in Figure 11.26.

The mass fraction distribution across the channel at the y coordinate halfway between pellet and outflow boundary confirms the $M = 3533$ mesh supports an engineering accurate monotone solution, Figure 11.27.

11.13 Chapter Summary

The theorization $\Rightarrow$ implementation WS^N $\Rightarrow$ GWS^h/mGWS^h + θTS process for designing approximate solution strategies for PDE + BCs systems in the computational engineering sciences is now completed in multidimensions including initial value character. Deriving a specific algorithm starts with addressing the appropriate coupled conservation principles DM, DP, and DE, both thermal and mechanical, as well as DM_α for mass transport, all augmented with closure models pertinent to *the physics*. Close

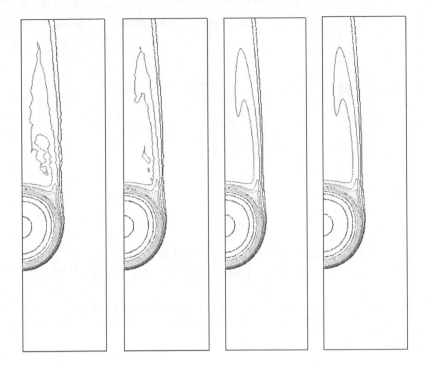

Figure 11.26   GWS^h $k=1$ NC basis algorithm steady solutions, pellet problem mass fraction distributions, Re = 75, Sc = 32, left to right: $M=$ 1961, 2206, 3533, 15,009 (a colored version of this figure is available at www.wiley.com/go/baker/finite)

attention is given to characterizing these PDE systems as (typically) EBV to ensure that encompassing BCs are well-posed.

The chapter structural mechanics lead off topic enabled examination of both DP and mechanical DE formulations, the latter closely associated with the variational calculus underlying structural mechanics FE theorization. The elasticity *divergence constraint*, clearly expressed in equations (11.10)–(11.11), lead to replacement of tensor calculus DP with the matrix calculus form (11.17), to enable use of the Green–Gauss divergence theorem. The dual approach of direct discrete implementation of mechanical DE and GWS^N for matrix DP clearly identified the latter BC integral functional.

The theorization complete, the tedious detail of forming theory stiffness and load matrices [STIFF] and {b} for the linear NC basis implementation exposed all associated matrix details for the closely coupled state variable. Comparison of the material property multipliers on [STIFF] and [DIFF], from heat transfer, confirmed the structures problem statement as highly diffusive. The plane stress computer lab well illustrated the requirement for accurate BC implementation.

The following thermal-fluid mechanics chapter topic enabled detailed DM, DP, DE manipulation to establish a well-posed PDE + BCs system prior to GWS^h + θTS

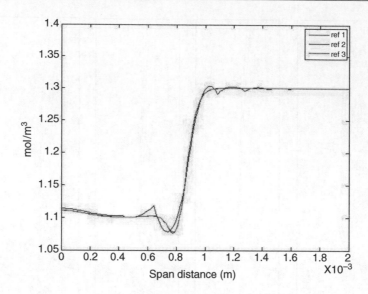

Figure 11.27 GWS^h $k = 1$ NC basis steady solution distribution profiles at $y = 5.0$, Re $= 75$, Sc $= 32$, $M = 2206$, 3533, 15,009 (a colored version of this figure is available at www.wiley. com/go/baker/finite)

identification. The new multimember state variable formulation rigorously enforced the *divergence constraint*, in the process identifying a Robin BC, typically replaced in CFD practice with a *discrete* DOF Dirichlet approximation. The generated DP curl-nonlinearity was precisely transformed to an element *hypermatrix* statement via direct use of calculus with vector field theory.

The topic computer labs amply illustrate the requirement for close attention to the homogeneous Neumann outflow BC. The isothermal validation problem identified the consequence of assuming reduced dimensionality. The latter thermal specification required *solution-adapted* mesh refinement to control inadequate resolution-generated dispersion error.

A brief introduction to the variational calculus introduced the *Lagrangian*, leading to a DP formulation for normal mode analyses in mechanical vibration. The resultant singular GWS^N matrix statement enabled exposing the process for generation of harmonic solutions. The associated computer lab detailed vibration mode frequency adjustment via geometry alteration.

The terminal engineering sciences topic is a multiphysics statement in solid–fluid mass transfer, a process intrinsic to production processing in chemical engineering. The non-D parameters intrinsic to fluid mechanics performance again play a major role in performance diagnostics, and identifying well posed BCs again became important. The requirement for solution-adapted mesh refinement again supported identification of mesh M necessary to ensure an engineering accurate solution.

Exercises

11.1 From first principles, derive a linear elasticity plane stress/strain PDE system, (11.10) or (11.11), hence verify the vector PDE (11.12).

11.2 Detail construction of the matrix PDE GWS^N for plane stress/strain, equation (11.15).

11.3 Verify discrete extremization of the FE implementation $\prod_e$ (11.23). Then compare term by term with the matrix GWS^N, (11.15), after discrete implementation.

11.4 Verify construction of [**B**], (11.29), for basis $\{N_1(\zeta_\alpha)\}$.

11.5 Compute several terms in the matrix product (11.31).

11.6 Compare select matrix elements of [STIFF]$_e$, (11.36), with [DIFF]$_e$, (9.35).

11.7 Expanding the curl in equation (11.41) via its determinant definition, confirm that equation (11.39) is identically satisfied.

11.8 Substituting equation (11.41) into equation (11.42) confirm equation (11.44).

11.9 Generate the TS on streamfuction (11.47), using equation (11.46), then modify it to the Robin BC (11.48).

11.10 Confirm the two lower order TS expressions in equation (11.49).

11.11 Write the template statements for the Newton algorithm (11.54) for the FE algorithm statements (11.52) and (11.55).

11.12 Confirm (11.67) via the Euler–Lagrange equation definition.

Computer Labs

The COMSOL .mph files required for conducting the computer labs discussed in this chapter may be downloaded from their user community site www.comsol.com/community/exchange/?page=2.

11.1 In structural mechanics, the linear, elastic, homogeneous media plane stress/-strain assumption leads to the mechanical energy DE amenable to GWS^h FE implementations. The design study seeks to adjust the shape of an initially-circular hole in a plate, under tension to moderate the maximum local stress concentration. Execute the base mesh case, then a solution-adapted regular mesh refinement to estimate mesh adequacy. Recognizing the wall BC divergence constraint issue, alter the BCs as directed and repeat the mesh refinement. Then adjust the shape of the circular hole, keeping the area constant, such that the extremum von Mises stress is minimized for the *entire* plate.

11.2 The classic Navier–Stokes validation is flow in a duct with sudden enlargement in cross-section, the step-wall diffuser. In 2-D, the NS DM and DP transformation to streamfunction–vorticity enables a mathematically well-posed PDE + BCs system. The design goal of this lab is to establish the primary recirculation bubble intercept with the lower wall for $100 \leq \mathrm{Re} \leq 600$. The attachment coordinate is identified by the wall vorticity distribution DOF sign change.

11.3 Heat-transfer devices in practical geometries typically involve fluid flowing over a tube bank. Buoyancy effects can exert a profound impact on this process. Following the text discussion, execute the COMSOL file case for the base Re and Gr definitions. Then determine the inflow data yielding Gr/Re^2 definitions for heat-transfer mode alteration. Note the issue with outflow BC for the larger Re specification, hence fix the problem. Compute the average Nu associated with each heat transfer mode.

11.4 Multidimensional structural mechanical vibration analyses are usually cast in normal mode form, hence DP is transformed to an eigenvalue statement. For the given L-shaped membrane, execute the COMSOL base case solving for the first ten eigenvalues and eigenmodes. Execute a regular mesh-refinement study to assess natural frequency prediction accuracy. Then place a fillet in the membrane concave corner, regenerate the mesh and corresponding solutions, and compare eigenvalues/eigenmodes to the base case. Draw conclusions on the action of the fillet in modifying the lowest few normal mode shapes and frequencies.

11.5 This computer lab requires setup and execution of dual COMSOL modules for a multiphysics simulation of solid–fluid mass transfer/transport. Access the .mph file and execute the base case specification. Then adjust the solid effective diffusion coefficient D to alter the solid diffusion level, hence also the Schmidt number. Run a series of computer studies varying these data over the ranges $1E\text{-}06 \leq D \leq 9$ $E\text{-}06$, $5 \leq Re \leq 72$, hence $4 \leq Sc \leq 31$ and observe solution distinctions. Perform a solution adapted regular mesh refinement for the largest ReSc specification being particularly attentive to control of dispersion error oscillations just downstream of the pellet major diameter.

References

[1] COMSOL, http://comsol.com.

[2] Shames, I.H. (1964) *Mechanics of Deformable Solids*, Prentice-Hall, Englewood Cliffs, NJ.

[3] Carey, G.F. and Oden, J.T. (1984) *Finite Elements, a Second Course*, Prentice-Hall, Englewood Cliffs, NJ.

[4] Cook, R.D., Malkus, D.S., Plesha, M.E., and Witt, R.J. (2002) *Concepts and Applications of Finite Element Analysis*, Wiley, New York, NY.

[5] Baker, A.J. (2013) *Optimal Modified Continuous Galerkin CFD*, Wiley, London.

[6] Gresho, P.M. and Sani, R.L. (1998) *Incompressible Flow and the Finite Element Method*, Wiley, London.

[7] Williams, P.T. and Baker, A.J. (1996) Incompressible computational fluid dynamics and the continuity constraint method for the 3-D Navier–Stokes equations. *J. Numer. Heat Tr., Part B, Fund.*, **29**, 137–273.

[8] Armaly, B.F., Durst, F., Pereira, J.C.F., and Schonung, B. (1983) Experimental and theoretical investigation of backward-facing step flow. *J. Fluid Mechanics*, **127**, 473–496.

[9] Williams, P.T. and Baker, A.J. (1997) Numerical simulations of laminar flow over a 3D back-ward-facing step. *J. Numer. Methods Fluids*, **24**, 1–25.

[10] Incropera, F.P. and Dewitt, D.P. (2002) *Fundamentals of Heat and Mass Transfer*, Wiley, New York, NY.

[11] Weinstock, R. (1952) *The Calculus of Variations*, McGraw-Hill, New York, NY.

12

Conclusion

This completes the designed exposure to the formulation process, underlying theory, and computational practice *substance* of FE completeness degree k trial space basis implemented weak form algorithms in the *computational engineering sciences*.

The mathematical summary of the theory/practice sequence $\text{WF} \Rightarrow \text{WS}^N \Rightarrow \text{GWS}^N + \theta\text{TS} \Rightarrow \text{GWS}^h/m\text{GWS}^h + \theta\text{TS} \Rightarrow \{\text{WS}\}_e$ is

$$\textit{approximation:} \quad q(\mathbf{x}, t) \approx q^N(\mathbf{x}, t) \equiv q^h(\mathbf{x}, t) \equiv \cup_e q_e(\mathbf{x}, t)$$

$$\textit{FE basis implementation:} \quad q_e(\mathbf{x}, t) \equiv \{N_k(\zeta_\alpha, \boldsymbol{\eta})\}^T \{Q(t)\}_e$$

$$\textit{error extremization:} \quad m\text{GWS}^N = \int_\Omega \Psi_\beta(\mathbf{x}) L^m(q^N) d\tau \equiv \{0\}, \forall \beta$$

$$\textit{matrix statement:} \quad m\text{GWS}^h + \theta\text{TS} \Rightarrow [\text{JAC}]\{\delta Q\} = -\{\text{F}(\{Q\})\}_n$$

$$\textit{assembly:} \quad [\text{JAC}] = S_e([\text{JAC}]_e),$$
$$\{\text{F}(\{Q\})\} = S_e(\{\text{F}(\{Q\})\}_e)$$

$$\textit{asymptotic convergence:} \quad \left\| e^h(n\Delta t) \right\|_E \leq C h^{2\gamma} \|\text{data}\|_{L2}^2 + C_t \Delta t^{f(\theta)} \|q_0\|_E$$
$$\gamma = \min(1; \, k+1-m, \, r-m \, \text{for Pa}^{-1} > 0), f(\theta) = (2,3)$$

$$\textit{error spectra:} \quad U_\kappa(\lambda), G(\lambda) \Rightarrow f(\kappa, k, h, \Delta t, \theta, \alpha, \beta, \gamma)$$

Hopefully at this terminal point the reader fully appreciates that $\text{WF} \Rightarrow m\text{GWS}^N + \theta\text{TS}$ is *totally analytical*, developed in the continuum, hence *absolutely independent* of the choice of discrete implementation. The text preference to implement the theory using FE trial space basis functions reflects on the precision of calculus and vector field theory, used without exception, to generate the matrix *computable form*. It should likewise be obvious that alternative discrete implementations of the theory exist and can be made, as illustrated by the brief $\text{FD}/\text{FV}^h/\text{GWS}^h$ comparison.

The theory and mathematical foundations developed herein are full dimensional. The computer lab experiences have been restricted to $n = 1,2$ dimensional problem statements to keep this aspect of text content pedagogically tractable. Full dimensional compute capability is readily available in modern, commercial FE-implemented *problem solving environments* in the computational engineering sciences.

The professional or student interested in $mGWS^h + \theta TS$ application to genuine 3-D PDE + BCs + IC statements in the nonlinear, multiphysics computational engineering sciences has hopefully gained a *sense of comfort* with the rigorous theoretical foundation *and* the error prediction practice of solution-adapted mesh refinement. That accomplished, template organization of theory conversion to code practice precisely addresses the nuances of the decisions required. That these might be accomplished is the prime reason for authoring this text!

Index

Accuracy, 8, 28, 29, 33, 44–45, 51, 56, 60,
62, 67, 68, 71, 84, 85, 93, 111, 114,
127, 128, 132, 141–3, 149–52, 160,
170, 177, 188, 189, 192, 194, 200,
211, 242, 247
  boundary flux, 45
  and convergence, 47, 67, 127–30,
141–2, 157
  engineering, 84, 93, 142, 200, 226, 236,
240, 242, 247
  Gauss quadrature, 135
  optimal, 143
  order of, 60, 149, 150, 160, 194
  TS modified, 171, 190, 213
Adiabatic, 128, 201
Algebra,
  equation system, 3, 6, 24, 26, 27, 30,
160, 186
  linear, 8, 10, 128, 183, 203, 213
Approximate factorization, 207
Approximation,
  finite difference (FD), 30–32, 45, 150,
151
  finite element (FE), 34, 185, 219
  finite volume (FV), 154
  frequencies, 98
  global, 59
Artificial diffusion, 88, 167, 169, 171,
173, 177, 191, 193, 194, 212

Asymptotic
  convergence, 45, 51, 56, 60, 62, 63, 64,
81–83, 89, 90, 97, 109–14, 163, 179,
188, 192, 199, 225
  error estimate, 28, 29, 59–64, 82, 109,
129, 141, 152, 154, 163, 164, 175,
192, 212, 213
Axisymmetric, 101, 105, 106, 129, 177

Basis, finite element,
  bilinear, 134, 138
  biquadratic, 209
  cubic, 63, 78
  linear, 40, 42, 51, 54, 118, 235
  p-element, 68
  quadratic, 78
Boundary conditions (BCs),
  convection, 49, 51, 106, 108, 117, 136,
167, 183, 192
  Dirichlet, 24, 25, 27, 35, 43, 45, 49, 52,
67, 117, 136, 165, 184, 199, 203, 219,
231
  flux, 47, 129, 152, 219
  Neumann, 35, 101, 186, 196, 219, 220,
224, 226, 230, 235, 236, 241, 256
  radiation, 109, 117, 136, 139, 177
  Robin, 49, 70, 106, 127, 136, 148,
150, 177, 184, 186, 230, 231, 234,
252, 256

Finite Elements ⇔ Computational Engineering Sciences, First Edition. A. J. Baker.
© 2012 John Wiley & Sons, Ltd. Published 2012 by John Wiley & Sons, Ltd.